Spectral and Scattering Theory

Spectral and Scattering Theory

Edited by
Alexander G. Ramm
Kansas State University
Manhattan, Kansas

Plenum Press • New York and London

Library of Congress Cataloging-in-Publication Data

On file

Proceedings of sessions from the First Congress of the International Society for Analysis, Applications, and Computing, held June 2 – 6, 1997, at the University of Delaware, Newark, Delaware

ISBN 0-306-45829-2

© 1998 Plenum Press, New York
A Division of Plenum Publishing Corporation
233 Spring Street, New York, N.Y. 10013

http://www.plenum.com

10 9 8 7 6 5 4 3 2 1

All rights reserved

No part of this book may be reproduced, stored in a retrieval system, or transmitted in any form or by any means, electronic, mechanical, photocopying, microfilming, recording, or otherwise, without written permission from the Publisher

Printed in the United States of America

PREFACE

In this volume selected papers delivered at the special session on "Spectral and scattering theory" are published.

This session was organized by A. G. Ramm at the first international congress of ISAAC (International Society for Analysis, Applications and Computing) which was held at the University of Delaware, June 3–7, 1997. The papers in this volume deal with a wide variety of problems including some nonlinear problems (Schechter, Trenogin), control theory (Shubov), fundamental problems of physics (Kitada), spectral and scattering theory in waveguides and shallow ocean (Ramm and Makrakis), inverse scattering with incomplete data (Ramm), spectral theory for Sturm–Liouville operators with singular coefficients (Yurko) and with energy-dependent coefficients (Aktosun, Klaus, and van der Mee), spectral theory of Schrödinger operators with periodic coefficients (Kuchment, Vainberg), resolvent estimates for Schrödinger-type and Maxwell's operators (Ben-Artzi and Nemirovsky), Schrödinger operators with von Neumann–Wigner type potentials (Rejto and Taboada), principal eigenvalues for indefinite-weight elliptic operators (Pinchover), and symmetric solutions of Ginzburg–Landau equations (Gustafson).

These papers will be of interest to a wide audience including mathematicians, physicists, and theoretically oriented engineers.

A. G. Ramm
Manhattan, KS

CONTENTS

1. Wave Scattering in 1-D Nonconservative Media 1
   Tuncay Aktosun, Martin Klaus, and Cornelis van der Mee

2. Resolvent Estimates for Schrödinger-type and Maxwell Equations with
   Applications 19
   Matania Ben-Artzi and Jonathan Nemirovsky

3. Symmetric Solutions of Ginzburg–Landau Equations 33
   S. Gustafson

4. Quantum Mechanics and Relativity: Their Unification by Local Time 39
   Hitoshi Kitada

5. On Embedded Eigenvalues of Perturbed Periodic Schrödinger Operators 67
   Peter Kuchment and Boris Vainberg

6. On Principal Eigenvalues for Indefinite-Weight Elliptic Problems 77
   Yehuda Pinchover

7. Scattering by Obstacles in Acoustic Waveguides 89
   A. G. Ramm and G. N. Makrakis

8. Recovery of Compactly Supported Spherically Symmetric Potentials from the
   Phase Shift of the s-Wave 111
   A. G. Ramm

9. A Turning Point Problem Arising in Connection with a Limiting Absorption
   Principle for Schrödinger Operators with Generalized Von Neumann–Wigner
   Potentials 131
   Peter Rejto and Mario Taboada

10. Eigenvalue Problems for Semilinear Equations 157
    Martin Schechter

11. Spectral Operators Generated by 3-Dimensional Damped Wave Equation and
    Applications to Control Theory 177
    Marianna A. Shubov

12. Invertibility of Nonlinear Operators and Parameter Continuation Method 189
    Vladilen A. Trenogin

13. Sturm–Liouville Differential Operators with Singularities 199
    V. Yurko

Index 207

WAVE SCATTERING IN 1-D NONCONSERVATIVE MEDIA

Tuncay Aktosun,[1] Martin Klaus,[2] and Cornelis van der Mee[3]

[1]Department of Mathematics
North Dakota State University
Fargo, ND
[2]Department of Mathematics
Virginia Polytechnic Institute and State University
Blacksburg, VA
[3]Department of Mathematics
University of Cagliari
Cagliari, Italy

ABSTRACT

In this review paper, the generalized Schrödinger equation $d^2\psi/dx^2 + k^2\psi = [ikP(x) + Q(x)]\psi$ is considered, where $P(x)$ and $Q(x)$ are real, integrable potentials with finite first moments. The scattering solutions and the bound state solutions are studied, the scattering coefficients and their small-k and large-k asymptotics are analyzed. Unless $P(x) \leq 0$, it is shown that there may be bound states at complex energies, degenerate bound states, and singularities of the transmission coefficient for real k. Some illustrative examples are provided.

1. INTRODUCTION

In this paper we are interested in analyzing the scattering problem for

$$-\psi^{+\prime\prime}(k,x) + [ikP(x) + Q(x)]\psi^+(k,x) = k^2\psi^+(k,x), \qquad x \in \mathbf{R}, \tag{1.1}$$

where $\mathbf{R}$ is the real line, the prime denotes the derivative with respect to x, k is a complex parameter, and $P(x)$ and $Q(x)$ are real-valued functions such that $P, Q \in L_1^1(\mathbf{R})$. By $L_1^1(\mathbf{R})$ we denote the Lebesgue-measurable functions $f(x)$ such that $||f||_{1,1}$ is finite, where $||f||_{1,1} = \int_{-\infty}^{\infty} dx\,(1+|x|)\,|f(x)|$. Most of our results are valid under the weaker assumptions $P, Q \in L^1(\mathbf{R})$, and the first moments are needed only when we consider (1.1) near or at $k=0$. The reason for us to use the superscript plus in (1.1) will be apparent in Section 3: Associated with (1.1) is the related equation (3.1), where the superscript minus appears.

There are several reasons why the analysis of (1.1) is important. In quantum mechanics, (1.1) describes the behavior of a particle of momentum k and energy k^2 interacting with the energy-dependent potential $ikP(x)+Q(x)$. In this case $\psi^+(k,x)$ corresponds to the wavefunction. Moreover, in the frequency domain, (1.1) describes the wave propagation in a one-dimensional medium where $P(x)$ represents energy absorption or generation in the medium and $Q(x)$ is the restoring force density. In this context, the time-domain analog of (1.1) is given by

$$\frac{\partial^2 u}{\partial x^2} - \frac{\partial^2 u}{\partial t^2} - P(x)\frac{\partial u}{\partial t} = Q(x)u, \qquad t,x \in \mathbf{R},$$

where the wave speed is equal to 1. When $P(x) \leq 0$, there is net absorption; however, unless otherwise stated we will not put any restriction on the sign of $P(x)$.

The scattering states of (1.1) correspond to its solutions behaving like e^{ikx} or e^{-ikx} as $x \to \pm\infty$ for $k \in \mathbf{R}$. As indicated in Section 2, such solutions can be extended continuously and analytically in k to the upper-half complex plane $\mathbf{C}^+$. If at a certain k-value in $\mathbf{C}^+$ such solutions decay exponentially as $x \to \pm\infty$, we obtain a bound state. A bound state of (1.1) corresponds to a nontrivial solution belonging to $L^2(\mathbf{R})$.

When $P(x) \equiv 0$, (1.1) is reduced to the usual Schrödinger equation

$$-\psi^{[0]''}(k,x) + Q(x)\psi^{[0]}(k,x) = k^2 \psi^{[0]}(k,x), \qquad x \in \mathbf{R}. \qquad (1.2)$$

When $Q \in L^1(\mathbf{R})$, the scattering theory for (1.2) is well understood. Let us use $N(P,Q)$ to denote the number of bound states (including multiplicities) of (1.1); hence $N(0,Q)$ denotes the number of bound states of (1.2). When $Q \in L^1(\mathbf{R})$ there may be infinitely many bound states of (1.2) all having negative energies accumulating at zero, but each bound state is simple. When $N(0,Q)$ is finite, we let $k = i\kappa_j$ for $j = 1,\cdots,\mathcal{N}$ denote the corresponding momentum values with $0 < \kappa_1 < \cdots < \kappa_{\mathcal{N}}$. If $Q \in L^1_1(\mathbf{R})$, it is assured that $N(0,Q)$ is finite. Since $P(x)$ is real, unless $P(x) \equiv 0$, (1.1) is a non-self-adjoint equation and there may be complex eigenvalues, i.e. values of k^2 at which there exist solutions belonging to $L^2(\mathbf{R})$. Furthermore, (1.1) may also have eigenvalues that are not simple. Some examples of complex or multiple eigenvalues will be given in Section 8.

This paper is organized as follows. In Section 2 we consider the Jost solutions of (1.1), i.e. certain scattering solutions in terms of which the scattering coefficients are defined. In Section 3 we study the scattering coefficients and some of their properties. Section 4 is about the large-k and small-k asymptotics of the scattering coefficients. In Section 5 the bound states of (1.1) are considered, some estimates are obtained on the number of bound states, and a Levinson theorem is presented relating the number of bound states to the argument of the transmission coefficient. In Section 6 we show that the negative-energy bound states of (1.1) can be analyzed in terms of the eigenvalue branches of the operator $\mathcal{O}_\beta$ defined in (6.1). Section 7 is devoted to the connection between the zeros of the Jost solutions of (1.1) and the bound states. In Section 8 we present various examples illustrating the theory contained in the prior sections. Finally, in Section 9 we conclude with a brief summary of some inverse scattering problems for (1.1).

Since this is a review paper, no proofs are included. For proofs and further details, we refer the reader to [1] and the references therein.

2. SCATTERING SOLUTIONS

Among the scattering solutions of (1.1), we have the Jost solution from the left $f_l^+(k,x)$ and the Jost solution from the right $f_r^+(k,x)$ satisfying the boundary conditions

$$f_l^+(k,x) = e^{ikx} + o(1), \qquad f_l^{+\prime}(k,x) = ike^{ikx} + o(1), \qquad x \to +\infty, \tag{2.1}$$

$$f_r^+(k,x) = e^{-ikx} + o(1), \qquad f_r^{+\prime}(k,x) = -ike^{-ikx} + o(1), \qquad x \to -\infty. \tag{2.2}$$

These solutions satisfy

$$f_l^+(k,x) = \frac{1}{T^+(k)} e^{ikx} + \frac{L^+(k)}{T^+(k)} e^{-ikx} + o(1), \qquad x \to -\infty, \tag{2.3}$$

$$f_r^+(k,x) = \frac{1}{T^+(k)} e^{-ikx} + \frac{R^+(k)}{T^+(k)} e^{ikx} + o(1), \qquad x \to +\infty, \tag{2.4}$$

where $T^+(k)$ is the transmission coefficient, and $R^+(k)$ and $L^+(k)$ are the reflection coefficients from the right and from the left, respectively. Let $[F;G] = FG' - F'G$ denote the Wronskian. Note that the same transmission coefficient appears in both (2.3) and (2.4); this is because $[f_l^+(k,x); f_r^+(k,x)]$ is independent of x, and the values of that Wronskian at $x = \pm\infty$ show that the transmission coefficient from the left is the same as the transmission coefficient from the right.

Define the Faddeev functions from the left $m_l^+(k,x)$ and from the right $m_r^+(k,x)$ as

$$m_l^+(k,x) = e^{-ikx} f_l^+(k,x), \qquad m_r^+(k,x) = e^{ikx} f_r^+(k,x).$$

By $\overline{\mathbf{C}^+}$ we denote $\mathbf{C}^+ \cup \mathbf{R}$. In the next theorem we show that, for each fixed x, the Jost solutions can be analytically extended to $\mathbf{C}^+$.

Theorem 2.1. *Assume $P, Q \in L^1(\mathbf{R})$. Then, for each $x \in \mathbf{R}$, the functions $m_l^+(k,x)$, $m_r^+(k,x)$, $m_l^{+\prime}(k,x)$, and $m_r^{+\prime}(k,x)$ are analytic in $\mathbf{C}^+$ and continuous in $\overline{\mathbf{C}^+} \setminus \{0\}$. Similarly, for each $x \in \mathbf{R}$, the functions $f_l^+(k,x), f_r^+(k,x), f_l^{+\prime}(k,x)$, and $f_r^{+\prime}(k,x)$ are analytic in $\mathbf{C}^+$ and continuous in $\overline{\mathbf{C}^+} \setminus \{0\}$. Moreover, we have*

$$|m_l^+(k,x)| \leq Ce^{C/|k|}, \qquad |m_r^+(k,x)| \leq Ce^{C/|k|}, \qquad k \in \overline{\mathbf{C}^+} \setminus \{0\},$$

$$|m_l^{+\prime}(k,x)| \leq C(1+|k|)e^{C/|k|}, \qquad |m_r^{+\prime}(k,x)| \leq C(1+|k|)e^{C/|k|}, \qquad k \in \overline{\mathbf{C}^+} \setminus \{0\},$$

where C is a constant independent of x and k. If we further assume $Q \in L_1^1(\mathbf{R})$, then the Faddeev functions, the Jost solutions, and their x-derivatives are continuous at $k = 0$ as well. In this case, for $k \in \overline{\mathbf{C}^+}$ we have

$$|m_l^+(k,x)| \leq C[1 + \max\{0, -x\}], \qquad |m_r^+(k,x)| \leq C[1 + \max\{0, x\}],$$

$$|m_l^{+\prime}(k,x)| \leq C[1+|k|][1+\max\{0,-x\}], \qquad |m_r^{+\prime}(k,x)| \leq C[1+|k|][1+\max\{0,x\}].$$

Theorem 2.2. *Assume* $P, Q \in L^1(\mathbf{R})$. *For each* $k \in \overline{\mathbf{C}^+}$, *the quantities* $m_l^+(k,x)$, $m_r^+(k,x)$, $m_l^{+\prime}(k,x)$, *and* $m_r^{+\prime}(k,x)$ *are bounded and continuous, and we have*

$$m_l^+(k,x) = \begin{cases} 1+o(1), & x \to +\infty, \\ \frac{1}{T^+(k)}+o(1), & x \to -\infty, \end{cases}$$

$$m_r^+(k,x) = \begin{cases} \frac{1}{T^+(k)}+o(1), & x \to +\infty, \\ 1+o(1), & x \to -\infty, \end{cases}$$

$$m_l^{+\prime}(k,x) = o(1), \qquad m_r^{+\prime}(k,x) = O(1), \qquad x \to +\infty,$$

$$m_l^{+\prime}(k,x) = O(1), \qquad m_r^{+\prime}(k,x) = o(1), \qquad x \to -\infty.$$

Proposition 2.3. *Assume* $P \in L^1(\mathbf{R})$ *and* $Q \in L_1^1(\mathbf{R})$. *Then,* $f_l^+(0,x)$ *and* $f_r^+(0,x)$ *are determined by* $Q(x)$ *alone, and we have*

$$f_l^+(0,x) = f_l^{[0]}(0,x), \qquad f_r^+(0,x) = f_r^{[0]}(0,x). \tag{2.5}$$

Generically $f_l^{[0]}(0,x)$ and $f_r^{[0]}(0,x)$ are linearly independent, but in the exceptional case these two functions are linearly dependent [2–4]. From Proposition 2.3 we see that $Q(x)$ alone determines whether we are in the generic or exceptional case. In the exceptional case, let us define

$$\gamma = \frac{f_l^{[0]}(0,x)}{f_r^{[0]}(0,x)}. \tag{2.6}$$

Then γ is a real, nonzero constant determined by $Q(x)$ alone, and we have $\gamma = f_l^{[0]}(0, -\infty)$ and $\gamma = 1/f_r^{[0]}(0, +\infty)$.

The transformation $k \mapsto -\overline{k}$ in the complex plane is a reflection with respect to the imaginary axis, where the overline denotes complex conjugation. Under this transformation, we have $ik \mapsto \overline{ik}$ and

$$f_l^\pm(-\overline{k},x) = \overline{f_l^\pm(k,x)}, \qquad f_r^\pm(-\overline{k},x) = \overline{f_r^\pm(k,x)}, \qquad k \in \overline{\mathbf{C}^+}. \tag{2.7}$$

Hence, for real k, we get

$$f_l^\pm(-k,x) = \overline{f_l^\pm(k,x)}, \qquad f_r^\pm(-k,x) = \overline{f_r^\pm(k,x)}, \qquad k \in \mathbf{R}.$$

3. SCATTERING COEFFICIENTS

The transmission coefficient $T^+(k)$ given in (2.3) and (2.4) can also be defined in terms of a Wronskian of the Jost solutions of (1.1). However, this is not true for the reflection coefficients. In order to write the reflection coefficients of (1.1) in terms of Wronskians of the Jost solutions, we also need to consider the differential equation

$$-\psi^{-\prime\prime}(k,x) + [-ikP(x) + Q(x)]\psi^-(k,x) = k^2 \psi^-(k,x), \qquad x \in \mathbf{R}. \tag{3.1}$$

Notice that (3.1) is obtained from (1.1) by changing the sign of $P(x)$. Let $f_l^-(k,x)$ and $f_r^-(k,x)$ denote the Jost solutions of (3.1) from the left and from the right, respectively, satisfying the boundary conditions (2.1) and (2.2), respectively. As in (2.3) and (2.4), in terms of the spatial asymptotics of these Jost solutions we can define the transmission coefficient $T^-(k)$, the reflection coefficient from the right $R^-(k)$, and the reflection coefficient from the left $L^-(k)$. In terms of the Jost solutions of (1.1) and (3.1), we have

$$[f_l^\pm(k,x);f_r^\pm(k,x)] = -\frac{2ik}{T^\pm(k)}, \qquad k \in \overline{\mathbf{C}^+}, \tag{3.2}$$

$$[f_l^\pm(k,x);f_r^\mp(-k,x)] = \frac{2ikL^\pm(k)}{T^\pm(k)} = -\frac{2ikR^\mp(-k)}{T^\mp(-k)}, \qquad k \in \mathbf{R}, \tag{3.3}$$

$$[f_r^\pm(k,x);f_l^\mp(-k,x)] = -\frac{2ikR^\pm(k)}{T^\pm(k)} = \frac{2ikL^\mp(-k)}{T^\mp(-k)}, \qquad k \in \mathbf{R}. \tag{3.4}$$

The scattering matrices $\mathbf{S}^+(k)$ associated with (1.1) and $\mathbf{S}^-(k)$ associated with (3.1) are defined as

$$\mathbf{S}^\pm(k) = \begin{bmatrix} T^\pm(k) & R^\pm(k) \\ L^\pm(k) & T^\pm(k) \end{bmatrix}.$$

Let $\mathbf{S}^{[0]}(k)$ denote the scattering matrix associated with (1.2):

$$\mathbf{S}^{[0]}(k) = \begin{bmatrix} T^{[0]}(k) & R^{[0]}(k) \\ L^{[0]}(k) & T^{[0]}(k) \end{bmatrix},$$

where $T^{[0]}(k)$ is the transmission coefficient, and $R^{[0]}(k)$ and $L^{[0]}(k)$ are the reflection coefficients from the right and from the left, respectively. When $Q \in L_1^1(\mathbf{R})$, it is known that $\mathbf{S}^{[0]}(k)$ exists and is continuous for $k \in \mathbf{R}$. From (2.7) and (3.2) we obtain

$$\frac{k}{\overline{T^\pm(-\bar{k})}} = \frac{k}{T^\pm(k)}, \qquad k \in \overline{\mathbf{C}^+}.$$

Although the quantities given on the right-hand sides of (3.2)-(3.4) exist and are continuous in their respective domains, the scattering coefficients $T^\pm(k), R^\pm(k)$, and $L^\pm(k)$ do not necessarily exist or are not necessarily continuous for all $k \in \mathbf{R}$. When they exist, we have

$$\mathbf{S}^\pm(-k) = \overline{\mathbf{S}^\pm(k)}, \qquad k \in \mathbf{R}.$$

Using the Jost solutions of (1.1) and (3.1), we obtain the Wronskian relations

$$[f_l^\pm(k,x);f_l^\mp(-k,x)] = -2ik = -2ik\frac{1 - L^\pm(k)L^\mp(-k)}{T^\pm(k)\,T^\mp(-k)},$$

$$[f_r^\pm(k,x);f_r^\mp(-k,x)] = 2ik = 2ik\frac{1 - R^\pm(k)R^\mp(-k)}{T^\pm(k)\,T^\mp(-k)}.$$

Contrary to the unitarity of $\mathbf{S}^{[0]}(k)$, the matrices $\mathbf{S}^\pm(k)$ are in general not unitary, but instead, for $k \in \mathbf{R}$ except at the singularities of $\mathbf{S}^\pm(k)$ as indicated in Theorem 3.1, we have

$$\mathbf{S}^\pm(k)\,\overline{\mathbf{S}^\mp(k)}^t = \mathbf{I},$$

where $\mathbf{I}$ is the 2×2 unit matrix and the superscript t denotes the matrix transpose. For such k we also have

$$\det \mathbf{S}^\pm(k) = T^\pm(k)^2 - L^\pm(k)\,R^\pm(k) = \frac{T^\pm(k)}{T^\mp(-k)}.$$

Theorem 3.1. *Assume $P, Q \in L^1(\mathbf{R})$. Then:*

(a) *The functions $1/T^\pm(k)$ are analytic in $\mathbf{C}^+$ and continuous in $\overline{\mathbf{C}^+} \setminus \{0\}$, their zeros in $\mathbf{C}^+$ are all isolated and can only accumulate on the real axis. The transmission coefficients $T^\pm(k)$ cannot have any zeros in $\overline{\mathbf{C}^+} \setminus \{0\}$.*

(b) *The quantities $L^\pm(k)/T^\pm(k)$ and $R^\pm(k)/T^\pm(k)$ are continuous on $\mathbf{R} \setminus \{0\}$.*

(c) *For fixed $k_0 \in \mathbf{R} \setminus \{0\}$, the quantities $1/T^+(k_0)$ and $R^+(k_0)/T^+(k_0)$ cannot be zero simultaneously; similarly, $1/T^+(k_0)$ and $L^+(k_0)/T^+(k_0)$ cannot be zero simultaneously. If $1/T^+(k_0) = 0$ for some $k_0 \in \mathbf{R} \setminus \{0\}$, then the quantities $R^\pm(k_0)/T^\pm(k_0)$, $L^\pm(k_0)/T^\pm(k_0)$, $R^\pm(-k_0)/T^\pm(-k_0)$, $L^\pm(-k_0)/T^\pm(-k_0)$ are all nonzero.*

(d) *$R^+(k)$ is continuous for $k \in \mathbf{R} \setminus \{0\}$ if and only if $1/T^+(k)$ does not vanish for $k \in \mathbf{R} \setminus \{0\}$. Similarly, $L^+(k)$ is continuous for $k \in \mathbf{R} \setminus \{0\}$ if and only if $1/T^+(k)$ does not vanish for $k \in \mathbf{R} \setminus \{0\}$.*

(e) *For $k \in \mathbf{R} \setminus \{0\}$, the quantity $T^+(k)$ is continuous if and only if $R^+(k)$ is continuous; equivalently, $T^+(k)$ is continuous if and only if $L^+(k)$ is continuous.*

(f) *If, in addition, $Q \in L^1_1(\mathbf{R})$, then $k/[(k+i)\,T^\pm(k)]$ are continuous and bounded in $\overline{\mathbf{C}^+}$, and $kL^\pm(k)/T^\pm(k)$ and $kR^\pm(k)/T^\pm(k)$ are continuous and bounded on $\mathbf{R}$.*

Proposition 3.2. *Assume $P(x) \leq 0$ and $P, Q \in L^1(\mathbf{R})$. Then $\mathbf{S}^+(k)$ is continuous on $\mathbf{R} \setminus \{0\}$. If we further assume $Q \in L^1_1(\mathbf{R})$, then $\mathbf{S}^+(k)$ is continuous on $\mathbf{R}$.*

Theorem 3.3. *Assume $P, Q \in L^1(\mathbf{R})$. The scattering coefficients satisfy*

$$\frac{1}{|T^\pm(k)|^2} = 1 + \left|\frac{L^\pm(k)}{T^\pm(k)}\right|^2 \mp \int_{-\infty}^{\infty} dx\,|f_l^\pm(k,x)|^2 P(x), \quad k \in \mathbf{R} \setminus \{0\},$$

$$\frac{1}{|T^\pm(k)|^2} = 1 + \left|\frac{R^\pm(k)}{T^\pm(k)}\right|^2 \mp \int_{-\infty}^{\infty} dx\,|f_r^\pm(k,x)|^2 P(x), \quad k \in \mathbf{R} \setminus \{0\}.$$

Hence, if $P(x) \leq 0$, $1/T^+(k)$ cannot have any zeros for $k \in \mathbf{R}$, and we have

$$|T^+(k)|^2 + |L^+(k)|^2 \leq 1, \quad |T^+(k)|^2 + |R^+(k)|^2 \leq 1, \quad k \in \mathbf{R}.$$

If $1/T^+(k)$ does not have any zeros for $k \in \mathbf{R}$ and $P(x) \geq 0$, then we have

$$|T^+(k)|^2 + |L^+(k)|^2 \geq 1, \quad |T^+(k)|^2 + |R^+(k)|^2 \geq 1, \quad k \in \mathbf{R}.$$

Corollary 3.4. *Assume $P, Q \in L^1(\mathbf{R})$ and $P(x) \leq 0$. Then, for $k \in \mathbf{R} \setminus \{0\}$, we have $\frac{1}{|T^+(k)|^2} \geq 1$, and hence $1/T^+(k)$ cannot vanish on $\mathbf{R}$. Moreover, for $k \in \mathbf{R} \setminus \{0\}$, we have $\frac{1}{|T^+(k)|} \geq \frac{1}{|T^-(k)|}$.*

4. ASYMPTOTICS OF SCATTERING COEFFICIENTS

The large-k asymptotics of $\mathbf{S}^{\pm}(k)$ are summarized in the following theorem.

Theorem 4.1. *Assume* $P, Q \in L^1(\mathbf{R})$. *Then*

$$\frac{1}{T^{\pm}(k)} \exp\left(\pm \frac{1}{2} \int_{-\infty}^{\infty} dx\, P(x)\right) = 1 + o(1), \qquad k \to \infty \text{ in } \overline{\mathbf{C}^+},$$

$$\frac{R^{\pm}(k)}{T^{\pm}(k)} = o(1), \qquad \frac{L^{\pm}(k)}{T^{\pm}(k)} = o(1), \qquad k \to \pm\infty.$$

Corollary 4.2. *Assume* $P, Q \in L^1(\mathbf{R})$. *If* $1/T^{\pm}(k)$ *does not vanish for* $k \in \mathbf{R}$, *then its number of zeros in* $\mathbf{C}^+$ *is finite. This occurs, in particular, if* $P(x) \leq 0$.

Next, we analyze the small-k asymptotics of $\mathbf{S}^{\pm}(k)$ in the generic and exceptional cases separately. In the exceptional case, we will see that $\mathbf{S}^+(0)$ is not determined by $Q(x)$ alone and obtain $\mathbf{S}^+(0)$ explicitly in terms of $P(x)$ and $Q(x)$.

Theorem 4.3. *Assume* $P \in L^1(\mathbf{R})$ *and* $Q \in L^1_1(\mathbf{R})$ *and suppose that we are in the generic case. Then* $R^{\pm}(0) = L^{\pm}(0) = -1$, $T^{\pm}(k)$ *vanish linearly as* $k \to 0$ *in* $\overline{\mathbf{C}^+}$, *and*

$$\lim_{k \to 0} \frac{2ik}{T^+(k)} = \lim_{k \to 0} \frac{2ik}{T^-(k)} = \lim_{k \to 0} \frac{2ik}{T^{[0]}(k)}.$$

Furthermore, $\det \mathbf{S}^{\pm}(0) = -1$, *and we have*

$$T^{\pm}(k) = \frac{2ik}{\int_{-\infty}^{\infty} dy\, Q(y) f_l^{[0]}(0,y)} + o(k), \qquad k \to 0 \text{ in } \overline{\mathbf{C}^+}.$$

Theorem 4.4. *In the exceptional case, under the assumptions* $P, Q \in L^1_1(\mathbf{R})$, *we have*

$$T^{\pm}(0) = \frac{2\gamma}{\gamma^2 + 1 \mp \int_{-\infty}^{\infty} dx\, P(x) f_l^{[0]}(0,x)^2}, \tag{4.1}$$

$$L^{\pm}(0) = \frac{\gamma^2 - 1 \pm \int_{-\infty}^{\infty} dx\, P(x) f_l^{[0]}(0,x)^2}{\gamma^2 + 1 \mp \int_{-\infty}^{\infty} dx\, P(x) f_l^{[0]}(0,x)^2},$$

$$R^{\pm}(0) = \frac{1 - \gamma^2 \pm \int_{-\infty}^{\infty} dx\, P(x) f_l^{[0]}(0,x)^2}{\gamma^2 + 1 \mp \int_{-\infty}^{\infty} dx\, P(x) f_l^{[0]}(0,x)^2},$$

where γ *is the constant defined in* (2.6).

Theorem 4.5. *Let* $Q \in L^1_1(\mathbf{R})$, *and assume that* $P \in L^1(\mathbf{R})$ *in the generic case and* $P \in L^1_1(\mathbf{R})$ *in the exceptional case. Then:*

(a) If any one of $1/T^+(k)$, $R^+(k)/T^+(k)$, $L^+(k)/T^+(k)$ is continuous at $k=0$, then all three are continuous at $k=0$. Moreover, either $1/T^+(k)$ and $1/T^-(k)$ are both continuous at $k=0$ or both discontinuous at $k=0$.

(b) In the generic case the six quantities $1/T^\pm(k)$, $R^\pm(k)/T^\pm(k)$, and $L^\pm(k)/T^\pm(k)$ are all discontinuous at $k=0$; in the exceptional case, these quantities are all continuous at $k=0$.

(c) In the exceptional case $1/T^+(k)$ vanishes at $k=0$ if and only if $\int_{-\infty}^{\infty} dx P(x) f_l^{[0]}(0,x)^2$ is equal to γ^2+1, where γ is the constant defined in (2.6). In the generic case, $k/T^+(k)$ has a nonzero limit as $k \to 0$.

(d) Either the three quantities $T^+(k)$, $R^+(k)$, $L^+(k)$ are all continuous on $\mathbf{R}$, or they are all discontinuous on $\mathbf{R}$.

In the special situation when $Q(x)=0$, we have $f_l^{[0]}(0,x) = f_r^{[0]}(0,x) = 1$ and hence $\gamma = 1$. This corresponds to the exceptional case. Using these values in (4.1) we see that

$$\frac{1}{T^+(0)} = 1 - \frac{1}{2}\int_{-\infty}^{\infty} dx P(x).$$

Hence, if $\int_{-\infty}^{\infty} dx P(x) = 2$, no matter how smooth $P(x)$ is, we have $\frac{1}{T^+(0)} = 0$ and $\frac{L^+(0)}{T^+(0)} = \frac{R^+(0)}{T^+(0)} = 1$. In this case $\mathbf{S}^+(0)$ is clearly undefined.

5. BOUND STATES

Although $T^{[0]}(k)$ cannot have any singularities when $k \in \mathbf{R}$, we cannot rule out singularities of $T^+(k)$ when $k \in \mathbf{R}$ unless $P(x) \leq 0$, as we have seen at the end of Section 4. Some other examples of such singularities will be presented in Section 8.

Theorem 5.1. Assume $P, Q \in L^1(\mathbf{R})$. The zeros of $1/T^+(k)$ on the real axis do not correspond to the bound states of (1.1). Each zero of $1/T^+(k)$ in $\mathbf{C}^+$ corresponds to a bound state of (1.1). Conversely, if (1.1) has a bound state at some $k_0 \in \mathbf{C}^+$, it is necessary that $1/T^+(k_0) = 0$.

Let us define

$$P_{\min} = \operatorname*{ess\,inf}_{x \in \mathbf{R}} P(x), \qquad P_{\max} = \operatorname*{ess\,sup}_{x \in \mathbf{R}} P(x), \qquad Q_{\min} = \operatorname*{ess\,inf}_{x \in \mathbf{R}} Q(x), \qquad (5.1)$$

$$\beta^* = P_{\max}/2 + \sqrt{P_{\max}^2/4 - Q_{\min}}.$$

Note that, if $P, Q \in L^1(\mathbf{R})$, it follows that $P_{\max} \geq 0$ with equality holding if and only if $P(x) \leq 0$, that $Q_{\min} \leq 0$ with equality holding if and only if $Q(x) \geq 0$, and that $P_{\min} \leq 0$ with equality holding if and only if $P(x) \geq 0$. Furthermore, $\beta^* \geq P_{\max}$ with equality holding if and only if $Q(x) \geq 0$.

Theorem 5.2. Assume $P, Q \in L^1(\mathbf{R})$, $P(x) \not\equiv 0$, and $P_{\max}$ is finite. Then the zeros of $1/T^+(k)$ for $P_{\max}/2 \leq \operatorname{Im} k < \beta^*$ can only occur on the imaginary axis, and all such zeros are simple. If, in addition, $Q_{\min}$ is finite, then there are no zeros of $1/T^+(k)$ in the region $\{k \in \mathbf{C}^+ : (\operatorname{Im} k)^2 - (\operatorname{Re} k)^2 - (\operatorname{Im} k)P_{\max} \geq -Q_{\min}\}$. Consequently, $1/T^+(k)$ has no zeros in $\mathbf{C}^+$ satisfying $\operatorname{Im} k \geq \beta^*$.

Theorem 5.3. *Assume $Q(x) = 0$ and $P \in L^1(\mathbf{R})$. If $\int_{-\infty}^{\infty} dx\, P(x) > 2$, then (1.1) has at least one bound state at $k = i\beta$ for some positive β. If $\int_{-\infty}^{\infty} dx\, |P(x)| \leq 2$, then $1/T^+(k)$ has no zeros in $\mathbf{C}^+$.*

Theorem 5.4. *Assume that $P \in L^1(\mathbf{R})$ and $Q \in L_1^1(\mathbf{R})$. At the k-values in $\mathbf{C}^+$ satisfying $\int_{-\infty}^{\infty} dx\, |ikP(x) + Q(x)| \leq 2|k|$, there are no zeros of $1/T^+(k)$. Moreover, there are no zeros of $1/T^+(k)$ in $\mathbf{C}^+ \setminus \{i\kappa_1, \cdots, i\kappa_N\}$ satisfying $|T^{[0]}(k)|\, \|P\|_{1,1} < 2e^{-\|Q\|_{1,1}}$, where κ_j correspond to the bound states of (1.2).*

Next we analyze the change in $N(P,Q)$, the number of bound states of (1.1), when we perturb $P(x)$ or $Q(x)$. In the next two theorems we write $T^+(k;P,Q)$ for the transmission coefficient of (1.1) to emphasize its dependence on $P(x)$ and $Q(x)$. By $\|f\|_1$ we denote the norm on $L^1(\mathbf{R})$, i.e. $\|f\|_1 = \int_{-\infty}^{\infty} dx\, |f(x)|$.

Theorem 5.5. *Assume $P_1, P_2 \in L^1(\mathbf{R})$ and $Q_1, Q_2 \in L_1^1(\mathbf{R})$, and suppose $1/T^+(k;P_1,Q_1)$ does not have any real zeros and $Q_1(x)$ is a generic potential. If $\|P_1 - P_2\|_1 + \|Q_1 - Q_2\|_{1,1}$ is small enough, then*

(a) *$1/T^+(k;P_2,Q_2)$ does not have any real zeros.*

(b) *$N(P_2,Q_2) = N(P_1,Q_1)$.*

(c) *If all zeros of $1/T^+(k;P_1,Q_1)$ are simple and purely imaginary, then the zeros of $1/T^+(k;P_2,Q_2)$ are also simple and purely imaginary.*

Theorem 5.6. *Assume $P_1, P_2, Q \in L_1^1(\mathbf{R})$, $1/T^+(k;P_1,Q)$ does not have any real zeros, and $Q(x)$ is an exceptional potential. If $\|P_1 - P_2\|_{1,1}$ is small enough, then*

(a) *$1/T^+(k;P_2,Q)$ does not have any real zeros.*

(b) *$N(P_2,Q) = N(P_1,Q)$.*

(c) *If all zeros of $1/T^+(k;P_1,Q)$ are simple and purely imaginary, then the zeros of $1/T^+(k;P_2,Q)$ are also simple and purely imaginary.*

When $P(x) \leq 0$, we can say more about the bound states of (1.1). From Theorem 5.2 we get the following:

Corollary 5.7. *Assume $P(x) \leq 0$ and $P, Q \in L^1(\mathbf{R})$. Then, the poles of $T^+(k)$ in $\mathbf{C}^+$ are all purely imaginary and simple. In addition, assume that $Q_{\min}$ defined in (5.1) is finite; then there are no zeros of $1/T^+(k)$ in $\mathbf{C}^+$ for $\operatorname{Im} k \geq \sqrt{-Q_{\min}}$. In particular, if $P(x) \leq 0$ and $Q(x) \geq 0$, then $1/T^+(k)$ has no zeros in $\mathbf{C}^+$.*

When $P(x) \leq 0$, under additional assumptions on $P'(x)$, Pivovarchik has shown that [5] the number of bound states of the radial analog of (1.1) is independent of $P(x)$ and that [6–8] the bound states can only occur when k is located on the positive imaginary axis in $\mathbf{C}^+$ and each bound state is simple. The results were actually obtained for a class of abstract operator polynomials with the radial analog of (1.1) as an example. It is possible to obtain Pivovarchik's results on the full line and without assuming the differentiability of $P(x)$.

Theorem 5.8. *Assume $P, Q \in L^1(\mathbf{R})$ and $P(x) \leq 0$. If $N(0,Q) = +\infty$, then we also have $N(P,Q) = +\infty$. If $N(0,Q)$ is finite, then we have $N(P,Q) = N(0,Q)$. Thus, the number of bound states of (1.1) coincides with the number of bound states of (1.2).*

Theorem 5.9. *Assume $N(0,Q)$ is finite and nonzero, $P(x) \leq 0$, and suppose $P, Q \in L^1(\mathbf{R})$ and P_{min} is finite, where P_{min} is the constant defined in (5.1). Let $k = i\kappa_j$ correspond to the bound states of (1.2) for $j = 1, \cdots, \mathcal{N}$. Then, the zeros of $1/T^+(k)$ in $\mathbf{C}^+$ occur at $k = i\beta_j$ satisfying $\beta_* \leq \beta_j \leq \kappa_j$ for $j = 1, \cdots, \mathcal{N}$, where $\beta_* = P_{min}/2 + \sqrt{P_{min}^2/4 + \kappa_1^2}$. In particular, $\beta_1 \geq \beta_*$ and $\beta_{\mathcal{N}} \leq \kappa_{\mathcal{N}}$, with equalities holding if and only if $P(x) \equiv 0$.*

Theorem 5.10. *Assume $P, Q \in L^1(\mathbf{R})$, $P(x) \leq 0$, and $N(0,Q) = +\infty$, and let $\{\mathcal{E}_j\}$ and $\{\mathcal{E}_j^{[0]}\}$ for $j \geq 1$ denote the bound-state energies of (1.1) and (1.2), respectively, ordered such that $\mathcal{E}_j < \mathcal{E}_{j+1}$ and $\mathcal{E}_j^{[0]} < \mathcal{E}_{j+1}^{[0]}$. Then, we have $\mathcal{E}_j^{[0]} \leq \mathcal{E}_j < 0$ for $j \geq 1$, and hence the bound-state energies of (1.1) cannot occur below the lowest bound-state energy of (1.2).*

Recall that the Levinson theorem [9, 10] relates the number of bound states of the Schrödinger equation to the change in the phase of the transmission coefficient. Next we present an analog of the Levinson theorem for (1.1).

Theorem 5.11. *Assume that $P \in L^1(\mathbf{R})$ in the generic case and $P \in L_1^1(\mathbf{R})$ in the exceptional case and that $Q \in L_1^1(\mathbf{R})$, and suppose $1/T^+(k)$ does not have any real zeros. Then the number of bound states of (1.1) is given by*

$$N(P,Q) = \frac{d}{2} + \frac{1}{\pi} \arg T^+(0+),$$

where $d = 0$ in the exceptional case and $d = 1$ in the generic case, and $\arg T^+(k)$ denotes the continuous branch of the argument of $T^+(k)$ normalized so that $\arg T^+(+\infty) = 0$.

6. EIGENVALUE BRANCHES

In this section we consider the bound states of (1.1) when k is on the positive imaginary axis in $\mathbf{C}^+$; in other words, we consider the negative-energy bound states of (1.1). As indicated in Theorem 5.1, (1.1) cannot have any bound states at zero or positive energies. When $P(x) \leq 0$, as seen in Theorem 5.9, the bound states of (1.1) can only occur at negative energies. However, unless $P(x) \leq 0$, there may exist also bound states at complex energies, some examples of which will be given in Section 8.

The negative-energy bound states of (1.1) can be analyzed in terms of the eigenvalue curves of the differential operator $\mathcal{O}_\beta$ given by

$$\mathcal{O}_\beta = -d^2/dx^2 + Q(x) - \beta P(x). \tag{6.1}$$

Let us write (1.1) when $k = i\beta$ as a system of two simultaneous equations:

$$-\psi'' + V(\beta, x)\psi = E(\beta)\psi, \tag{6.2}$$

$$E(\beta) = -\beta^2, \tag{6.3}$$

where β is considered to be a parameter in the potential $V(\beta, x) = Q(x) - \beta P(x)$ of the Schrödinger equation (6.2), and $E(\beta)$ denotes the corresponding energy for each β. Each bound-state energy of (1.2) gives rise to an eigenvalue branch $E(\beta)$ of $\mathcal{O}_\beta$. Note that for each $\beta > 0$, a nontrivial solution of (1.1) belonging to $L^2(\mathbf{R})$ corresponds to an eigenvector of $\mathcal{O}_\beta$

with the eigenvalue E, which we write $E(\beta)$ to emphasize its dependence upon β. Thus, we see that the negative-energy bound states of (1.1) correspond to the eigenvalues $E(\beta)$ that intersect the parabola $E = -\beta^2$ in the (β, E)-plane. Assume that the eigenvalue branch $E(\beta)$ and the parabola $E = -\beta^2$ intersect at $(\beta_0, -\beta_0^2)$. Then, β_0 corresponds to a bound state of (1.1) with negative energy. Conversely, any negative bound-state energy of (1.1) can be identified with a simultaneous solution of (6.2) and (6.3). Hence, we can analyze the negative-energy bound states of (1.1) by analyzing the eigenvalue curves of $\mathcal{O}_\beta$ and their intersections with the parabola $E = -\beta^2$.

Associated with the eigenvalue $E_0(\beta)$ there exists [11] a real-valued, analytic eigenvector $\psi(\beta,x)$. Near $\beta = \beta_0$ we have the convergent expansions

$$E_0(\beta) = \sum_{n=0}^{\infty} a_n (\beta - \beta_0)^n, \qquad \psi(\beta,x) = \sum_{n=0}^{\infty} \psi_n(x)(\beta - \beta_0)^n, \qquad (6.4)$$

with $\psi_n \in L^2(\mathbf{R})$ for $n \geq 0$. One may choose $\psi_0(x) = f_l^+(i\beta_0, x)$. We can recursively determine a_n and $\psi_n(x)$. In fact, we have

$$a_0 = E_0(\beta_0), \qquad a_1 = -\frac{1}{\|\psi_0\|_2^2} \int_{-\infty}^{\infty} dx\, P(x)\, \psi_0(x)^2, \qquad (6.5)$$

$$a_n = -\frac{1}{\|\psi_0\|_2^2} \int_{-\infty}^{\infty} dx\, \psi_0(x) \left(P(x)\psi_{n-1}(x) + \sum_{j=1}^{n-1} a_j \psi_{n-j}(x) \right), \qquad n \geq 2.$$

For the lowest eigenvalue, one obtains

$$a_2 = -\frac{1}{\|\psi_0\|_2^2} \int_{-\infty}^{\infty} \frac{dx}{\psi_0(x)^2} \left(\int_{-\infty}^{x} dt\, \psi_0(t)^2 [P(t) + a_1] \right)^2.$$

Theorem 6.1. *Suppose $P,Q \in L^1(\mathbf{R})$. Then, the lowest eigenvalue branch satisfies $E''(\beta) \leq 0$ for $\beta > 0$ with equality holding if and only if $P(x) \equiv 0$.*

An eigenvalue curve $E(\beta)$ may cut the parabola $E = -\beta^2$ at two or more points, and if this happens each intersection gives rise to a negative-energy bound state of (1.1). Moreover, $E(\beta) + \beta^2$ may have double or higher-order zeros; then, the order of the zero of $E(\beta) + \beta^2$ is the same as the multiplicity of the corresponding bound state.

Theorem 6.2. *Suppose $P,Q \in L^1(\mathbf{R})$. Then, $1/T^+(i\beta)$ has a zero of order m at some positive β_0 if and only if the function $E_0(\beta) + \beta^2$ has a zero of order m at β_0, where $E_0(\beta)$ denotes the unique eigenvalue branch of the operator $\mathcal{O}_\beta$ satisfying $E_0(\beta) \to -\beta_0^2$ as $\beta \to \beta_0$. Moreover, if (1.2) has $N(0,Q)$ bound states, then (1.1) has at least $N(0,Q)$ bound states with negative energies.*

If β_0 corresponds to a zero of $E_0(\beta) + \beta^2$ of order m for some $m \geq 1$, then the coefficients a_n in (6.4) are determined for $n = 0, 1, \cdots, m-1$ by expanding $E_0(\beta) + \beta^2$ about β_0. Thus, for $m = 1$ we get $a_0 = -\beta_0^2$; for $m = 2$ we have $a_0 = -\beta_0^2$, $a_1 = -2\beta_0$; for $m = 3$ we get $a_0 = -\beta_0^2$, $a_1 = -2\beta_0$, $a_2 = -1$; for $m \geq 4$ we get $a_0 = -\beta_0^2$, $a_1 = -2\beta_0$, $a_2 = -1$, and $a_3 = \cdots = a_{m-1} = 0$.

If $P(x) \leq 0$, from (6.5) we see that $a_1 \geq 0$ for any positive β_0 with equality holding if and only if $P(x) \equiv 0$. Thus, when $P(x) \leq 0$ we have $E'(\beta) \geq 0$, and as a result each eigenvalue branch $E_j(\beta)$ is a nondecreasing function of β. Therefore, for $\beta > 0$, the graph of each eigenvalue branch $E_j(\beta)$ intersects the parabola $E = -\beta^2$ at exactly one point, say $(\beta_j, -\beta_j^2)$, and each $E_j(\beta)$ gives rise to exactly one solution of (6.3). Hence, there is a one-to-one correspondence between the bound states of (1.1) and the bound states of (1.2), and $E_j(\beta)$ satisfies $E_j(0) = -\kappa_j^2$. The number $N(P,Q)$ is equal to the number of intersections of the parabola in (6.3) with the eigenvalue branches $E_j(\beta)$ for $j \geq 1$. Since each of the $N(0,Q)$ branches is responsible for exactly one intersection, we conclude that $N(P,Q) = N(0,Q)$. Note that if $Q \in L^1(\mathbf{R})$ but $Q \notin L_1^1(\mathbf{R})$, it is possible that $N(0,Q) = +\infty$, but then we also have $N(P,Q) = +\infty$.

7. ZEROS OF JOST SOLUTIONS

In this section we study the zeros of the Jost solutions of (1.1) for fixed k and analyze the number of such zeros in relation to the bound states of (1.1) and (1.2).

Concerning the zeros of the Jost solutions of (1.2) when k is on the positive imaginary axis in $\mathbf{C}^+$, the following is already known [12, 13]:

Proposition 7.1. *Suppose $Q \in L^1(\mathbf{R})$ and $\beta > 0$. Then the number of zeros of $f_l^{[0]}(i\beta,x)$ is equal to the number of bound states of (1.2) with energies contained in the interval $(-\infty, -\beta^2)$. Suppose further that $Q \in L_1^1(\mathbf{R})$. Then, the number of zeros of $f_l^{[0]}(0,x)$ is equal to $N(0,Q)$.*

The next proposition concerns the zeros of the Jost solutions when k lies off the positive imaginary axis in $\mathbf{C}^+$.

Proposition 7.2. *Assume $P,Q \in L^1(\mathbf{R})$ and $k \in \mathbf{C}^+$. If $P(x) \leq 2\,\mathrm{Im}\,k$, then $f_l^+(k,x)$ and $f_r^+(k,x)$ cannot vanish for any $x \in \mathbf{R}$.*

When k is confined to the positive imaginary axis, one can analyze the zeros of the Jost solutions of (1.1) by using the methods [13, 14] developed for (1.2). At a fixed nonnegative β, one can show that $f_l^+(i\beta,x)$ and $f_r^+(i\beta,x)$ have the same number of zeros. These zeros are simple, and they are interlaced when $f_l^+(i\beta,x)$ and $f_r^+(i\beta,x)$ are linearly independent.

Theorem 7.3. *Suppose $P \in L^1(\mathbf{R})$ and $Q \in L_1^1(\mathbf{R})$, and assume that (1.1) has a bound state of multiplicity m at $k = i\beta_0$ for some positive β_0. Then, the number of zeros of $f_l^+(i\beta,x)$ behaves in the following manner as β is increased from $\beta_0 - \epsilon$ to $\beta_0 + \epsilon$ for sufficiently small and positive ϵ: If m is even, then the number of zeros is either constant throughout the interval $(\beta_0 - \epsilon, \beta_0 + \epsilon)$ or it is constant in $(\beta_0 - \epsilon, \beta_0) \cup (\beta_0, \beta_0 + \epsilon)$ but one less at β_0. If m is odd, then the number of zeros either increases or decreases by one as β crosses β_0. The number of zeros of $f_l^+(i\beta,x)$ can only change at β-values corresponding to the bound states of (1.1).*

When $P(x) \leq 0$, the number of zeros of the Jost solutions of (1.1) when k is on the positive imaginary axis in $\mathbf{C}^+$ is related to the bound states in a simple manner.

Theorem 7.4. *Assume that $P \in L^1(\mathbf{R})$, $Q \in L_1^1(\mathbf{R})$, and $P(x) \leq 0$. Then, for each $\beta \geq 0$, the functions $f_l^+(i\beta,x)$ and $f_r^+(i\beta,x)$ have the same number of zeros, and this number is equal to the number of bound states of (1.1) with energies contained in the interval $(-\infty, -\beta^2)$.*

For further results on the zeros of the Jost solutions of (1.1), we refer the reader to [1].

8. EXAMPLES

In this section we present explicitly solved examples illustrating the theory presented in the earlier sections. The numerical values in these examples were obtained by using the mathematical software Maple.

Our first example shows that if we relax the condition $P \in L^1(\mathbf{R})$, the scattering matrix $\mathbf{S}^+(k)$ may not exist at all.

Example 8.1. *Let $P(x)$ and $Q(x)$ have support in $(0,+\infty)$ and be given by*

$$Q(x) = \theta(x)\frac{2}{(1+x)^2}, \qquad P(x) = \theta(x)\frac{2}{1+x}, \tag{8.1}$$

where $\theta(x)$ is the Heaviside function; thus $P \notin L^1(\mathbf{R})$. Two linearly independent solutions of (1.1) are given by

$$\psi_1^+(k,x) = \theta(x)\frac{e^{-ikx}}{1+x} + \theta(-x)\left[e^{-ikx} - \frac{\sin kx}{k}\right],$$

$$\psi_2^+(k,x) = \theta(x)\left[x+1+\frac{i}{k}-\frac{1}{2k^2(1+x)}\right]e^{ikx} + \theta(-x)F(k,x),$$

where we have defined

$$F(k,x) = \left(\frac{1}{k}+\frac{1}{2k^3}\right)\sin kx + \left(1+\frac{i}{k}-\frac{1}{2k^2}\right)e^{ikx}.$$

Note that $\psi_1^+(k,x) \to 0$ and $\psi_2^+(k,x) = O(x)$ as $x \to +\infty$; hence, we cannot form a solution of (1.1) asymptotic to e^{ikx} as $x \to +\infty$. Although we can form a linear combination of $\psi_1^+(k,x)$ and $\psi_2^+(k,x)$ that is asymptotic to e^{-ikx} as $x \to -\infty$, the resulting function is not bounded as $x \to +\infty$. Thus, there are no scattering solutions and no scattering matrices corresponding to the potentials given in (8.1). Note that the scattering matrix $\mathbf{S}^{[0]}(k)$ corresponding to (1.2) with $Q(x)$ given in (8.1) is well defined, and we have

$$T^{[0]}(k) = \frac{2k^2}{2k^2+2ik-1}, \qquad L^{[0]}(k) = -R^{[0]}(k) = \frac{1}{2k^2+2ik-1}.$$

Contrary to the case $P(x) = 0$, the scattering matrix $\mathbf{S}^+(k)$ is in general not determined if one of the reflection coefficients and the bound state energies are known, as illustrated by the following example.

Example 8.2. *Assume $Q \in L^1_1(\mathbf{R})$ is an exceptional potential without bound states. Let*

$$P(x) = 2\frac{f_l^{[0]\prime}(0,x)}{f_l^{[0]}(0,x)}, \tag{8.2}$$

where $f_l^{[0]}(0,x)$ is the zero-energy Jost solution of (1.2); note that $f_l^{[0]}(0,x)$ is uniquely determined by $Q(x)$ alone and $P(x)$ given in (8.2) necessarily belongs to $L^1(\mathbf{R})$. The corresponding scattering matrices $\mathbf{S}^\pm(k)$ can be evaluated explicitly, and we have

$$T^+(k) = \frac{1}{\gamma}, \quad L^+(k) = 0, \quad R^+(k) = 2\int_{-\infty}^{\infty} dy\,\frac{f_l^{[0]\prime}(0,y)}{f_l^{[0]}(0,y)^3}e^{-2iky}, \tag{8.3}$$

$$T^-(k) = \gamma, \quad R^-(k) = 0, \quad L^-(k) = -2\gamma^2 \int_{-\infty}^{\infty} dy \frac{f_l^{[0]\prime}(0,y)}{f_l^{[0]}(0,y)^3} e^{2iky},$$

where γ is the constant defined in (2.6). As seen from (8.3), $T^+(k)$ and $L^+(k)$ cannot determine $R^+(k)$, and there are infinitely many $R^+(k)$ corresponding to these two scattering coefficients. Therefore, the coefficients $P(x)$ and $Q(x)$ cannot in general be determined from the scattering data consisting of the transmission coefficient and of only one of the reflection coefficients. Note also that $\mathbf{S}^+(k)$ is in general not determined by only one or two of its entries.

In the next example we show that $1/T^+(k)$ may have zeros on $\mathbf{R}$ or off the positive imaginary axis in $\mathbf{C}^+$. We also consider the zeros of $1/T^+(k)$ on the positive imaginary axis and illustrate the fact that unless $P(x) \le 0$, the number of negative-energy bound states of (1.1) may be more than the number of bound states of (1.2).

Example 8.3. For real parameters a and b, let

$$P(x) = \begin{cases} b, & x \in (0,1), \\ 0, & \text{elsewhere}, \end{cases} \qquad Q(x) = \begin{cases} a, & x \in (0,1), \\ 0, & \text{elsewhere}. \end{cases} \tag{8.4}$$

The resulting transmission coefficient can be obtained explicitly and we have

$$\frac{1}{T^+(k)} = e^{ik}\left[\cos s + \frac{k^2+s^2}{2iks}\sin s\right],$$

where we have defined $s = \sqrt{k^2 - ibk - a}$. Let us use an overline on the last digit to indicate a roundoff. When $a = -9.273\overline{8}$ and $b = 3.970\overline{8}$, we find simple zeros of $1/T^+(k)$ at $k = \pm 1$. When $a = 0$ we have

$$\frac{1}{T^+(0)} = 1 - \frac{b}{2}, \qquad \frac{L^+(0)}{T^+(0)} = \frac{R^+(0)}{T^+(0)} = \frac{b}{2},$$

and hence $1/T^+(0) = 0$ if $b = 2$. When $b = 0$ and $a < 0$, we have a square-well potential, and in this case (1.1) has $\mathcal{N}$ bound states such that

$$(\mathcal{N}-1)\pi < \sqrt{-a} \le \mathcal{N}\pi. \tag{8.5}$$

When $a = -100$ and $b = 0$, from (8.5) we see that we get four bound states of (1.1) at $k = i\kappa_j$ with

$$\kappa_1 = 1.9\overline{3}, \quad \kappa_2 = 6.4\overline{1}, \quad \kappa_3 = 8.5\overline{5}, \quad \kappa_4 = 9.6\overline{5}. \tag{8.6}$$

When $a = -100$ and $b = -10$, there are four bound states at $k = i\beta_j$, where

$$\beta_1 = 0.7\overline{6}, \quad \beta_2 = 3.5\overline{5}, \quad \beta_3 = 5.1\overline{1}, \quad \beta_4 = 5.9\overline{2}. \tag{8.7}$$

When $a = -100$ and $b = -100$, there are still four bound states with

$$\beta_1 = 0.1\overline{1}, \quad \beta_2 = 0.5\overline{8}, \quad \beta_3 = 0.8\overline{6}, \quad \beta_4 = 0.9\overline{7}. \tag{8.8}$$

From (8.6)-(8.8), we see that as b becomes more negative the bound-state energies are pushed toward zero. Now let us see what happens when $b > 0$. By Theorem 5.3, if $a = 0$ and $b > 2$, we must have a bound state at $k = i\beta$ for some positive β. Letting $a = 0$ and $b = 21/10$, we obtain a bound state at $k = 0.1\overline{5}i$; from the plot of $T^+(k)$ for $k \in (0, +\infty)$, we see that this is the only bound state. Note that when $b > 0$ we cannot exclude the possibility of bound states with k off the imaginary axis in $\mathbf{C}^+$. For example, when $a = -93/10$ and $b = 4$, we find bound states at $k = \pm 0.97\overline{64} + 0.02\overline{33}i$; in this case there are no other bound states. Choosing $a = 0$ and $b = 10$, we obtain over 200 bound states, only three of which correspond to the k-values on the positive imaginary axis with $k = i\beta_j$, where

$$\beta_1 = 2.1\overline{4}, \quad \beta_2 = 5.9\overline{6}, \quad \beta_3 = 9.2\overline{7}. \tag{8.9}$$

Choosing $a = 0$ and $b = 100$, when k is on the positive imaginary axis we obtain thirty-one bound states with

$$\begin{aligned}
&\beta_1 = 0.1\overline{0}, \quad &&\beta_2 = 0.4\overline{1}, \quad &&\beta_3 = 0.9\overline{3}, \quad &&\beta_4 = 1.6\overline{7}, \\
&\beta_5 = 2.6\overline{4}, \quad &&\beta_6 = 3.8\overline{5}, \quad &&\beta_7 = 5.3\overline{3}, \quad &&\beta_8 = 7.0\overline{9}, \\
&\beta_9 = 9.1\overline{9}, \quad &&\beta_{10} = 11.6\overline{9}, \quad &&\beta_{11} = 14.6\overline{3}, \quad &&\beta_{12} = 18.2\overline{0}, \\
&\beta_{13} = 22.6\overline{1}, \quad &&\beta_{14} = 28.4\overline{3}, \quad &&\beta_{15} = 37.6\overline{3}, \quad &&\beta_{16} = 60.4\overline{1}, \\
&\beta_{17} = 69.6\overline{9}, \quad &&\beta_{18} = 75.6\overline{0}, \quad &&\beta_{19} = 80.1\overline{1}, \quad &&\beta_{20} = 83.7\overline{7}, \\
&\beta_{21} = 86.8\overline{3}, \quad &&\beta_{22} = 89.4\overline{2}, \quad &&\beta_{23} = 91.6\overline{3}, \quad &&\beta_{24} = 93.5\overline{2}, \\
&\beta_{25} = 95.1\overline{2}, \quad &&\beta_{26} = 96.4\overline{6}, \quad &&\beta_{27} = 97.5\overline{7}, \quad &&\beta_{28} = 98.4\overline{6}, \\
&\beta_{29} = 99.1\overline{4}, \quad &&\beta_{30} = 99.6\overline{2}, \quad &&\beta_{31} = 99.9\overline{1},
\end{aligned}$$

and there are also many more bound states corresponding the k-values off the imaginary axis in $\mathbf{C}^+$. Note that the bound states may occur even when $a > 0$ and $b > 0$. For example, when $a = 1$ and $b = 10$, we obtain many bound states, four of which correspond to the k-values on the positive imaginary axis with $k = i\beta_j$, where

$$\beta_1 = 0.1\overline{3}, \quad \beta_2 = 2.5\overline{0}, \quad \beta_3 = 5.6\overline{3}, \quad \beta_4 = 9.1\overline{6}.$$

Next we present an example where $T^+(k)$ has a double pole on the positive imaginary axis in $\mathbf{C}^+$.

Example 8.4. Let

$$P(x) = \frac{4b\epsilon c e^{-2\epsilon|x|}}{1 + ce^{-2\epsilon|x|}}, \quad Q(x) = \frac{4\epsilon^2 c e^{-2\epsilon|x|}[-3b - 2 + b^2 c e^{-2\epsilon|x|}]}{(1 + ce^{-2\epsilon|x|})^2},$$

with $\epsilon > 0$, $c \in (-1, -5 + \sqrt{20})$, and $b \in \mathbf{R}$. The transmission coefficient can be evaluated explicitly, and we have

$$T^+(k) = \frac{k(k + i\epsilon)^2 (1 + c)^{2b}}{(k - k_0)(k - k_+)(k - k_-)}, \tag{8.10}$$

where we have defined

$$k_0 = i\frac{\epsilon}{1+c}[-1 + c + 2bc],$$

$$k_\pm = \frac{i\epsilon}{2(1+c)}\left[(-1 + c + 4bc) \pm \sqrt{1 + c^2 + 14c + 16bc}\right].$$

When $b = -(c^2 + 14c + 1)/(16c)$, we get $k_\pm = -i\epsilon(c^2 + 10c + 5)/[8(1+c)]$, and hence $T^+(k)$ given in (8.10) has a double pole on the positive imaginary axis. When $b = (1-c)/(4c)$, note that k_0 is located on the negative imaginary axis and that k_+ and k_- are symmetrically located on the real axis; thus, in this case $T^+(k)$ has poles on the real axis. When $b = -(5 + \sqrt{5})/10$ and $c = -5 + \sqrt{20}$, we get $k_+ = k_- = 0$, and hence $T^+(k)$ has a simple pole at $k = 0$.

The next example concerns the zeros of $f_l^+(i\beta,x)$ when $\beta > 0$.

Example 8.5. *Consider the same $P(x)$ and $Q(x)$ as studied in Example 8.3. For the various specific values of a and b listed in that example, the zeros of $1/T^+(i\beta)$ are all simple. Hence, as Theorem 7.3 states, we expect the number of zeros of $f_l^+(i\beta,x)$ and $f_r^+(i\beta,x)$ to change by one at the zeros of $1/T^+(i\beta)$ as β varies in $(0,+\infty)$. For example, when $a = 0$ and $b = 10$, using β_1, β_2, and β_3 given in (8.9), one finds that $f_l^+(i\beta,x)$ has one zero for $\beta \in [0,\beta_1)$, two zeros for $\beta \in (\beta_1,\beta_2)$, one zero for $\beta \in (\beta_2,\beta_3)$, and no zeros for $\beta \in (\beta_3,+\infty)$. When $a = 0$ and $b = 21/10$, one finds that $f_l^+(i\beta,x)$ has one zero for $\beta \in [0,\beta_1)$ and no zeros for $\beta \in (\beta_1,+\infty)$, where $\beta_1 = 0.1\overline{5}$. When $a = 0$ and $b = 100$, one finds that $f_l^+(i\beta,x)$ has no zeros for $\beta \in (\beta_{31},+\infty)$, one zero for $\beta \in [0,\beta_1)$ and one zero for $\beta \in (\beta_{30},\beta_{31})$, j zeros for $\beta \in (\beta_{j-1},\beta_j)$ and j zeros for $\beta \in (\beta_{31-j},\beta_{32-j})$ with $j = 2,3,\cdots,15$, and sixteen zeros for $\beta \in (\beta_{16},\beta_{17})$. On the other hand, for $a = 19.85\overline{2}$ and $b = 10$, there is one negative-energy bound state of (1.1) of multiplicity two occurring at $k = i\beta_1$ with $\beta_1 = 4.72\overline{4}$; in this case $f_l^+(i\beta,x)$ has no zeros for any $\beta \geq 0$.*

9. INVERSE PROBLEMS

Inverse problems related to (1.1) consist of the recovery of $P(x)$ or $Q(x)$ from an appropriate set of scattering data. One such inverse problem is to recover both $P(x)$ and $Q(x)$. In the radial case, when there are no bound states, Jaulent and Jean presented [15] an inversion method when $Q(x)$ is real and $P(x)$ is imaginary. They also extended their method to solve the inverse problem on the full line for real $Q(x)$ and imaginary $P(x)$ [16,17]. By this method, using the scattering data $\{R^+(k), R^-(k)\}$, one solves a pair of two coupled Marchenko integral equations, and these solutions are used in a first-order ordinary differential equation whose solution leads to $P(x)$. Jaulent [18] also extended this method to the case when $P(x)$ is real although complete details and proofs were not given. When $P(x)$ is purely imaginary and $\int_{-\infty}^{\infty} dz P(z) = 0$, Sattinger and Szmigielski [19] showed that one can simplify the method of Jaulent and Jean and recover $P(x)$ by solving an algebraic equation rather than a differential equation.

When $P(x)$ is purely imaginary, the methods available for self-adjoint differential operators can be employed to analyze the inverse scattering problem for (1.1); in this case the scattering matrices $\mathbf{S}^\pm(k)$ are unitary and the reflection coefficients cannot exceed one in absolute value. However, when $P(x)$ is real, the differential operator pertaining to (1.1) is no longer self-adjoint and the scattering matrices $\mathbf{S}^\pm(k)$ are no longer unitary. Consequently, the analysis of the direct and inverse scattering problems with real $P(x)$ is more difficult than with imaginary $P(x)$. As we have seen, for example, the non-self-adjointness of the differential operator in (1.1) may lead to singularities of the transmission coefficient on $\mathbf{R}$, and the reflection coefficients may not be bounded by one in absolute value and hence the Marchenko integral operators are in general not contractive. When $P(x) \leq 0$, some of the usual properties of the one-dimensional Schrödinger equation given in (1.2), such as the simplicity of the poles of the transmission coefficient, the confinement of these poles to the positive imaginary axis in $\mathbf{C}^+$, and the absence of singularities of the transmission coefficient at real-k values are still valid for (1.1). However, in the available inversion methods to recover $P(x)$ and $Q(x)$ one needs the scattering data associated with both (1.1) and (3.1). Hence, even when we study the inversion problem for absorptive media where one requires $P(x) \leq 0$, one may have to deal with bound-state scattering data for (3.1) which may involve complex-energy or degenerate bound states, unless the absorption is sufficiently weak.

Finally, let us mention the study by Kaup [20] on the direct and inverse scattering problem for

$$\phi'' + \left[k^2 + \frac{1}{4\beta^2}\right]\phi = [ikP(x) + Q(x)]\phi, \tag{9.1}$$

where β is a nonzero constant and $P, Q \in L_1^1(\mathbf{R})$. Under additional restrictions on $P(x)$, Tsutsumi [21] analyzed the scattering problem for (9.1) with $\beta = \frac{1}{2}$ by using a 2×2 matrix analog of (1.1) with k replaced by $\sqrt{k^2 + 1}$. Sattinger and Szmigielski [22] studied the direct and inverse scattering problem for (9.1) when $\beta = \frac{1}{2}$, $\int_{-\infty}^{\infty} dx P(x) = 0$, and $P(x)$ and $Q(x)$ are in the Schwartz space. The inverse scattering problem for (9.1) is used to solve an initial-value problem for a coupled system of two nonlinear evolution equations, and that inverse problem is analyzed by studying an associated Riemann–Hilbert problem [22]. We should emphasize that both the direct and inverse scattering problems for (1.1) are somewhat different from those for (9.1). The direct problem for (9.1) is analyzed in the complex-z plane using Kaup's transformations $k = \frac{1}{4}[2z - 1/(2\beta^2 z)]$ and $E = \sqrt{k^2 + 1} = \frac{1}{4}[2z + 1/(2\beta^2 z)]$.

ACKNOWLEDGMENT

The research leading to this article was supported in part by the National Science Foundation under grant DMS-9501053, and by CNR, MURST, and a University of Cagliari coordinated research grant.

REFERENCES

1. T. Aktosun, M. Klaus, and C. van der Mee, *Wave scattering with absorption*, J. Math. Phys., to appear.
2. L. D. Faddeev, *Properties of the S-matrix of the one-dimensional Schrödinger equation*, Amer. Math. Soc. Transl. **2**, 139–166 (1964) [Trudy Mat. Inst. Steklova **73**, 314–336 (1964) (Russian)].
3. P. Deift and E. Trubowitz, *Inverse scattering on the line*, Comm. Pure Appl. Math. **32**, 121–251 (1979).
4. K. Chadan and P. C. Sabatier, *Inverse problems in quantum scattering theory*, 2nd ed., Springer-Verlag, New York (1989).
5. V. N. Pivovarchik, *On the discrete spectrum of a boundary value problem*, Sib. Math. J. **31**, 853–856 (1990) [Sibirsk. Mat. Zh. **31**, 182–186 (1990) (Russian)].
6. V. N. Pivovarchik, *Eigenvalues of a certain quadratic pencil of operators*, Funct. Anal. Appl. **23**, 70–72 (1989) [Funkt. Anal. i ego Priloz. **23**, 80–81 (1989) (Russian)].
7. V. N. Pivovarchik, *On the total algebraic multiplicity of the spectrum in the right-half plane for a quadratic operator pencil*, St. Petersburg Math. J. **3**, 447–454 (1992) [Algebra i Analiz **3**, 223–230 (1991) (Russian)].
8. V. N. Pivovarchik, *On positive spectra of one class of polynomial operator pencils*, Integr. Eqs. Oper. Th. **19**, 314–326 (1994).
9. N. Levinson, *On the uniqueness of the potential in a Schrödinger equation for a given asymptotic phase*, Danske Vid. Selsk. Mat.-Fys. Medd. **25**, 1–29 (1949).
10. R. G. Newton, *Inverse scattering. I. One dimension*, J. Math. Phys. **21**, 493–505 (1980).
11. T. Kato, *Perturbation theory for linear operators*, 2nd ed., Springer, Berlin, 1976.
12. M. Reed and B. Simon, *Methods of modern mathematical physics. IV. Analysis of operators*, Academic Press, New York, 1978.
13. J. Weidmann, *Spectral theory of ordinary differential operators*, Lecture Notes in Mathematics, 1258, Springer, Berlin, 1987.
14. E. A. Coddington and N. Levinson, *Theory of ordinary differential equations*, McGraw-Hill, New York, 1955.

15. M. Jaulent and C. Jean, *The inverse s-wave scattering problem for a class of potentials depending on energy*, Comm. Math. Phys. **28**, 177–220 (1972).
16. M. Jaulent and C. Jean, *The inverse problem for the one-dimensional Schrödinger equation with an energy-dependent potential. I*, Ann. Inst. Henri Poincaré A **25**, 105–118 (1976).
17. M. Jaulent and C. Jean, *The inverse problem for the one-dimensional Schrödinger equation with an energy-dependent potential. II*, Ann. Inst. Henri Poincaré A **25**, 119–137 (1976).
18. M. Jaulent, *Inverse scattering problems in absorbing media*, J. Math. Phys. **17**, 1351–1360 (1976).
19. D. H. Sattinger and J. Szmigielski, *Energy dependent scattering theory*, Differ. Integral Eqs. **8**, 945–959 (1995).
20. J. Kaup, *A higher-order water-wave equation and the method for solving it*, Progr. Theor. Phys. **54**, 396–408 (1975).
21. M. Tsutsumi, *On the inverse scattering problem for the one-dimensional Schrödinger equation with an energy dependent potential*, J. Math. Anal. Appl. **83**, 316–350 (1981).
22. D. H. Sattinger and J. Szmigielski, *A Riemann–Hilbert problem for an energy dependent Schrödinger operator*, Inverse Problems **12**, 1003–1025 (1996).

RESOLVENT ESTIMATES FOR SCHRÖDINGER-TYPE AND MAXWELL EQUATIONS WITH APPLICATIONS

Matania Ben-Artzi* and Jonathan Nemirovsky

Institute of Mathematics
Hebrew University
Jerusalem 91904, Israel

It is the purpose of this paper to present some recent results concerning properties of solutions to time dependent Schrödinger-type equations of the form,

$$\frac{1}{i}\frac{du}{dt} = (H_0 + V)u,$$
$$u(0) = u_0 \in L^2(\mathbb{R}^n). \tag{0.1}$$

(where the perturbation V is a real potential $V(x)$), as well as Maxwell's equations for the case of layered anisotropic media, i.e.,

$$-i\frac{\partial}{\partial t}\mathbf{E}(\mathbf{x},t) = -i\varepsilon(x_1)^{-1} \cdot \mathbf{curl}(\mathbf{H}(\mathbf{x},t)) \quad -i\frac{\partial}{\partial t}\mathbf{H}(\mathbf{x},t) = i\mu(x_1)^{-1} \cdot \mathbf{curl}(\mathbf{E}(\mathbf{x},t)) \tag{0.2}$$

$$\mathbf{div}(\varepsilon(x_1) \cdot \mathbf{E}(\mathbf{x},t)) = 0 \quad \mathbf{div}(\mu(x_1) \cdot \mathbf{H}(\mathbf{x},t)) = 0, \tag{0.3}$$

where $\varepsilon(\cdot), \mu(\cdot)$ are are symmetric and positive definite measurable (3×3)-real valued matrices defined on $\mathbb{R}^1$, that are uniformly bounded from above and from below as follows,

$$0 < \varepsilon_m \cdot I \leq \varepsilon(x_1) \leq \varepsilon_M \cdot I \quad \text{and} \quad 0 < \mu_m \cdot I \leq \mu(x_1) \leq \mu_M \cdot I \quad (x_1 \in \mathbb{R}^1). \tag{0.4}$$

In this case $\begin{bmatrix}\mathbf{E}\\\mathbf{H}\end{bmatrix}$ is considered as an element in a "weighted" $(L^2(\mathbb{R}^3))^6$ space $\mathbb{H}$, equipped with the scalar product,

$$\left\langle \begin{bmatrix}\mathbf{E}_1\\\mathbf{H}_1\end{bmatrix}, \begin{bmatrix}\mathbf{E}_2\\\mathbf{H}_2\end{bmatrix} \right\rangle = \int_{\mathbb{R}^3} \mathbf{E}_1(\mathbf{x}) \cdot \varepsilon(x_1)\overline{\mathbf{E}_2(\mathbf{x})}d\mathbf{x} + \int_{\mathbb{R}^3} \mathbf{H}_1(\mathbf{x}) \cdot \mu(x_1)\overline{\mathbf{H}_2(\mathbf{x})}d\mathbf{x}.$$

---

*Partially supported by the Fund for Basic Research, Israel Academy of Sciences

Spectral and Scattering Theory, edited by Ramm,
Plenum Press, New York, 1998

1. RESOLVENT ESTIMATES, LOCAL SMOOTHING AND LONG-TIME DECAY FOR SCHRÖDINGER-TYPE OPERATORS H_0 AND $H_0 + V$

For the operator H_0, we take one of two cases.

A. $H_0 = f(-\Delta)$, where $f(\theta)$, $\theta \in \overline{\mathbb{R}_+} = [0, \infty)$, is a real-valued non-negative continuously differentiable function satisfying the following assumptions.

(1) The derivative $f'(\theta)$ is positive and uniformly Hölder continuous in $\overline{\mathbb{R}_+}$, i.e., for some $C > 0$, $0 < \alpha \leq 1$ and all $\theta_1, \theta_2 \in \overline{\mathbb{R}_+}$, $|\theta_1 - \theta_2| \leq 1$,

$$|f'(\theta_1) - f'(\theta_2)| \leq C|\theta_1 - \theta_2|^\alpha.$$

(2) $\lim_{\theta \to +\infty} f(\theta) = +\infty$.

Assume $f(0) = 0$, without loss of generality.

In particular, the free "relativistic Schrödinger operator" $H_0 = (I - \Delta)^{1/2}$ (see [11, 18]) is included, by taking $f(\theta) = (1+\theta)^{1/2}$. It is closely related, of course, to the Klein–Gordon equation [4] and to the Dirac operator [2], and our results apply to these cases as well.

B. $H_0 = -\Delta - x_1$, $x = (x_1, x_2, \ldots, x_n) \in \mathbb{R}^n$.

This is the "free Stark Hamiltonian" the quantum-mechanical Hamiltonian (free of charges) in the presence of a uniform electric field in the x_1-direction.

The spectrum $\sigma(H_0)$ in both cases is absolutely continuous. In Case A $\sigma(H_0) = \overline{\mathbb{R}_+}$ and in Case B $\sigma(H_0) = \mathbb{R}$.

The fundamental fact in both cases is the validity of a "Limiting Absorption Principle" (henceforth LAP). Roughly, it means that the resolvent operators $R_0(z) = (H_0 - z)^{-1}$, $\text{Im}\, z \neq 0$, possess limiting values, as $\text{Im}\, z \to 0$, in suitable operator topologies. The absolute continuity of the spectrum follows as a corollary [1].

To formulate the LAP explicitly, we introduce the weighted-L^2 spaces $L^{2,s}(\mathbb{R}^n)$, $s \in \mathbb{R}$, normed by,

$$\|f\|_s^2 := \int_{\mathbb{R}^n} (1 + |x|^2)^s |f(x)|^2 dx. \tag{1.1}$$

The norm in $L^2(\mathbb{R}^n)$ ($s = 0$) will be designated as $\|\cdot\|$.

For $s \geq 0$ we denote by $B_s = B(L^{2,s}(\mathbb{R}^n), L^{2,-s}(\mathbb{R}^n))$ the space of bounded linear operators from $L^{2,s}$ to its dual $L^{2,-s}$, equipped with the operator norm. Our LAP theorem reads as follows.

Theorem 1.1. *Let H_0 be as in Cases A or B. For every $s > \frac{1}{2}$, the operator-valued function*

$$z \to R_0(z) = (H_0 - z)^{-1}, \quad \text{Im}\, z > 0 \ (\text{resp. } \text{Im}\, z < 0),$$

can be extended continuously to the real axis (avoiding $\lambda = 0$ in Case A), in the uniform operator topology of B_s.

Furthermore, the limiting values

$$R_0^\pm(\lambda) = \lim_{\epsilon \to 0+} R_0(\lambda \pm i\epsilon) \tag{1.2}$$

are Hölder continuous in B_s.

In Case B, the free Stark Hamiltonian, the proof is well-known [3, 7, 20]. In fact, one can replace the weight function $(1+|x|^2)^{s/2}$ in (1.1) by a weight function depending only on x_1 (and even nonincreasing as $x_1 \to -\infty$). The reason for choosing the more restrictive weight (1.1) is in the need to gain decay for the limits (1.2) as $|\lambda| \to \infty$, as will be seen below.

For Case A, the theorem is of course classical for $H_0 = -\Delta$ (i.e., $f(\theta) = \theta$) [1]. Here is the proof for the general case.

Proof: It suffices to prove the statement for $0 = f(0) < \lambda < f(\infty) = \infty$.

Denote by $\{E_0(\lambda)\}$, $\{E_{H_0}(\lambda)\}$, the spectral families associated with $-\Delta$, H_0 respectively. In view of the monotonicity of f,

$$E_{H_0}(\lambda) = E_0\left(f^{-1}(\lambda)\right), \quad f(0) < \lambda < f(\infty). \tag{1.3}$$

As was shown in [7] the operator-valued function $E_0(\mu)$, $\mu > 0$, is weakly differentiable in $B(L^{2,s}, L^{2,-s})$, and its weak derivative $A_0(\mu) \in B(L^{2,s}, L^{2,-s})$ satisfies

$$[A_0(\mu)\varphi, \psi] = \frac{d}{d\mu}(E_0(\mu)\varphi, \psi) = \frac{1}{2\sqrt{\mu}} \int_{|\xi|^2 = \mu} \hat{\varphi}(\xi)\overline{\hat{\psi}(\xi)}\, d\sigma_\xi, \tag{1.4}$$

where $\varphi, \psi \in L^{2,s}$ and $[\,,\,]$ denotes the $(L^{2,-s}, L^{2,s})$ pairing. Hence, by (1.3) $E_{H_0}(\lambda)$ is also weakly differentiable in $B(L^{2,s}, L^{2,-s})$ and its weak derivative $A_{H_0}(\lambda) \in B(L^{2,s}, L^{2,-s})$ satisfies,

$$A_{H_0}(\lambda) = \left[f'\left(f^{-1}(\lambda)\right)\right]^{-1} \cdot A_0\left(f^{-1}(\lambda)\right), \quad f(0) < \lambda < f(\infty). \tag{1.5}$$

It follows from [7] that $A_0(\mu)$ is locally Hölder continuous in the uniform operator topology of $B(L^{2,s}, L^{2,-s})$ (with exponent depending only on s, n). Combining this with Assumption A we conclude from (1.5) that $A_{H_0}(\lambda)$ is locally Hölder continuous in $B(L^{2,s}, L^{2,-s})$. As shown in [7], this yields the statements of the theorem. □

It follows from the definition of $A_{H_0}(\lambda)$ that the resolvent $R_{H_0}(z) = (H_0 - z)^{-1}$, satisfies,

$$R_{H_0}(z) = \int \frac{1}{\lambda - z} A_{H_0}(\lambda) d\lambda, \quad \text{Im}\, z \neq 0. \tag{1.6}$$

The next step is to study the decay properties of $A_{H_0}(\lambda)$ as $|\lambda| \to \infty$. The considerations in [8] show that the gain in regularity is exactly that rate of decay. In addition, that same decay implies that the operator $(1+|x|^2)^{-s}(I+|H_0|)^\beta$ is H_0-smooth [17], for a certain constant $\beta > 0$. Following [9, 11] we can state the following theorem, where we assume $n \geq 3$.

Theorem 1.2. *Assume $s > 1$ in Case A and $s = \frac{3}{4}$ in Case B. Assume further, in Case A, that $f(\lambda) \sim \lambda^a$, $f'(\lambda) \sim \lambda^{a-1}$, $a \geq \frac{1}{2}$, as $\lambda \to +\infty (g(\lambda) \sim h(\lambda)$ if $C^{-1} \leq h(\lambda)/g(\lambda) \leq C$ for large λ).*

Take $\beta = \frac{2a-1}{4a}$ in Case A and $\beta = \frac{1}{4}$ in Case B.
Then, for every $u_0 \in L^2(\mathbb{R}^n)$,

$$\int_\mathbb{R} \int_{\mathbb{R}^n} (1+|x|^2)^{-s} |(I+|H_0|)^\beta \exp(itH_0) u_0(x)|^2 dx dt \leq C \|u_0\|^2. \tag{1.7}$$

An outline of the Proof:

If $u(0) = \varphi \in L^2(\mathbb{R}^n)$, then $u(t) = \exp(-itH_0)\varphi$ can be expressed as

$$u(t) = \int_{f(0)}^{f(\infty)} e^{-it\lambda} A_{H_0}(\lambda) \varphi\, d\lambda, \tag{1.8}$$

which must be interpreted in the appropriate weak sense.

We shall outline a proof for Case A for $a = \frac{1}{2}$ (for more details see [11]). The general case where $a \neq \frac{1}{2}$ is easily obtained with similar arguments.

We shall use the following trace lemma, the proof of which can be found in [10], Appendix. Here and below we let $H_s(\mathbb{R}^n)$ be the Sobolev space of order s, with norm $\|\cdot\|_{H_s}$.

Lemma 1.3 (Trace). *Let* $\hat{\varphi} \in H_s(\mathbb{R}^n)$, $n \geq 3$, $\frac{1}{2} < s < \frac{3}{2}$. *Then for any* $r > 0$,

$$\left(\int_{|\xi|=r} |\hat{\varphi}(\xi)|^2 \, d\sigma_r \right)^{\frac{1}{2}} \leq C \cdot \mathrm{Min}\left(r^{s-\frac{1}{2}}, 1\right) \cdot \|\hat{\varphi}\|_{H_s},$$

where C depends only on s, n.

Combining the trace lemma with (1.4)–(1.5) we obtain for $\lambda > f(0)$,

$$\varphi, \psi \in L^{2,s}(\mathbb{R}^n),$$

$$|[A_{H_0}(\lambda)\varphi, \psi]| \leq C \left[f'\left(f^{-1}(\lambda)\right) \right]^{-1} \cdot \left(f^{-1}(\lambda)\right)^{-\frac{1}{2}} \cdot \mathrm{Min}\left(f^{-1}(\lambda)^{s-\frac{1}{2}}, 1\right) \|\varphi\|_s \|\psi\|_s .$$

In particular, since $f'(\lambda) \sim C/\sqrt{\lambda}$ ($\lambda \to \infty$), we have the following bound.

$$\sup \left\{ \|A_{H_0}(\lambda)\|_{B(L^{2,1}, L^{2,-1})}, \lambda > f(0) \right\} < \infty. \tag{1.9}$$

Note that for $a = \frac{1}{2}$, $\beta = 0$. To prove (1.7), note that by an obvious density argument, it suffices to establish (1.7) for $\varphi \in C_0^\infty(\mathbb{R}^n)$. Let $w(x,t) \in C_0^\infty(\mathbb{R}^{n+1})$ and use (1.8) to write

$$\int_{\mathbb{R}} (u(t), w(\cdot, t)) \, dt = \int_{\mathbb{R}} \int_{f(0)}^{\infty} e^{-it\lambda} [A_{H_0}(\lambda)\varphi, w(\cdot, t)] \, d\lambda \, dt$$

$$= \int_{f(0)}^{\infty} \left[A_{H_0}(\lambda)\varphi, \int_{\mathbb{R}} e^{+it\lambda} w(\cdot, t) \, dt \right] d\lambda . \tag{1.10}$$

Denoting by $\tilde{w}(\cdot, \lambda) = \int_{\mathbb{R}} e^{+it\lambda} w(\cdot, t) \, dt$ the Fourier transform of $w(\cdot, t)$ with respect to t, we rewrite (1.10) as,

$$\int_{\mathbb{R}} (u(t), w(\cdot, t)) \, dt = \int_{f(0)}^{\infty} [A_{H_0}(\lambda)\varphi, \tilde{w}(\cdot, \lambda)] \, d\lambda . \tag{1.11}$$

Since $A_{H_0}(\lambda) = \frac{d}{d\lambda} E_{H_0}(\lambda)$, the bilinear form $[A_{H_0}(\lambda)\cdot, \cdot]$ is nonnegative and, for any

$$\psi \in L^{2,s}(\mathbb{R}^n), \quad \int_{f(0)}^{\infty} [A_{H_0}(\lambda)\psi, \psi] \, d\lambda = \|\psi\|^2 .$$

Using these considerations and the Cauchy–Schwartz inequality in (1.11) we obtain

$$\left| \int_{\mathbb{R}} (u(t), w(\cdot, t)) \, dt \right| \leq \left(\int_{f(0)}^{\infty} [A_{H_0}(\lambda)\varphi, \varphi] \, d\lambda \right)^{\frac{1}{2}} .$$

$$\cdot \left(\int_{f(0)}^{\infty} [A_{H_0}(\lambda) \tilde{w}(\cdot,\lambda), \tilde{w}(\cdot,\lambda)] \, d\lambda \right)^{\frac{1}{2}}$$

$$\leq C \|\varphi\| \cdot \left(\int_{f(0)}^{\infty} \|\tilde{w}(\cdot,\lambda)\|_1^2 \, d\lambda \right)^{\frac{1}{2}}, \tag{1.12}$$

where we have used (1.9) in the last step. The Plancherel theorem now yields,

$$\left| \int_{\mathbb{R}} (u(t), w(\cdot,t)) \, dt \right| \leq C \|\varphi\| \left(\int_{\mathbb{R}} \|w(\cdot,t)\|_1^2 \, dt \right)^{\frac{1}{2}},$$

which can be rewritten as,

$$\left| \int_{\mathbb{R}} \left((1+|x|^2)^{-\frac{1}{2}} u(t), w(\cdot,t) \right) dt \right| \leq C \|\varphi\| \left(\int_{\mathbb{R}^{n+1}} |w(x,t)|^2 \, dx \, dt \right)^{\frac{1}{2}},$$

and which is equivalent to (1.7) by duality. $\square$

Remark that it follows from (1.7) that the solution to the (unperturbed) equation $i^{-1} u_t = H_0 u$ is locally in the domain of $(I + |H_0|)^\beta$ ("local smoothing") for $a.e. t \in \mathbb{R}$ and the t-integrability implies the decay as $|t| \to \infty$. Note also that the case $a = \frac{1}{2}$ (the Klein–Gordon or "relativistic Schrödinger" equation) yields long-time decay, but no smoothing. The case $H_0 = -\Delta$ has been extensively studied [8, 10, 13, 14].

The above theorems have their counterparts in the presence of a nonvanishing potential $V(x)$. We shall state here only a result concerning the Stark Hamiltonian (Case B) and refer the reader to [11] for Case A.

Theorem 1.4 ([9]). *Let $V(x)$ be a real potential satisfying:*

1. $|V(x)| \leq C(1+|x|)^{-3/2}$, $x \in \mathbb{R}^n$.

2. *For some $a > 0$ the derivative $\frac{\partial}{\partial x_1} V(x)$ exists for $x_1 > a$ and satisfies,*

$$\sup_{x_1 > a} \left| \frac{\partial}{\partial x_1} V(x) \right| < 1.$$

Then $H = -\Delta - x_1 + V(x)$ is self adjoint on $D(-\Delta - x_1)$ and for every $\beta < \frac{1}{2}$ there exists a constant $C = C_\beta > 0$ such that, if $u_0 \in L^2(\mathbb{R}^n)$,

$$\int_{\mathbb{R}} \int_{\mathbb{R}^n} (1+|x|^2)^{-\frac{3}{4}} \left| (I+|H|)^{\frac{\beta}{2}} \exp(itH) u_0 \right|^2 dx \, dt \leq C \|u_0\|^2. \tag{1.13}$$

2. LIMITING ABSORPTION FOR MAXWELL EQUATIONS FOR STRATIFIED ANISOTROPIC MEDIA

We now turn to Maxwell's equations (0.2)-(0.3). The right-hand side of (0.2), is a self-adjoint operator H (cf. Lemma 2.2 in [12], [17]), which is defined by,

$$H \begin{bmatrix} \mathbf{E} \\ \mathbf{H} \end{bmatrix} = \begin{bmatrix} -i\varepsilon(x_1)^{-1} \nabla \times \mathbf{H} \\ i\mu(x_1)^{-1} \nabla \times \mathbf{E} \end{bmatrix}.$$

The "$\nabla \times$" here and below should be interpreted in the sense of distributions. The domain of H is given by,

$$D(H) = \left\{ \begin{bmatrix} \mathbf{E} \\ \mathbf{H} \end{bmatrix} \in \mathbb{H} \,\middle|\, \nabla \times \mathbf{E} \in (L^2(\mathbb{R}^3))^3, \, \nabla \times \mathbf{H} \in (L^2(\mathbb{R}^3))^3 \right\}.$$

Consider $\hat{W}$ defined by,

$$\hat{W} = \left\{ \begin{bmatrix} \mathbf{E} \\ \mathbf{H} \end{bmatrix} \in (L^2(\mathbb{R}^3))^6 \,\middle|\, \nabla \cdot (\varepsilon(x_1)\mathbf{E}(\mathbf{x})) = 0, \, \nabla \cdot (\mu(x_1)\mathbf{H}(\mathbf{x})) = 0 \right\}$$

where again, "$\nabla \cdot$" should be interpreted in the sense of distributions.

Recall that $\hat{W}$ is a reducing subspace for H and recall that $\hat{W} \perp \ker H$ (in the Hilbert space $\mathbb{H}$). In the following we shall denote by $H_{\hat{W}}$ the part of H in $\hat{W}$, i.e., $H = H_{\hat{W}} \oplus 0$. The operator $H_{\hat{W}}$ is self-adjoint in the Hilbert space $\hat{W}$.

Our main objective now is a presentation of a LAP result for $H_{\hat{W}}$ (in a recent work [16]). The LAP was established under the following assumptions on $\varepsilon(\cdot)$ and $\mu(\cdot)$.

Assumption (on $\varepsilon(\cdot)$ and $\mu(\cdot)$).

1. $\varepsilon(\cdot)$ and $\mu(\cdot)$ are piecewise continuous on $\mathbb{R}^1$ with finitely many jump discontinuities.

2. $\varepsilon(\cdot)$ and $\mu(\cdot)$ are scalar for $|x_1| > x_c$ for some fixed $x_c > 0$, namely,

$$\varepsilon(x_1) = \varepsilon_\pm I, \quad \pm x_1 > x_c$$
$$\mu(x_1) = \mu_\pm I, \quad \pm x_1 > x_c.$$

3.
$$\frac{1}{\varepsilon_- \mu_-} \neq \frac{1}{\varepsilon_+ \mu_+}.$$

Without loss of generality we take,

$$\frac{1}{\varepsilon_- \mu_-} > \frac{1}{\varepsilon_+ \mu_+}.$$

We shall denote by $\hat{W}^s$ ($s \in \mathbb{R}$) the Hilbert space of elements in

$$\left\{ \begin{bmatrix} \mathbf{E} \\ \mathbf{H} \end{bmatrix} \in (L^2(\mathbb{R}^3, (1+x_1^2)^s(1+x_2^2+x_3^2)^s d\mathbf{x}))^6 \,\middle|\, \nabla \cdot (\varepsilon(x_1)\mathbf{E}) = 0 \text{ and } \nabla \cdot (\mu(x_1)\mathbf{E}) = 0 \right\}$$

endowed with the scalar product

$$\left\langle \begin{bmatrix} \mathbf{E}_1 \\ \mathbf{H}_1 \end{bmatrix}, \begin{bmatrix} \mathbf{E}_2 \\ \mathbf{H}_2 \end{bmatrix} \right\rangle = \int_{\mathbb{R}^3} \mathbf{E}_1(\mathbf{x}) \cdot \varepsilon(x_1) \overline{\mathbf{E}_2(\mathbf{x})} (1+x_1^2)^s (1+x_2^2+x_3^2)^s d\mathbf{x} +$$

$$+ \int_{\mathbb{R}^3} \mathbf{H}_1(\mathbf{x}) \cdot \mu(x_1) \overline{\mathbf{H}_2(\mathbf{x})} (1+x_1^2)^s (1+x_2^2+x_3^2)^s d\mathbf{x}.$$

Note that for $s > 0$, $\hat{W}^s$ is dense in $\hat{W}$ because $\hat{W}$ is the closure of

$$\left\{ \begin{bmatrix} \mathbf{E}(\cdot) \\ \mathbf{H}(\cdot) \end{bmatrix} \in \varepsilon(x_1)^{-1}(C_0^\infty(\mathbb{R}^3))^3 \oplus \mu(x_1)^{-1}(C_0^\infty(\mathbb{R}^3))^3 \,\middle|\, \mathbf{\nabla} \cdot (\varepsilon(x_1)\mathbf{E}) = \mathbf{\nabla} \cdot (\mu(x_1)\mathbf{H}) = 0 \right\}$$

in $\mathbb{H}$ (cf. e.g. [15]).

We shall say that $\varepsilon(\cdot)$ and $\mu(\cdot)$ are of the form of a *generalized Šolc filter* if $\varepsilon_{12}(x_1) = \varepsilon_{13}(x_1) = \varepsilon_{21}(x_1) = \varepsilon_{12}(x_1) = 0$ and $\mu_{12}(x_1) = \mu_{13}(x_1) = \mu_{21}(x_1) = \mu_{12}(x_1) = 0$. For further details regarding Šolc filters see e.g. [19,21,22].

Our LAP theorem for the case of the Maxwell equations reads as follows.

Theorem 2.1. *Let*

$$R_{H_{\hat{W}}}(z) = \left(H_{\hat{W}} - z\right)^{-1} \quad \text{for} \quad \operatorname{Im} z \neq 0.$$

Fix $s > \frac{1}{2}$ and let $\lambda \in \mathbb{R} \setminus \{0\}$. Suppose that $\varepsilon(\cdot)$ and $\mu(\cdot)$ are of the form of a generalized Šolc filter. Then the limits

$$R^{\pm}_{H_{\hat{W}}}(\lambda) = \lim_{\kappa \to 0^+} R_{H_{\hat{W}}}(\lambda \pm i\kappa)$$

exist in the uniform operator topology of

$$B\left(\hat{W}^s(\mathbb{R}^3), \hat{W}^{-s}(\mathbb{R}^3)\right).$$

Furthermore, the convergence is uniform for λ in compact subsets of $\mathbb{R} \setminus \{0\}$, and the limits $R^{\pm}_{H_{\hat{W}}}(\lambda)$ are locally Hölder continuous (in the uniform operator topology) with exponent α which depends on s.

Remark 2.1. The assumption that $\varepsilon(\cdot)$ and $\mu(\cdot)$ are of the form of a generalized Šolc filter is redundant if one can prove there are no critical points on the guided-wave dispersion surfaces. This happens in particular if there is no wave-guiding.

The proof of Theorem 2.1 is based on a new technique can be used for establishing LAP for square roots of operators (cf. [16], Th. 2.1). It turns out that in many cases it is easier to prove a LAP result for H^2 rather than proving it for H (for example H^2 might be an elliptic operator). Note that in a general situation $H \neq \sqrt{H^2}$ (the positive square root of H^2), for example in the case of Maxwell's equations (or in the case of Dirac's operator) the generator H is not a function of H^2 and therefore one cannot apply Theorem 1.1.

In [16] the LAP for $H_{\hat{W}}$ is proved with the following strategy.

1. Establish a LAP for $H_{\hat{W}}^2$.

2. Prove that $H_{\hat{W}}$ and $H_{\hat{W}}^2$ are closable in $\hat{W}^{-s}$ with $D(\overline{H_{\hat{W}}^2}) \subset D(\overline{H_{\hat{W}}})$ (the closures are in the graph norm of $\hat{W}^{-s} \times \hat{W}^{-s}$). This statement is easily obtained with the theory of strictly m-sectorial forms.

3. Apply Theorem 3.1 (see Section 3 below).

The proof of LAP for $H_{\hat{W}}^2$ (step 1.) is an extension of the techniques used in [5] for the (scalar) acoustic wave equation. In the case of *anisotropic* media the Maxwell equations do not decouple into scalar equations. In fact, due to the anisotropy it is even impossible to use the celebrated T.E. – T.M. decomposition. The details of the LAP proof for $H_{\hat{W}}^2$ will be published in [16]. In Section 4 we will describe the spectral properties of $H_{\hat{W}}^2$ that are essential for the proof of LAP.

3. LIMITING ABSORPTION PRINCIPLE FOR SQUARE ROOTS

In this section we derive a general theorem concerning a LAP for H using the validity of such a principle for H^2. The operators H, H^2 and the Hilbert space $\mathbb{H}$ are not necessarily related to Maxwell's equations. In fact, throughout this section, we shall follow the general abstract conventions used in [7].

Let H be a self adjoint operator defined in a separable Hilbert space $\mathbb{H}$ and let $R_H(z) = (H-z)^{-1}$ (resp. $R_{H^2}(z) = (H^2-z)^{-1}$) be the resolvent operator associated with H (resp. H^2). Let $\mathbb{X}$ and $\mathbb{Y}$ be a Banach spaces such that $\mathbb{X}$ is densely and continuously embedded in $\mathbb{H}$ (i.e., $i : \mathbb{X} \hookrightarrow \mathbb{H}$) and such that $\mathbb{H}$ is densely and continuously embedded in $\mathbb{Y}$ (i.e., $j : \mathbb{H} \hookrightarrow \mathbb{Y}$).

We shall assume the following,

1. The operators H and H^2 (defined on $jD(H), jD(H^2)$) are closable in $\mathbb{Y}$. We shall denote by $\overline{H}, \overline{H^2}$ their closures in $\mathbb{Y}$ and by $D(\overline{H}), D(\overline{H^2})$ their respective domains.

2. The domain $D(\overline{H^2})$ is contained in $D(\overline{H})$ (i.e., $D(\overline{H^2}) \subset D(\overline{H})$).

With this setting we have the following result,

Theorem 3.1. *Let $\lambda_2 > \lambda_1 > 0$. Suppose that the resolvent operator $R_{H^2}(z)$, has a continuous extension to $\mathbb{C}^+ \cup (\lambda_1^2, \lambda_2^2)$ (resp. $\mathbb{C}^- \cup (\lambda_1^2, \lambda_2^2)$) with respect to the uniform operator topology $\mathbb{B}(\mathbb{X}, \mathbb{Y})$. Let $R^{\pm}_{H^2}(\nu) = \lim_{\kappa \to 0^+} (H^2 - \nu \pm i\kappa)^{-1}$ be the corresponding limiting values.*

Then, $R_H(z)$ has a continuous extension to $\mathbb{C}^+ \cup (\lambda_1, \lambda_2)$ (resp. $\mathbb{C}^- \cup (\lambda_1, \lambda_2)$) with respect to the uniform operator topology $\mathbb{B}(\mathbb{X}, \mathbb{Y})$ and the limiting values (for $(\lambda \in (\lambda_1, \lambda_2))$) are given by,

$$R^{\pm}_H(\lambda) = \lim_{\substack{\pm \mathrm{Im} z > 0 \\ z \to \lambda}} (H-z)^{-1} = (\overline{H}+\lambda) \lim_{\substack{\pm \mathrm{Im} z > 0 \\ z \to \lambda}} (H^2-z^2)^{-1} = (\overline{H}+\lambda) R^{\pm}_{H^2}(\lambda^2). \quad (3.1)$$

Furthermore, if $\nu \to R^{\pm}_{H^2}(\nu)$ are locally Hölder continuous (in $\mathbb{B}(\mathbb{X}, \mathbb{Y})$) with respect to $\nu \in (\lambda_1^2, \lambda_2^2)$ then the limiting values $R^{\pm}_H(\lambda) = \lim_{\kappa \to 0^+} (H - \lambda \pm i\kappa)^{-1}$ are also locally Hölder continuous in $\mathbb{B}(\mathbb{X}, \mathbb{Y})$.

(Observe that the right-hand side of (3.1) is well defined since it follows from the above setting that, $\mathrm{Ran}(R^{\pm}_{H^2}(\lambda^2)) \subset D(\overline{H})$ for every $\lambda \in (\lambda_1, \lambda_2)$ (see also [7], Th. 2.4])).

Remark 3.1. Note that if $(\lambda_1, \lambda_2) \subset \mathbb{R}^-$ then the same result holds with the following changes: $R_H(z)$ is continuous in $\mathbb{C}^+ \cup (\lambda_1, \lambda_2)$ (resp. $\mathbb{C}^- \cup (\lambda_1, \lambda_2)$), when, $R_{H^2}(z)$ is continuous in $\mathbb{C}^- \cup (\lambda_1, \lambda_2)$ (resp. $\mathbb{C}^+ \cup (\lambda_1, \lambda_2)$) and the limiting values are given by,

$$R^{\pm}_H(\lambda) = (\overline{H}+\lambda) R^{\mp}_{H^2}(\lambda^2) \quad (\lambda \in (\lambda_1, \lambda_2)).$$

Proof of Theorem 3.1:

We shall prove the result for the case $(\lambda_1, \lambda_2) \subset \mathbb{R}^+$. The proof for the case $(\lambda_1, \lambda_2) \subset \mathbb{R}^-$ is similar.

Let us define the Banach space $\mathbb{Y}_{H^2}$ (as in [7]) as $D(\overline{H^2})$ equipped with the graph norm,

$$\|\varphi\|^2_{\mathbb{Y}_{H^2}} = \|\varphi\|^2_{\mathbb{Y}} + \|\overline{H^2}\varphi\|^2_{\mathbb{Y}}.$$

Observe that for every $\nu \in (\lambda_1^2, \lambda_2^2)$ and $x \in \mathbb{X}$,

$$\lim_{\substack{\operatorname{Im} z > 0 \\ z \to \nu}} R_{H^2}(z)x = R_{H^2}^+(\nu)x,$$

and

$$\lim_{\substack{\operatorname{Im} z > 0 \\ z \to \nu}} \overline{H^2} R_{H^2}(z)x = x + \nu R_{H^2}^+(\nu)x.$$

Since $\overline{H^2}$ is closed in $\mathbb{Y}$ we have

$$\overline{H^2} R_{H^2}^+(\nu) = I + \nu R_{H^2}^+(\nu). \tag{3.2}$$

It follows that $R_{H^2}(z)$ has a continuous extension to $\mathbb{C}^+ \cup (\lambda_1^2, \lambda_2^2)$ (resp. $\mathbb{C}^- \cup (\lambda_1^2, \lambda_2^2)$) with respect to the uniform operator topology $\mathbb{B}(\mathbb{X}, \mathbb{Y}_{H^2})$.

Next, consider $\overline{H}$ as an operator defined in $\mathbb{Y}_{H^2}$. By the definition of $\mathbb{Y}_{H^2}$ and since $D(\overline{H^2}) \subset D(\overline{H})$ it follows by the closed graph theorem, that $\overline{H} : \mathbb{Y}_{H^2} \to \mathbb{Y}$ is a bounded operator.

Writing,

$$(H-z)^{-1} = (H+z)(H^2 - z^2)^{-1} = (H+z) R_{H^2}(z^2) \qquad (z \in \mathbb{C}^\pm)$$

and noting the above continuity of $R_{H^2}(z^2)$ when $z \to \lambda + i0$ ($\lambda_1 < \lambda < \lambda_2$) we conclude that $R_H(z)$ can be extended continuously to $\mathbb{C}^+ \cup (\lambda_1, \lambda_2)$ (resp. $\mathbb{C}^- \cup (\lambda_1, \lambda_2)$) with respect to the uniform operator topology $\mathbb{B}(\mathbb{X}, \mathbb{Y})$.

Thus, the following limits exist (in the uniform operator topology $\mathbb{B}(\mathbb{X}, \mathbb{Y})$),

$$R_H^\pm(\lambda) := \lim_{\substack{\pm \operatorname{Im} z > 0 \\ z \to \lambda}} R_H(z) = (\overline{H} + \lambda) R_{H^2}^\pm(\lambda^2) \qquad (\lambda \in (\lambda_1, \lambda_2)). \tag{3.3}$$

Observe that the assumption $(\lambda_1, \lambda_2) \subset \mathbb{R}^+$ was used in (3.3).

Finally, given our setting, observe that in view of (3.2) it follows that if $R_{H^2}^\pm(\nu)$ are locally Hölder continuous with respect to the topology of $\mathbb{B}(\mathbb{X}, \mathbb{Y})$ then $R_{H^2}^\pm(\nu) \in \mathbb{B}(\mathbb{X}, \mathbb{Y}_{H^2})$ and the family $R_{H^2}^\pm(\nu)$ is locally Hölder continuous with respect to the uniform operator topology of $\mathbb{B}(\mathbb{X}, \mathbb{Y}_{H^2})$. Hence, the local Hölder continuity of $R_H^\pm(\lambda)$ follows from the local Hölder continuity of $R_{H^2}^\pm(\lambda^2)$ (in $\mathbb{B}(\mathbb{X}, \mathbb{Y}_{H^2})$) and from (3.1). □

Remark 3.2. One can verify that the reason for excluding the point $\lambda = 0$ in the proof of LAP for the square root H, lies in the fact that $\lambda = 0$ is a critical point of the polynomial $p(\lambda) = \lambda^2$ (i.e., $\left|\frac{\partial p}{\partial \lambda}\right|_{\lambda = \lambda_{\text{critical}}} = 0$).

4. SPECTRAL ANALYSIS OF $H_{\hat{W}}^2$ IN THE CASE OF STRATIFIED ANISOTROPIC MEDIA

In this section we shall describe spectral properties of $H_{\hat{W}}^2$ that are essential for the proof of LAP.

In the case of Maxwell's equations the operator $H_{\hat{W}}^2$ decomposes in the following manner.

$$H_{\hat{W}}^2 \begin{bmatrix} \mathbf{E} \\ \mathbf{H} \end{bmatrix} = \begin{bmatrix} \varepsilon(x_1)^{-1} \boldsymbol{\nabla} \times \left(\mu(x_1)^{-1} \boldsymbol{\nabla} \times \mathbf{E} \right) \\ \mu(x_1)^{-1} \boldsymbol{\nabla} \times \left(\varepsilon(x_1)^{-1} \boldsymbol{\nabla} \times \mathbf{H} \right) \end{bmatrix}.$$

Thus we can write $H_{\hat{W}}^2 = L \oplus G$ where,

$$LE = \varepsilon(x_1)^{-1} \nabla \times \left(\mu(x_1)^{-1} \nabla \times E \right),$$

and G is defined similarly replacing $\varepsilon(\cdot)$ by $\mu(\cdot)$ (and vice versa). L is a self adjoint operator in a Hilbert space (that we shall denote by $\mathbb{H}_\varepsilon$) of elements in $(L^2(\mathbb{R}^3))^3$ equipped with the scalar product,

$$\langle E_1, E_2 \rangle_{\mathbb{H}_\varepsilon} = \int_{\mathbb{R}^3} E_1(x) \cdot \varepsilon(x_1) \overline{E_2(x)} dx.$$

Note that L represents an elliptic 3×3 system, for which the LAP is easier to prove. Let $F_{2,3}$ be the (partial) Fourier–Plancherel transform defined by,

$$(F_{2,3} u)(x_1, p_2, p_3) = \frac{1}{2\pi} \int_{\mathbb{R}^2} dx_2 dx_3 \exp(-ip_2 x_2 - ip_3 x_3) u(x_1, x_2, x_3)$$

$$\text{for} \quad \left(u \in \left(C_0^\infty(\mathbb{R}^3) \right)^3 \right).$$

The operator $F_{2,3}$ is extended in a standard way to a unitary transformation from $\mathbb{H}_\varepsilon$ onto $\int_{\mathbb{R}^2} \oplus \mathbb{H}_{\varepsilon; p_2, p_3} dp_2 dp_3$ where $\mathbb{H}_{\varepsilon; p_2, p_3} \equiv \mathbb{H}_{\varepsilon; 0, 0}$ (independently of (p_2, p_3)) is a weighted $(L^2(\mathbb{R}^1))^3$ space with inner product,

$$\langle E_1, E_2 \rangle_{\mathbb{H}_{\varepsilon; p_2, p_3}} = \int_{\mathbb{R}} E_1(x_1) \cdot \varepsilon(x_1) \overline{E_2(x_1)} dx_1.$$

By setting,

$$\hat{L} = F_{2,3} L F_{2,3}^{-1} \qquad D(\hat{L}) = F_{2,3} D(L).$$

We obtain a representation of L as a direct integral of the form,

$$\hat{L} = \int_{\mathbb{R}^2} \oplus L_{p_2, p_3} dp_2 dp_3 \quad \text{in} \quad \int_{\mathbb{R}^2} \oplus \mathbb{H}_{\varepsilon; p_2, p_3} dp_2 dp_3,$$

where, $\{L_{p_2, p_3}\}$ is a family of self-adjoint operators defined in the fibers $\{\mathbb{H}_{\varepsilon; p_2, p_3}\}$ as follows (essentially, replacing ∇ by (∂_1, ip_2, ip_3)). $E \in D(L_{p_2, p_3})$ with $L_{p_2, p_3} E = v$, if and only if,

$$u := -\mu^{-1}(-i\partial_1, p_2, p_3) \times E, \quad \text{is in } (L^2(\mathbb{R}^1))^3,$$

and

$$v := \varepsilon^{-1}(-i\partial_1, p_2, p_3) \times u, \quad \text{is in } (L^2(\mathbb{R}^1))^3,$$

(where, the ∂_1 derivative is in the sense of distributions).

Given a self-adjoint operator T, let $\sigma_p(T)$ (resp. $\sigma_c(T)$) denote the point (resp. continuous) spectrum of T. Let us now discuss the structure of $\sigma_p(L_{p_2, p_3})$ in terms of p_2 and p_3. When $\varepsilon(\cdot)$ and $\mu(\cdot)$ are scalar, $\sigma_p(L_{p_2, p_3})$ depends only on $p_2^2 + p_3^2$. However, under general anisotropic conditions, $\sigma_p(L_{p_2, p_3})$ is not axisymmetric in p_2 and p_3. At any rate, in view of Assumption 2, and (0.4) we obtain the following result (see [16]).

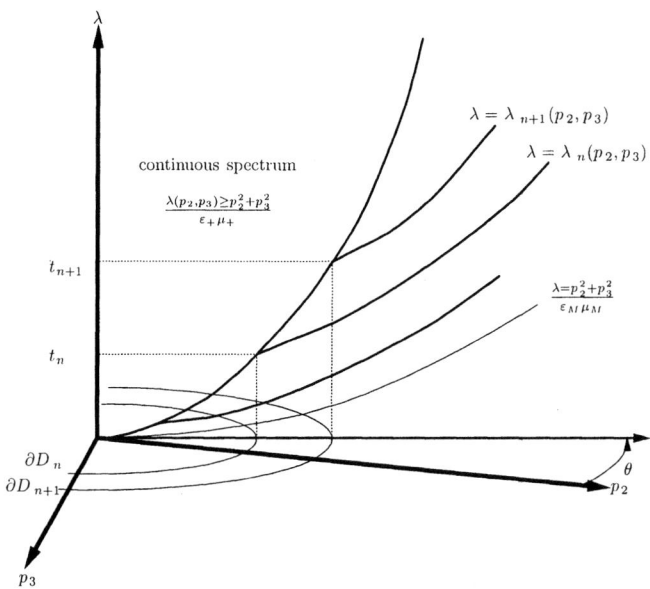

Figure 1.

Lemma 4.1. *For every $(p_2,p_3) \in \mathbb{R}^2$ we have,*

$$\sigma_c(L_{p_2,p_3}) = \left[\frac{p_2^2 + p_3^2}{\varepsilon_+ \mu_+}, \infty\right)$$

$$\sigma_p(L_{p_2,p_3}) \subset \left(\frac{p_2^2 + p_3^2}{\varepsilon_M \mu_M}, \frac{p_2^2 + p_3^2}{\varepsilon_+ \mu_+}\right) \cup \{0\}.$$

In addition, $\sigma_p(L_{p_2,p_3}) \setminus \{0\}$ is finite in $\left[\frac{p_2^2+p_3^2}{\varepsilon_M \mu_M}, \frac{p_2^2+p_3^2}{\varepsilon_+ \mu_+}\right]$ and the multiplicity of each eigenvalue in $\sigma_p(L_{p_2,p_3}) \setminus \{0\}$ is at most two.

In Figure 1 we depict the spectrum of L_{p_2,p_3} along a ray in the (p_2,p_3) plane.

Using Assumption 2 and (0.4) we obtain the following lemma whose proof is given in [16].

Lemma 4.2. *The positive eigenvalues and the corresponding eigenvectors of L_{p_2,p_3} depend smoothly on (p_2,p_3). More specifically, the number of positive eigenvalues is locally constant. For each point $(\tilde{p}_2,\tilde{p}_3) \in \mathbb{R}^2$ with $\sigma_p(L_{\tilde{p}_2,\tilde{p}_3}) \neq \{0\}$ there is a neighborhood U of $(\tilde{p}_2,\tilde{p}_3)$ and an enumeration of the eigenvalues by C^∞ functions,*

$$\lambda_n : U \to \mathbb{R} \qquad 1 \leq n \leq N \qquad (N = N(U)).$$

The corresponding eigenvectors $\mathbf{u}_n(\cdot,p_2,p_3) \in \mathbb{H}_{\varepsilon;o,o}$ form an orthonormal set for each fixed (p_2,p_3) and are differentiable in U as functions of (p_2,p_3).

Thus, locally, the eigenvalues of L_{p_2,p_3} form a finite number of smooth surfaces.

Loosely speaking, it turns out that if

$$\nabla_{p_2,p_3} \lambda_n(p_2,p_3) \neq 0 \quad \text{for} \quad (p_2,p_3) \in \mathbb{R}^2 \setminus \{(0,0)\}$$

then the derivative of the spectral measure of L is Hölder continuous. On the other hand, critical points of $\lambda_n(p_2,p_3)$ (i.e., points $(p_2,p_3) \in \mathbb{R}^2$ where $\nabla_{p_2,p_3} \lambda_n(p_2,p_3) = 0$) can cause "spectral concentration" points in the spectrum of L. When this phenomenon occurs the derivative of the spectral measure of L might not be Hölder continuous. Since the Hölder continuity of the derivative of the spectral measure of L is essential for our proof of a LAP, we will not be able to prove a LAP at such points. In some situations (see the Lemma below) it is possible to prove that there are no critical points. However in the general anisotropic situation we cannot rule out the possibility of critical points.

Lemma 4.3. *Suppose that in addition to Assumption 2, $\varepsilon(\cdot)$ and $\mu(\cdot)$ are of the form of a generalized Šolc filter (i.e., $\varepsilon_{12}(x_1) = \varepsilon_{13}(x_1) = \varepsilon_{21}(x_1) = \varepsilon_{12}(x_1) = 0$ and $\mu_{12}(x_1) = \mu_{13}(x_1) = \mu_{21}(x_1) = \mu_{12}(x_1) = 0$). Then*

$$4\lambda_n(p_2,p_3) \geq (p_2,p_3) \cdot \nabla_{p_2,p_3} \lambda_n(p_2,p_3) > 0 \quad \text{for} \quad (p_2,p_3) \neq 0.$$

It turns out that for our proof of the LAP (see [16]) we also need some information about the behavior of the surfaces $\lambda_n(p_2,p_3)$ near the thresholds. So we add the following assumptions on $\sigma_p(L_{p_2,p_3})$.

Assumption 1. Let each of the functions $\lambda_n(p_2,p_3)$, be defined in a maximal open domain $D_n \subset \mathbb{R}^2 \setminus \{(0,0)\}$. We assume that $\lambda_n(p_2,p_3)$ and $\nabla \lambda_n(p_2,p_3)$ can be extended continuously to $D_n \cup \partial D_n$. In addition we also assume that ∂D_n is connected and that for each $(p_2,p_3) \in \partial D_n$ we have, $\lambda_n(p_2,p_3) = \frac{p_2^2 + p_3^2}{\varepsilon_+ \mu_+}$.

Geometrically speaking it means that the boundary of the surface $\lambda_n(p_2,p_3)$ lies on the "bottom surface" of the continuous spectrum see Figure 1.

Using Lemma 1.3, Lemma 4.1 and Assumption 1 one obtains the following (see [16]).

Lemma 4.4. *The boundaries ∂D_n are circles, i.e.,*

$$\textit{for each} \quad (p_2,p_3) \in \partial D_n, \quad \lambda_n(p_2,p_3) = \frac{p_2^2 + p_3^2}{\varepsilon_+ \mu_+} = t_n = \text{const}.$$

Furthermore,

$$(p_2,p_3) \cdot \nabla \lambda_n(p_2,p_3) = 2 \frac{p_2^2 + p_3^2}{\varepsilon_+ \mu_+} \quad ((p_2,p_3) \in \partial D_n).$$

Definition 4.1. The numbers t_n, $(n = 1,2,3,\dots)$ defined above are called the thresholds of L.

The fact that the thresholds are circles is important for the proof of the LAP because in this way one obtains proper estimates for the thresholds (see [16]).

REFERENCES

1. S. Agmon, Spectral properties of Schrödinger operators and scattering theory, Ann. Scuola Norm. Sup. Pisa **2** (1975), 151–218.

2. E. Balslev and B. Helffer, Limiting absorption principle and resonances for the Dirac operator, Advances in applied mathematics, (1992), 13,(2), 186-215,
3. M. Ben-Artzi, Unitary equivalence and scattering theory for Stark-like Hamiltonians, J. Math. Phys. **25** (1984), 951–964.
4. M. Ben-Artzi, Regularity and smoothing for some equations of evolution, in "Nonlinear Partial Differential Equations and Their Applications," H. Brezis, J. L. Lions (Eds.), Vol. II, pp. 1–12, Pittman Publ. 1994.
5. M. Ben-Artzi and Y. Dermenjian and J. C. Guillot, Acoustic waves in perturbed stratified fluids: A spectral theory, Comm. in PDE,(1989), 14,(4), 479-517.
6. M. Ben-Artzi and A. Devinatz, Resolvent estimates for a sum of tensor products with applications to the spectral theory of differential operators, J. d'Analyse Math. **43** (1984), 215–250.
7. M. Ben-Artzi and A. Devinatz, "The limiting absorption principle for partial differential operators," Memoirs Amer. Math. Soc. #364, 1987.
8. M. Ben-Artzi and A. Devinatz, Local smoothing and convergence properties for Schrödinger type equations, J. Func. Anal. **101** (1991), 231–254.
9. M. Ben-Artzi and A. Devinatz, Regularity and decay of solutions to the Stark evolution equation, J. Func. Anal. (to appear).
10. M. Ben-Artzi and S. Klainerman, Decay and regularity for the Schrödinger equation, J. d'Analyse Math. **58** (1992), 25–37.
11. M. Ben-Artzi and J. Nemirovsky, Remarks on relativistic Schrödinger operators and their extensions, Ann. Inst. H. Poincaré-Phys. Théorique, 67 (**1**), (1997), 29-39.
12. M. Sh. Birman and M. Z. Solomyak, L_2 theory of the Maxwell operator in arbitrary domains,Usp. Math. Nauk., **6**, 42 (1987), 61-76.
13. P. Constantin and J. C. Saut, Local smoothing properties of dispersive equations, J. Amer. Math. Soc. **1** (1988), 413–439.
14. C. E. Kenig, G. Ponce and L. Vega, Oscillatory integrals and regularity of dispersive equations, Indiana U. Math. J. **40** (1991), 33–69.
15. O. A. Ladyzhenskaya, "The mathematical theory of viscous incompressible flow", (1969), 23-31, Gordon and Breach Science publishers.
16. J. Nemirovsky, Limiting Absorption Principle and Spectral Analysis for Electro-magnetic Waves in Stratified Anisotropic Media, Ph. D. thesis (Hebrew University).
17. M. Reed and B. Simon, "Methods of Modern Mathematical Physics III — Scattering theory," Academic Press, 1972.
18. T. Umeda, Radiation conditions and resolvent estimates for relativistic Schrödinger operators, Ann. Inst. H. Poincaré-Phys. Théorique **63** (1995), 277–296.
19. I. Šolc and Českoslov, Časopis pro Fysiku, (1953), 3, p. 366.
20. K. Yajima, Spectral and scattering theory for Schrödinger operators with Stark-effect, J. Fac. Sci. Univ. Tokyo, Sec. 1A, **26** (1979), 377–389.
21. A. Yariv and P. Yeh, "Optical Waves in Crystals", (1984), Wiley-Interscience Publication.
22. P. Yeh, "Optical Waves in Layered Media", 1988, Wiley-Interscience Publication.

SYMMETRIC SOLUTIONS OF GINZBURG–LANDAU EQUATIONS*

S. Gustafson

Department of Mathematics
University of Toronto
Toronto, Canada

ABSTRACT

We discuss the existence and stability of spherically-symmetric solutions of the Ginzburg–Landau equation in all spatial dimensions. We briefly review known results along these lines for dimension two, and we show that in higher dimensions there is a stable symmetric solution of topological degree ± 1.

1. INTRODUCTION

The (time-independent) Ginzburg–Landau equation for vector fields $\psi : \mathbf{R}^d \to \mathbf{R}^d$ is

$$-\Delta \psi + (|\psi|^2 - 1)\psi = 0 \qquad (1.1)$$

together with the boundary condition

$$\lim_{|x| \to \infty} |\psi(x)| \to 1. \qquad (1.2)$$

Equation (1.1) (and time-dependent versions of it) arise in the study of superconductors and superfluids, and as a basic model in quantum field theory. They are among the simplest nonlinear equations, but their solutions display complex behavior, in which topology plays a fundamental role.

We note several properties of equation (1.1). Firstly, it is formally the Euler–Lagrange equation for the Ginzburg–Landau energy functional

$$E(\psi) = \frac{1}{2} \int_{\mathbf{R}^d} \{|\nabla \psi|^2 + \frac{1}{2}(|\psi|^2 - 1)^2\}. \qquad (1.3)$$

---

*Research on this paper was supported by NSERC under Grant NA7901.

Spectral and Scattering Theory, edited by Ramm,
Plenum Press, New York, 1998

Secondly, the functional (1.3) is invariant with respect to coordinate translations and rotations, as well as rotations of the target space (gauge rotations). Hence, applying one of these transformations to a solution of equation (1.1) results in another solution. Finally, for functions ψ satisfying the boundary condition (1.2), we can define the degree of ψ, $\deg(\psi)$, as the integer degree of the map

$$\left.\frac{\psi}{|\psi|}\right|_{|x|=R} : \mathbf{S}^{d-1} \to \mathbf{S}^{d-1}$$

for sufficiently large R.

We seek the simplest possible solutions of equation (1.1), with the idea that more complex solutions of corresponding time-independent equations (see section 2.2) can be built up from these. To this end, in this paper we study *spherically-symmetric* solutions of equation (1.1), by which we mean solutions, ψ, satisfying

$$\psi(gx) = \rho(g)\psi(x) \qquad (1.4)$$

for all $x \in \mathbf{R}^d$ and $g \in SO(d)$, and where $\rho : SO(d) \to SO(d)$ is a homomorphism (i.e., ρ is a d-dimensional real representation of $SO(d)$). Such a ψ is invariant under coordinate rotations, when compensated for by gauge transformations (see, e.g., [17], for a discussion of such functions).

We are also interested in finding stable solutions. In this setting, stable means linearly stable in the following sense. If ψ solves equation (1.1), then the linearized operator, L_ψ, around ψ is given by

$$L_\psi \xi = -\Delta \xi + (|\psi|^2 - 1)\xi + 2(\psi \cdot \xi)\psi$$

(note L_ψ is the Hessian of the energy E at ψ). We say ψ is *stable* if $L_\psi \geq 0$ on $L^2(\mathbf{R}^d; \mathbf{R}^d)$, and any zero-eigenvalues of L_ψ arise from the symmetry breaking of ψ (if A generates a one-parameter group of symmetries of equation (1.1) then $L_\psi A\psi = 0$). Formally, stability in this sense corresponds to strict local minimization (up to symmetries) of the energy.

In sections 2.1, 2.2 we give a brief review of results on symmetric solutions and their stability in the well-studied $d = 2$ case. In section 3 we investigate the existence of symmetric solutions for $d \geq 3$; we conclude that the only such solutions are of degree $0, \pm 1$. In section 4 we outline a proof of the stability of this degree ± 1 solution for $d = 2, 3$ (see theorem 4.1). This work draws heavily on the work of Y. Ovchinnikov and I. M. Sigal ([13]).

2. VORTICES IN TWO DIMENSIONS

2.1. The time-independent equation

When $d = 2$, equation (1.1) has well-known symmetric solutions of the form.

$$\psi_n(r, \theta) = f_n(r)e^{in\theta}$$

for each $n \in \mathbf{Z}$ (here (r, θ) are polar coordinates on $\mathbf{R}^2$ and we have identified the target space with $\mathbf{C}$). f_n satisfies $f_n(0) = 0$, $f_n(\infty) = 1$, and $f_n' > 0$. ψ_n is called an *n-vortex*, and is symmetric in the sense of (1.4) with respect to the homomorphism $\rho_n : SO(2) \to SO(2)$

given by $\rho_n : \alpha \mapsto n\alpha$ (where α represents a rotation by the angle α). Clearly, $\deg(\psi_n) = n$. Establishing the existence of a unique solution of this form amounts to solving an ODE for f_n. This can be done using a shooting method (see [3, 4, 7]). In [13], the existence of the n-vortices is established by minimizing a renormalized version of the Ginzburg–Landau energy (1.3) among functions of the form $f(x)e^{in\theta}$ with f real (for $|n| \geq 1$, $E(\psi_n) = \infty$).

The $0, \pm 1$-vortices are known to be stable (in the above sense), and the higher-degree vortices unstable ([10, 13]). An outline of the stability proof for the 1-vortex is given in section 4. In [11], it is shown that the $0, \pm 1$-vortices are (up to symmetries) the only stable solutions.

For a detailed study of equation (1.1) on bounded domains, see [1]. See [8] for a study of vortices in gauge theory.

2.2. Dynamics

The vortices can be viewed as static solutions of time-dependent equations. Of interest are the nonlinear heat equation (NLHE),

$$-\dot\psi = -\Delta\psi + (|\psi|^2 - 1)\psi$$

and the nonlinear Schrödinger equation (NLSE),

$$i\dot\psi = -\Delta\psi + (|\psi|^2 - 1)\psi.$$

For existence and uniqueness results for (NLHE), see, e.g., [2, 5, 18].

The fact that the $0, \pm 1$-vortices are linearly stable suggests that they are dynamically stable as solutions of these evolution equations. Some results in this direction are available for (NLHE). The stability of 0-vortices (constants) as solutions of (NLHE) is proved in [2]. In ([18], see also [6]), the dynamic stability of n-vortices (for all n) with respect to radially-symmetric perturbations is established for (NLHE).

The study of the general dynamics for these equations was initiated in [12] where an asymptotic 'particle + field' description of the dynamics is proposed. Here multi-scale matched asymptotic expansions are used to derive leading-order equations for the dynamics of slow-moving, widely separated vortices (i.e., localized regions where the field has a zero, and resembles a ± 1-vortex). The dynamic stability of ± 1-vortices is a basic assumption of such analysis.

For (NLHE), this program is continued in [9, 15], for example. The basic idea is that in the leading order, the positions of localized vortices evolve via a gradient flow. In this picture, vortices with local topological degrees of similar signs repel, and those with local topological degrees of opposite signs attract.

In [14], for example, this analysis is carried out for (NLSE). In this case, the leading order behavior of the vortex positions, $a_i(t)$ (with local degrees n_i), is that of a Hamiltonian flow with Hamiltonian

$$H(\{a_i\}) = -\pi \sum_{i \neq j} n_i n_j \log|a_i - a_j|.$$

3. SYMMETRIC SOLUTIONS FOR $D \geq 3$

In this section we determine that analogues of n-vortices exist only for $n = 0, \pm 1$ when $d \geq 3$.

Obviously the constant (of modulus 1) solutions exist in any dimension and have degree 0. The analogues of the ± 1-vortices in higher dimensions are solutions of the form

$$\psi_{\pm 1}(x) = \pm f(|x|)\hat{x} \tag{3.1}$$

where $\hat{x} = x/r$ and $f \geq 0$. Such functions have degree ± 1.

The existence and uniqueness of a solution of this form follows from the same ODE shooting method which gives existence of vortices in two dimensions. Just as for the n-vortices, f satisfies $f(0) = 0, f(\infty) = 1$, and $f' > 0$.

In fact, up to symmetries, there are no higher-degree spherically-symmetric solutions in the strict sense of (1.4). This follows immediately from the fact that the only d-dimensional real representations of $SO(d)$ for $d \geq 3$ are the trivial one (with homomorphism $\rho_0 : g \mapsto 1$ for all $g \in SO(d)$) and the natural action of $SO(d)$ on $\mathbf{R}^d$ (with homomorphism $\rho_1 : g \mapsto g$ for all $g \in SO(d)$). It is easy to show that any function which is spherically-symmetric with respect to ρ_1 is of the form (3.1) up to a gauge rotation. Any function which is spherically-symmetric with respect to the trivial representation depends only on the radial coordinate, $r = |x|$, and so has degree 0 (e.g., constants).

4. STABILITY OF THE DEGREE-ONE SOLUTION

In this section we outline the proof of

Theorem 4.1. *The degree-one solution, ψ_1, is stable*

for dimensions 2 and 3 (for $d = 2$, ψ_1 is the 1-vortex; in this case, see [13] for details of the proof; see also remark 4.1)

We recall that our object is to prove that the linearized operator,

$$L\xi = -\Delta \xi + (f^2 - 1)\xi + 2f^2(\hat{x} \cdot \xi)\hat{x}$$

is non-negative on $L^2(\mathbf{R}^d; \mathbf{R}^d)$, and that the only zero-eigenvalues of L correspond to symmetry zero-modes.

We proceed as follows. First, we construct an orthogonal decomposition of $L^2(\mathbf{S}^{d-1}; \mathbf{R}^d)$ into small subspaces which are invariant for the operators Δ_Ω (the spherical Laplacian) and $P : \xi \mapsto (\xi \cdot \hat{x})\hat{x}$ (which appear as the 'spherical parts' of L). Now because the radial and spherical variables in L separate ($L = -\Delta_r - (1/r^2)\Delta_\Omega + f(r)^2 - 1 + 2f(r)^2 P$), the action of L on $\xi \in L^2(\mathbf{R}^d; \mathbf{R}^d)$ is equivalent to the action of a block-diagonal operator on $L^2((0, \infty); r^{d-1} dr)$. That is, we have a direct sum decomposition of L into small matrices of ordinary differential operators in the radial variable, r. We then show each block is non-negative.

To decompose $L^2(\mathbf{S}^{d-1}; \mathbf{R}^d)$, we observe that L commutes with the image of $SO(d)$ under the representation, ρ, on $L^2(\mathbf{S}^{d-1}; \mathbf{R}^d)$ (or $L^2(\mathbf{R}^d; \mathbf{R}^d)$), given by

$$(\rho(g)\xi)(x) = g\xi(g^{-1}x)$$

for $\xi \in L^2$, $g \in SO(d)$. Therefore, L commutes with the images under ρ of the infinitesimal generators of 1-parameter subgroups in $SO(d)$ (i.e., Lie algebra elements).

For $d = 2$, the image under ρ of the generator of rotations in $SO(2)$ is

$$-\partial_\theta + \begin{pmatrix} 0 & -1 \\ 1 & 0 \end{pmatrix}$$

(where θ is the coordinate on S^1). The eigenspaces of this operator on $L^2(S^1;\mathbf{R}^2)$ are easily found, and are two-dimensional. These eigenspaces are invariant for L, and so we obtain a decomposition of L (on $L^2(\mathbf{R}^2;\mathbf{R}^2)$ now) into 2×2 blocks of ordinary differential operators in the radial variable.

For $d = 3$, we determine the images under ρ of the generators of rotations around each of the three axes. They are $L_i = (1/2)\epsilon_{ijk}(-x_j\partial_k + x_k\partial_j) + M_i$ for $i = 1,2,3$ where the 3×3 matrix M_i is $(M_i)_{jk} = \epsilon_{ijk}$ (ϵ_{ijk} is the totally antisymmetric tensor). As in the angular momentum calculation from quantum mechanics, L_3 and L^2 commute. A somewhat involved calculation determines the simultaneous eigenspaces of L_3 and L^2 on $L^2(S^2;\mathbf{R}^3)$ (which, again, are L-invariant). They are three-dimensional. In fact, each of these eigenspaces splits into smaller L-invariant pieces (a 1-dimensional piece and a 2-dimensional piece). This gives a decomposition of L (on $L^2(\mathbf{R}^3;\mathbf{R}^3)$) into 2×2 and 1×1 blocks of ordinary differential operators in the radial variable.

It remains to show non-negativity of each block on $L^2((0,\infty);r^{d-1}dr)$. For this, we take advantage of the symmetry zero-modes.

For both $d = 2$ and $d = 3$, the lowest 1×1 block (all the other 1×1 operators are greater) is the operator

$$l = -\Delta_r + \frac{d-1}{r^2} + f^2 - 1.$$

From the ODE satisfied by f, it is immediate that $lf = 0$. In fact, this zero-mode corresponds to the fact that ψ_1 breaks rotational symmetry. Using the facts that $f > 0$, $f \notin L^2$, a Perron–Frobenius type argument allows us to conclude that 0 is the bottom of the spectrum of l, and l has no 0-eigenvalue (see [13] for details).

The lowest 2×2 operator, $\tilde{l}$, in both cases has (up to orthogonal transformation)

$$\tilde{f} = \begin{pmatrix} f' \\ \sqrt{d-1}\frac{f}{r} \end{pmatrix}$$

as a zero-mode. This corresponds to the translational symmetry breaking of ψ_1. Again, $\tilde{f} > 0$, $\tilde{f} \notin L^2$, allow us to conclude that $\tilde{l} \geq 0$ with no zero-eigenvalue. We sketch the proof of this, following [13]. The technique extends somewhat standard results (e.g., [16]).

When $d = 2$, the matrix operator in question is $L = L_0 - V$ where

$$L_0 = \begin{pmatrix} -\Delta_r + \frac{4}{r^2} + f^2 - 1 & 0 \\ 0 & -\Delta_r + \frac{4}{r^2} + 3f^2 - 1 \end{pmatrix}$$

and

$$V = \frac{2}{r^2}\begin{pmatrix} 1 & 1 \\ 1 & 1 \end{pmatrix}.$$

One constructs the operator family

$$K(\lambda) = V^{1/2}R(\lambda)V^{1/2}$$

for $\lambda < 0$ where $R(\lambda) = (L_0 - \lambda)^{-1}$. The previous step implies $L_0 \geq 0$ with no 0-eigenvalue. $K(\lambda)$ can be extended to $\lambda = 0$. One can show that $\lambda \leq 0$ is the lowest eigenvalue of L if and only if 1 is the largest eigenvalue of $K(\lambda)$. Now, $R(\lambda)$ is positivity improving for $\lambda < 0$. So the fact that $V^{1/2}$ has positive entries implies $K(\lambda)$ is also positivity improving (this can also be extended to $\lambda = 0$). Now one uses this property (as in [16]) to prove that the highest eigenvalue of $K(\lambda)$ is simple, with strictly (componentwise) positive eigenfunction. Passing to L, we conclude the lowest eigenvalue of L has the same property. If L were to have a negative lowest eigenvalue, its (positive) eigenfunction would decay exponentially, and would be orthogonal to the positive function $\tilde{f}$, an impossibility. A further argument rules out the possibility of a genuine 0-eigenvalue.

This completes the outline of the stability proof. □

Remark 4.1. This result can be extended to all dimensions with a more careful argument. In particular, the unexpected splitting of $L^2(\mathbf{S}^{d-1}; \mathbf{R}^d)$ into 1- and 2-dimensional L-invariant pieces (rather that 3-dimensional ones), which we observed in the $d = 3$ case, occurs in general. In this way, the analysis carries through for all d.

ACKNOWLEDGMENT

The author would like to thank I. M. Sigal for suggesting the problems studied herein, and for many helpful discussions.

REFERENCES

1. F. Bethuel, H. Brezis, F. Hélein, *Ginzburg–Landau Vortices*, Birkhäuser: Basel (1994).
2. P. Bauman, C. Chen, D. Phillips, P. Sternberg, "Vortex annihilation in nonlinear heat flow for Ginzburg–Landau systems" *European J. of Applied Math* **6** (1995), 115–126.
3. Y. Chen, C. Elliott, T. Qi, "Shooting method for vortex solutions of a complex-valued Ginzburg–Landau equation" *Proc. of the Royal Society of Edinburgh* **124A** (1994) 1075–1068.
4. P. Fife, L. Peletier, "On the location of defects in stationary solutions of the Ginzburg–Landau equation on $\mathbf{R}^2$" *Quart. Appl. Math* **54** (1996) 85–104.
5. J. Ginibre, G. Velo, "The Cauchy problem for the complex Ginzburg–Landau equation I: compactness methods," *Physica D* **95** (1996) 191–228.
6. S. Gustafson, "Stability of vortex solutions of the Ginzburg–Landau equation" to appear in *PDEs and their Applications: CRM Proceeding and Lecture Notes*.
7. M. Hervé, R. Hervé, "Étude qualitative des solutions réelles d'une équation différentielle liée a l'équation de Ginzburg–Landau" *Ann Inst. Henri Poincaré, Analyse non linéaire* **11** no.4 (1994) 427–440.
8. A. Jaffe, C. Taubes, *Vortices and Monopoles*, Birkhäuser: Basel (1980).
9. F. H. Lin, "Some dynamical properties of Ginzburg–Landau vortices," *Comm. Pure Appl. Math* **49** (1996) 323–359.
10. P. Mironescu, "On the stability of radial solutions of the Ginzburg–Landau equation" *Journal of Functional Analysis* **130** (1995) 334–344.
11. P. Mironescu, "Les minimiseurs locaux pour l'équation de Ginzburg–Landau sont à symétrie radiale" *C. R. Acad. Sci. Paris, Ser. I, Math.* **323** no.6 (1996) 593–598.
12. J. Neu, "Vortices in complex scalar fields" *Physics D: Nonlinear Phenomena* **43** (1990) 385–406.
13. Y. Ovchinnikov, I. M. Sigal, "The Ginzburg–Landau equation I: static vortices," preprint.
14. Y. Ovchinnikov, I. M. Sigal, "The Ginzburg–Landau equation III: vortex dynamics," preprint.
15. J. Rubinstein, P. Sternberg, "On the slow motion of vortices in the Ginzburg–Landau heat flow" *SIAM J. Math. Anal.* **26** (1995) 1452–1466.
16. M. Reed, B. Simon, *Methods of Modern Mathematical Physics IV* Academic Press: New York (1979).
17. A. Schwarz, *Quantum Field Theory and Topology*, Springer-Verlag: Berlin (1993).
18. M. Weinstein, J. Xin, "Dynamic stability of the vortex solutions of the Ginzburg–Landau and nonlinear Schrödinger equations," *Comm. Math. Phys.* **180** (1996) 389–428.

4

QUANTUM MECHANICS AND RELATIVITY: THEIR UNIFICATION BY LOCAL TIME

Hitoshi Kitada

Department of Mathematical Sciences
University of Tokyo
Komaba, Meguro, Tokyo 153, Japan
E-mail: kitada@ms.u-tokyo.ac.jp

ABSTRACT

In a framework of a stationary universe, time is defined as a local and quantum-mechanical notion in the sense that it is defined for each local and quantum-mechanical system consisting of a finite number of particles. In this context, the total universe consisting of an infinite number of particles has no time associated, and quantum mechanics and general theory of relativity are united consistently. Relativistic Hamiltonians including gravitation are derived as a consequence of our treatment of observation. Related open problems in mathematical physics are presented.

1. INTRODUCTION

Physics is a work to explain phenomena, i.e. a job to give a description of visible events. Insofar as we understand physics as such activities, it is neither surprise nor ridiculous thing if one takes other ways in explaining phenomena than the present physics: The problem of combining relativity and quantum theories, which has been an old and difficult one, might be able to be considered from a different viewpoint than the present trends where relativity theory is tried to be quantized or quantum mechanics is tried to be modified relativistically. It is enough if one can explain relativistic quantum-mechanical phenomena or observations of them, in a systematic way. The purpose of the present paper is to give an attempt in this direction to explain relativistic quantum-mechanical phenomena. To make clear the contrast of our approach to the current physics, we briefly review the problems of physics in relation with relativity and quantum theories.

As is well-known, the solution $\psi = \psi(t) = \psi(x,t)$ ($x \in \mathbb{R}^{3N}$) of the Schrödinger equation for $N(\geq 1)$ particles, numbered as $\ell = 1, 2, \cdots, N$, with positions $x_\ell = (x_{\ell 1}, x_{\ell 2}, x_{\ell 3}) \in \mathbb{R}^3_{x_\ell}$ and masses $m_\ell > 0$:

$$\frac{1}{i}\frac{d\psi}{dt}(t) + H\psi(t) = 0, \quad \psi(0) = \phi, \quad \phi \in \mathcal{D}(H) \subset L^2(\mathbb{R}^{3N}),$$

$$H = -\sum_{\ell=1}^{N}\sum_{k=1}^{3}\frac{1}{2m_\ell}\frac{\partial^2}{\partial x_{\ell k}^2} + \sum_{1 \le i < j \le N} V_{ij}(x_i - x_j), \quad (1.1)$$

is invariant with respect to the Galilei transformation:

$$x'_\ell = x_\ell - vt, \quad (\ell = 1, 2, \cdots, N)$$
$$t' = t$$

up to a factor of absolute value 1:

$$\exp\left[i\sum_{\ell=1}^{N}\left(\frac{1}{2}m_\ell v^2 t - m_\ell v \cdot x_\ell\right)\right],$$

consistently with Born's interpretation [1]. Here we adopted the unit system such that $\hbar = \frac{h}{2\pi} = 1$; $v = (v_1, v_2, v_3) \in \mathbb{R}^3$ denotes the velocity between two inertial frames of reference; and $v \cdot x_\ell = \sum_{k=1}^{3} v_k x_{\ell k}$ is the inner product of v and x_ℓ. This implies that the Schrödinger equation is not invariant under Lorentz transformation: $x^{\mu'} = a^\mu_{\ \nu} x^\nu$ from $\mathbb{R}_t \times \mathbb{R}^3_{x_\ell}$ to $\mathbb{R}_{t'} \times \mathbb{R}^3_{x'_\ell}$ with $x^{0'} = ct'$, $x^{k'} = x'_{\ell k}$, $x^0 = ct$, and $x^k = x_{\ell k}$. Here c is the speed of light in vacuum and the coefficients $a^\mu_{\ \nu}$ are independent of ℓ and determined by the following condition with some extra informations:

$$(x^{1'})^2 + (x^{2'})^2 + (x^{3'})^2 - (x^{0'})^2 = (x^1)^2 + (x^2)^2 + (x^3)^2 - (x^0)^2.$$

Therefore the quantum mechanics which is described by Schrödinger equation is understood, in the current physical context, as incompatible with special theory of relativity. Even if the free energy part in (1.1) (i.e. the sum of the differential operators in (1.1)) is replaced by a Lorentz invariant one as we will do in (2.16) of Subsection 2, the Schrödinger equation is not Lorentz invariant. This fact is known as that the instantaneous force among particles, which depends only on the locations of those particles, is not relativistic, i.e. is not Lorentz invariant.

One of the features of the Schrödinger equation is that it yields the stability of matter (of the first kind, see [2]), which is violated in the classical framework of Maxwell's equations and Rutherford model of atoms, where atoms collapse by the continuous radiation of light from the electrons around the nucleus according to classical electromagnetism so that the electrons fall into the nucleus. However the Schrödinger operator H defined in (1.1) is bounded from below by some constant $-L > -\infty$ in the sense that L^2-inner product (Hf, f) satisfies

$$(Hf, f) \ge -L(f, f)$$

for any f belonging to the domain $\mathcal{D}(H)$ of H under a suitable assumption on the pair potentials V_{ij}. This means that the total energy of the quantum system does not decrease below $-L$, therefore the system does not collapse. In this respect, quantum mechanics remedies the difficulty of classical theory, while it is not Lorentz invariant.

In 1928, Dirac [3] introduced a system of equations, which is invariant under Lorentz transformation, and could explain some of the relativistic quantum-mechanical phenomena.

However, Dirac operator is not bounded from below, and Dirac equation does not imply the stability of matter unlike the Schrödinger equation.

Dirac thus proposed an idea that the vacuum is filled with electrons with negative energy so that the electrons around the nucleus cannot fall into the negative energy anymore by the Pauli exclusion principle, which explains the stability of matter. However, if one has to consider plural kinds of elementary particles at a time, one has to introduce the vacuum which is filled with those plural kinds of particles, and the vacuum can depend on the number of the kinds of particles which one takes into account. The vacuum then may not be determined to be unique. Further if one has to include Bosons into consideration, the Pauli exclusion principle does not hold and the stability of matter does not follow. In this sense, the idea of "'filling the negative energy sea,' unfortunately," "is ambiguous in the many-body case," as Lieb writes in [2], p. 33.

Quantum field theory is introduced (see [4] for a review) to overcome this difficulty as well as to explain the annihilation–creation phenomena of particles, which are familiar in elementary particle physics. Quantum field theory is a theory of infinite degrees of freedom. In the case of the Schrödinger equation (1.1), the degree of freedom is $3N$, the number of coordinates $x_{11}, x_{12}, x_{13}, \cdots, x_{N1}, x_{N2}$, and x_{N3} of N particles. Contrary to this, quantum field theory deals with the infinite number of particles, which makes it possible to discuss creation–annihilation processes inside the theory. However, since it deals with infinite number of freedom, even at the first step of the definition of the Hamiltonian of the system obtained by second quantization, there is a difficulty, the difficulty of divergence. This sort of difficulty appears at almost every stage of the development of the theory, and physicists had to find clever ways to avoid the difficulties at each step after the theory was introduced. Mathematically, the difficulty of divergence has not been overcome yet at all. Physicists however noticed that if one could get finite quantities in a systematic way by extracting some infinite quantities from the divergent quantities, then those finite quantities might express the reality. Actually in their explanation of Lamb shift, they seemed to have succeeded going in this way and to have been able to give predictions outstandingly close to experiments. However, the calculation done is up to the 6th or 8th order of a series giving Lamb shift or anomalous magnetic moment of electrons [5]. Dyson noticed [6] that the series has symptom to diverge to infinity.

The procedure mentioned in the above to yield finite quantities from infinite ones is called process of "renormalization," and still forms active areas of researches in theoretical physics. In the mathematical attempt, called "axiomatic quantum field theory," which was planned to clarify the meaning of quantum field theory and construct the theory consistently, it is known that in some mathematical but important examples (see, e.g., [7]), renormalizability conditions and the axioms of quantum field theory yield that the theory must not involve interaction terms inside the theory, i.e., the theory is void as a physics.

These are the situation currently understood as an incompatibility problem between quantum theory and special theory of relativity. In the case of general theory of relativity and quantum mechanics, the situation seems similar or no better (see, e.g., [8] for a review of the current approaches). The traditional attempts toward the unification of quantum theory and general relativity, like quantum gravity, superstring theory, and so on, are trying to find a way to unify them in a single layered theory where these two difficult theories should admit each other.

We present below an attempt in a different direction, where general theory of relativity and quantum mechanics are considered as independent aspects of nature, but as playing complementary roles to each other. Our approach may be called a two-layered theory, where these two theories have their own residences and they interfere only when observation is done.

A procedure which describes the interference between them at observation will be our basis of explanation of relativistic quantum-mechanical phenomena.

Our spirit behind the procedure we will introduce below for that interference is that what is intrinsic is the quantum-mechanical aspect of nature, while relativity plays a role of glasses to see nature. This attitude is contrary to the one adopted by current physics, which in origin comes from the spirit of Einstein [9]:

> Thus, according to the general theory of relativity, gravitation occupies an exceptional position with regard to other forces, particularly the electromagnetic forces, since the ten functions representing the gravitational field at the same time define the metrical properties of the space measured.

His position is that the metrical properties of space–time are intrinsic for nature, and space–time is a vessel of nature, into which other forces should be incorporated. In the framework of classical theory, electromagnetic forces can be treated in this direction in the sense that the equation for electromagnetic fields can be written as a tensor equation. In the framework of quantum theory, the characteristic of traditional approach is to treat gravity as a one which should be quantized, and the inclusion of other forces is a problem which is treated only after gravity is quantized successfully. In such attempts to quantize gravity or general theory of relativity, the canonical formalism of general theory of relativity is assumed usually, and this means that one has to introduce some global time coordinate which is common throughout the total universe. This itself produces a problem incompatible with the spirit of general theory of relativity that time is a local notion. If one would admit of introducing such a global time, it is difficult to reformulate general theory of relativity into canonical formalism even if gravitation is weak (see, e.g., [8]), and the quantization of gravity or space–time remains as a difficult problem even if one would defer to the global time.

To overcome these difficulties, we introduce a notion of local time t_L which is proper to each local system L consisting of a finite number of quantum-mechanical particles. Our local time is a *quantum-mechanical* notion inasmuch as it is defined in each quantum-mechanical system as a parameter t_L in the exponent of the propagator $\exp(-it_L H_L)$ describing the propagation of the local system L. It is a *local* notion defined for each local system L with a local Hamiltonian H_L, and this will enable us to regard the time t_L as a *classical general relativistic* local time, proper to the *center of mass* of the local system L, by identifying the classical particles with those centers of mass of local systems. We will show that these classical local times proper to the centers of mass of local systems constitute general relativistic notion of local times, compatible with the quantum mechanics inside each local system. The proof is, in part, a recall of the inclusion/exclusion assumption which has been adopted in physics, that the time t_B of a bigger system L_B which includes a smaller system L_S dominates the time t_S of L_S, i.e. the assumption that t_S must be equal to t_B if the system L_B includes L_S. Apart from this traditional position on which physics has been founded, we retrieve the independence of each local system and its time coordinate among local systems, and liberate them from the bondage of inclusion/exclusion relation, which has been implicitly assumed for systems of physical particles. Geometrically expressed, our position may be formulated as a vector bundle with base space X representing the Riemannian manifold consisting of classical particles, identified with the centers of mass of local systems, and with the local system L which obeys the quantum mechanics on its own geometry being associated as a fibre to each point $x \in X$, which is identified with the center of mass of the local system L.

There is a theory by Prugovečki [10] successful, in a sense, in quantizing general relativity, where he modifies quantum mechanics and general relativity so that the usual results are obtained as limits of his theory. His approach looks similar to ours in that he associates a *Lorentzian* quantum-mechanical world to each point of a Riemannian manifold as a fibre, regarding the total universe as a vector or fibre bundle equipped with connections compatible with the Riemannian metric of the base Riemannian space. Our approach differs from his in the following points:

1. Each quantum-mechanical world associated to a point of a Riemannian manifold is *Euclidean*;

2. We do not introduce any connections among those Euclidean quantum mechanics;

3. We treat electromagnetic forces and gravity on the same level in our explanation of observation, under the assumption that gravitation is weak; and

4. Quantum mechanics and general theory of relativity are intact in our formulation.

The explanation of observation stated in the third item is our point and is realized by a procedure which yields a quantized Hamiltonian including gravity and electric forces on the same level.

In our explanation of observation, we appeal to a procedure which transforms quantum-mechanical quantities into the classical quantities which obey the relativistic change of coordinates among the Euclidean quantum-mechanical worlds, which we call local systems. The quantum-mechanical world associated to each point of a Riemannian manifold has no relation with the Riemannian metric of the base space of our vector bundle, for we do not define any connections among the quantum-mechanical worlds. The procedure which transforms quantum-mechanical quantities into classical quantities is consistent with the two aspects of nature, i.e. with the quantum-mechanical aspect inside local systems and the general relativistic aspect outside local systems, because the results obtained by transformations are just concerned with observed facts. The relativity appears in our theory as "glasses," which deform quantum-mechanical quantities into classical relativistic quantities at each step of quantum-mechanical evolution, so that the resultant classical quantities accord with the observation. The intrinsic for our theory is quantum mechanics inside local systems, and relativity modifies quantum-mechanical calculations to accord with observations.

Summing up, our point is in the liberation of local systems from the inclusion/exclusion relation which has been an implicit assumption of physics. Instead of the inclusion/exclusion relation, we introduce a relation which transforms the quantum-mechanical values to classical relativistic values, as a procedure describing the interference between the two aspects of nature, the general relativistic aspect and the quantum-mechanical aspect.

In Part 2 of the paper, we give a presentation of our theory without using the notion of vector bundle. We first recall in Section 2.1 the basic notions related with the definition of local times from [11]. This notion of local times is a quantum-mechanical one defined in each local system consisting of a finite number of quantum-mechanical particles. In the sense that the local time is a local notion, it serves an ingredient which adheres the two layers: general theory of relativity and quantum mechanics. A result in many body quantum scattering is used to assign the usual meaning of time to our notion of local times. In Section 2.2, we review the proof of the consistency of the notion of local times with general theory of relativity. We give, in Section 2.3, a procedure of interpreting observation of quantum-mechanical process through the glasses of the relativity, yielding a relativistic quantum-mechanical Hamiltonian which explains gravitation and electric forces in quantum-mechanical way. In Part 3, we treat two examples following the spirit of Part 2. We present some open problems related with our formulation of physics in Part 4.

2. LOCAL TIME AND OBSERVATION

2.1. Local Time

To state our definition of quantum-mechanical local times, we begin with introducing a stationary universe ϕ. What we adopt here for the universe may be called a closed universe, within which is all and which has a definite property specified by a certain quantum-mechanical condition.

Let $\mathcal{H}$ be a separable Hilbert space, and set

$$\mathcal{U} = \{\phi\} = \bigoplus_{n=0}^{\infty} \left(\bigoplus_{\ell=0}^{\infty} \mathcal{H}^n \right) \quad (\mathcal{H}^n = \underbrace{\mathcal{H} \otimes \cdots \otimes \mathcal{H}}_{n \text{ factors}}).$$

$\mathcal{U}$ is called a Hilbert space of possible universes. An element ϕ of $\mathcal{U}$ is called a universe and is of the form of an infinite matrix $(\phi_{n\ell})$ with components $\phi_{n\ell} \in \mathcal{H}^n$. $\phi = 0$ means $\phi_{n\ell} = 0$ for all n, ℓ.

Let $\mathcal{O} = \{A\}$ be the totality of the selfadjoint operators A in $\mathcal{U}$ of the form $A\phi = (A_{n\ell}\phi_{n\ell})$ for $\phi = (\phi_{n\ell}) \in \mathcal{D}(A) \subset \mathcal{U}$, where each component $A_{n\ell}$ is a selfadjoint operator in $\mathcal{H}^n$. We assume the following condition for our universe ϕ.

Axiom 1. *There is a selfadjoint operator $H \in \mathcal{O}$ in $\mathcal{U}$ such that for some $\phi \in \mathcal{U} - \{0\}$ and $\lambda \in \mathbb{R}$*

$$H\phi = \lambda\phi \tag{2.1}$$

in the following sense: Let F_n be a finite subset of $\mathbb{N} = \{1, 2, \cdots\}$ with $\sharp(F_n)(=$ the number of elements in $F_n) = n$ and let $\{F_n^\ell\}_{\ell=0}^{\infty}$ be a countable set of such F_n. Then the formula (2.1) in the above means that there are integral sequences $\{n_k\}_{k=1}^{\infty}$ and $\{\ell_k\}_{k=1}^{\infty}$ and a real sequence $\{\lambda_{n_k \ell_k}\}_{k=1}^{\infty}$ such that $F_{n_k}^{\ell_k} \subset F_{n_{k+1}}^{\ell_{k+1}}$; $\bigcup_{k=1}^{\infty} F_{n_k}^{\ell_k} = \mathbb{N}$;

$$H_{n_k \ell_k} \phi_{n_k \ell_k} = \lambda_{n_k \ell_k} \phi_{n_k \ell_k}, \cdot \phi_{n_k \ell_k} \neq 0, \quad k = 1, 2, 3, \cdots; \tag{2.2}$$

and

$$\lambda_{n_k \ell_k} \to \lambda \quad \text{as} \quad k \to \infty.$$

H is an infinite matrix $(H_{n\ell})$ of selfadjoint operators $H_{n\ell}$ in $\mathcal{H}^n$. Axiom 1 asserts that this matrix converges in the sense of (2.1) on our universe ϕ. We remark that our universe ϕ is not determined uniquely by this condition.

The universe as a state ϕ is a whole, within which is all. As such a whole, the state ϕ can follow the two ways: The one is that ϕ develops along a global time T in the grand universe $\mathcal{U}$ under a propagation $\exp(-iTH)$, and another is that ϕ is a bound state of H. If there were such a global time T as in the first case, all phenomena had to develop along that global time T, and the locality of time would be lost. We could then *not* construct a notion of local times compatible with general theory of relativity. The only one possibility is therefore to adopt the stationary universe ϕ of Axiom 1.

The following axiom asserts the existence of configuration and momentum operators and that the canonical commutation relation between them holds. This is a basis of our definition of time, where configuration and momentum are given first, and then local times are defined in each local system of finite number of quantum-mechanical particles.

Axiom 2. *Let $n \geq 1$ and F_{n+1} be a finite subset of $\mathbb{N} = \{1, 2, \cdots\}$ with $\sharp(F_{n+1}) = n+1$. Then for any $j \in F_{n+1}$, there are selfadjoint operators $X_j = (X_{j1}, X_{j2}, X_{j3})$ and $P_j = (P_{j1}, P_{j2}, P_{j3})$ in $\mathcal{H}^n$, and constants $m_j > 0$ such that*

$$[X_{j\ell}, X_{km}] = 0, \quad [P_{j\ell}, P_{km}] = 0, \quad [X_{j\ell}, P_{km}] = i\delta_{jk}\delta_{\ell m},$$

$$\sum_{j \in F_{n+1}} m_j X_j = 0, \quad \sum_{j \in F_{n+1}} P_j = 0.$$

The Stone–von Neumann theorem and Axiom 2 specify the space dimension (see [12], p. 452) as 3 dimension. We identify $\mathcal{H}^n$ with $L^2(\mathbb{R}^{3n})$ in the following.

What we want to mean by the (n,ℓ)-th component $H_{n\ell}$ $(n, \ell \geq 0)$ of $H = (H_{n\ell})$ in Axiom 1 is the usual $N = n+1$ body Hamiltonian with center of mass removed in accordance with the requirement $\sum_{j \in F_{n+1}} m_j X_j = 0$ in Axiom 2. For the local Hamiltonian $H_{n\ell}$ we thus make the following postulate.

Axiom 3. *The component Hamiltonian $H_{n\ell}$ $(\ell \geq 0)$ of H in Axiom 1 is of the form*

$$H_{n\ell} = H_{n\ell 0} + V_{n\ell}, \quad V_{n\ell} = \sum_{\substack{\alpha = (i,j) \\ 1 \leq i < j < \infty, i, j \in F_N^\ell}} V_\alpha(x_\alpha)$$

on $C_0^\infty(\mathbb{R}^{3n})$, where $x_\alpha = x_i - x_j$ with x_i being the position vector of the i-th particle, and $V_\alpha(x_\alpha)$ is a real-valued measurable function of $x_\alpha \in \mathbb{R}^3$ which is $H_{n\ell 0}$-bounded with $H_{n\ell 0}$-bound of $V_{n\ell}$ less than 1. $H_{n\ell 0} = H_{(N-1)\ell 0}$ is the free Hamiltonian of the N-particle system, whose concrete form is similar to the interaction-free part of H in (1.1) of the introduction.

This axiom implies that $H_{n\ell} = H_{(N-1)\ell}$ is uniquely extended to a selfadjoint operator bounded from below in $\mathcal{H}^n = \mathcal{H}^{N-1} = L^2(\mathbb{R}^{3(N-1)})$ by the Kato–Rellich theorem.

We do not include vector potentials in the Hamiltonian $H_{n\ell}$ of Axiom 3, for we take the position that what is elementary is the electronic charge, and the magnetic forces are the consequence of the motions of charges. Thus when we restrict our attention to a system consisting of the N number of particles, the vector potential is redundant to our argument. It would be, however, a good approximation to introduce vector potentials, when we consider a subsystem of a bigger system, and we concentrate on the analysis of the behavior of that subsystem inside the bigger system.

Let P_H denote the orthogonal projection onto the space of bound states for a selfadjoint operator H. We recall that a state orthogonal to the space of bound states is called a scattering state. Let $\phi = (\phi_{n\ell})$ with $\phi_{n\ell} = \phi_{n\ell}(x_1, \cdots, x_n) \in L^2(\mathbb{R}^{3n})$ be the universe in Axiom 1, and let $\{n_k\}$ and $\{\ell_k\}$ be the sequences specified there. Let $x^{(n,\ell)}$ denote the relative coordinates of $n+1$ particles in F_{n+1}^ℓ.

Definition 2.1.

1. We define $\mathcal{H}_{n\ell}$ as the sub-Hilbert space of $\mathcal{H}^n$ generated by the functions $\phi_{n_k \ell_k}(x^{(n,\ell)}, y)$ of $x^{(n,\ell)} \in \mathbb{R}^{3n}$ with regarding $y \in \mathbb{R}^{3(n_k - n)}$ as a parameter, where k moves over a set $\{k \mid n_k \geq n, F_{n+1}^\ell \subset F_{n_k+1}^{\ell_k}, k \in \mathbb{N}\}$.

2. $\mathcal{H}_{n\ell}$ is called a *local universe* of ϕ.

3. $\mathcal{H}_{n\ell}$ is said to be non-trivial if $(I - P_{H_{n\ell}})\mathcal{H}_{n\ell} \neq \{0\}$.

The total universe ϕ is a single element in $\mathcal{U}$. The local universe $\mathcal{H}_{n\ell}$ may be richer and may have elements more than one. This is because we consider the subsystems of the universe consisting of a finite number of particles. These subsystems receive the influence from the other particles of infinite number outside the subsystems, and may vary to constitute a non-trivial subspace $\mathcal{H}_{n\ell}$.

To state this mathematically, let us assume that the pair potentials are of the Coulomb type $V_{ij}(x_{ij}) = c_{ij}|x_{ij}|^{-1}$ ($c_{ij} \in \mathbb{R}$), which are the typical examples we had in mind in Axiom 3. Consider, e.g., a system $H_{3\ell}$ consisting of four particles, the one of which has positive charge, and other three have negative charge. Then this system tends to scatter, i.e. it is probable that this system is in a scattering state with respect to the Hamiltonian $H_{3\ell}$ (see, e.g., [13], p. 50) for a theorem asserting the absence of eigenvalues for a similar case). Add one particle with positive charge to this system $H_{3\ell}$ to constitute a system $H_{n_k\ell_k}$ ($n_k = 4$). Then this new system may be in an eigenstate $\phi_{n_k\ell_k} = \phi_{4\ell_k}$ with respect to the extended Hamiltonian $H_{n_k\ell_k} = H_{4\ell_k}$ for *some* eigenvalue $\lambda_{n_k\ell_k} = \lambda_{4\ell_k}$ so that it satisfies the condition (2.2) in the above for a k, while the restriction $\phi_{4\ell_k}(x^{(3,\ell)}, y)$ to $\mathbb{R}^9_{x^{(3,\ell)}}$, with $y \in \mathbb{R}^3$ arbitrary but fixed, of the bound state $\phi_{4\ell_k}$ of $H_{4\ell_k}$ is a scattering state of the original system $H_{3\ell}$. Here $y \in \mathbb{R}^3$ is the intercluster coordinates between the added particle of positive charge and the center of mass of the four particles in the system $H_{3\ell}$. Namely, the extended system $H_{4\ell}$ is in a bound state $\phi_{4\ell_k}$, while the restriction $\phi_{4\ell_k}(x^{(3,\ell)}, y)$ moves over the scattering states of $H_{3\ell}$ belonging to the Hilbert space $L^2(\mathbb{R}^9_{x^{(3,\ell)}})$ of the state vectors for the system $H_{3\ell}$, and constitutes a nontrivial subspace $\mathcal{H}_{3\ell}$ of $\mathcal{H}^3$ when y varies.

Definition 2.2.

1. The restriction of H to $\mathcal{H}_{n\ell}$ is also denoted by the same notation $H_{n\ell}$ as the (n, ℓ)-th component of H.

2. We call the pair $(H_{n\ell}, \mathcal{H}_{n\ell})$ a local system.

3. The unitary group $e^{-itH_{n\ell}}$ ($t \in \mathbb{R}^1$) on $\mathcal{H}_{n\ell}$ is called the *proper clock* of the local system $(H_{n\ell}, \mathcal{H}_{n\ell})$, if $\mathcal{H}_{n\ell}$ is non-trivial: $(I - P_{H_{n\ell}})\mathcal{H}_{n\ell} \neq \{0\}$. (Note that the clock is defined only for $N = n + 1 \geq 2$, since $H_{0\ell} = 0$ and $P_{H_{0\ell}} = I$.)

4. The universe ϕ is called *rich* if $\mathcal{H}_{n\ell}$ equals $\mathcal{H}^n = L^2(\mathbb{R}^{3n})$ for all $n \geq 1$, $\ell \geq 0$. For a rich universe ϕ, $H_{n\ell}$ equals the (n, ℓ)-th component of H.

Definition 2.3.

1. The parameter t in the exponent of the proper clock $e^{-itH_{n\ell}} = e^{-itH_{(N-1)\ell}}$ of a local system $(H_{n\ell}, \mathcal{H}_{n\ell})$ is called the (quantum-mechanical) *proper time* or *local time* of the local system $(H_{n\ell}, \mathcal{H}_{n\ell})$, if $(I - P_{H_{n\ell}})\mathcal{H}_{n\ell} \neq \{0\}$.

2. This time t is denoted by $t_{(H_{n\ell}, \mathcal{H}_{n\ell})}$ indicating the local system under consideration.

This definition is a one reverse to the usual definition of the motion or dynamics of the N-body quantum systems, where the time t is given *a priori* and then the motion of the particles is defined by $e^{-itH_{(N-1)\ell}}f$ for a given initial state f of the system.

Time is thus defined only for local systems $(H_{n\ell}, \mathcal{H}_{n\ell})$ and is determined by the associated proper clock $e^{-itH_{n\ell}}$. Therefore there are infinitely many number of times $t = t_{(H_{n\ell}, \mathcal{H}_{n\ell})}$ each of which is proper to the local system $(H_{n\ell}, \mathcal{H}_{n\ell})$. In this sense time is a local notion. There is no time for the total universe ϕ in Axiom 1, which is a bound state of the total Hamiltonian H in the sense specified by the condition (2.1) of Axiom 1.

To see the meaning of our definition of time, we quote a theorem from [14]. To state the theorem we make some notational preparation concerning the local system $(H_{n\ell}, \mathcal{H}_{n\ell})$, assuming that the universe ϕ is rich: Let $b = (C_1, \cdots, C_{\sharp(b)})$ be a decomposition of the set $\{1,2,\cdots,N\}$ ($N = n+1$) into $\sharp(b)$ number of disjoint subsets $C_1, \cdots, C_{\sharp(b)}$ of $\{1,2,\cdots,N\}$. b is called a cluster decomposition. $H_b = H_{n\ell,b} = H_{n\ell} - I_b = H_{n\ell}^b + T_{n\ell,b} = H^b + T_b$ is the truncated Hamiltonian for the cluster decomposition b with $1 \leq \sharp(b) \leq N$, where I_b is the sum of the intercluster interactions between various two different clusters in b, and T_b is the sum of the intercluster free energies among various clusters in b. $x_b \in \mathbb{R}^{3(\sharp(b)-1)}$ is the intercluster coordinates among the centers of mass of the clusters in b, while $x^b \in \mathbb{R}^{3(N-\sharp(b))}$ denotes the intracluster coordinates inside the clusters of b so that $x \in \mathbb{R}^{3n} = \mathbb{R}^{3(N-1)}$ is expressed as $x = (x_b, x^b)$. Note that x^b is decomposed as $x^b = (x_1^b, \cdots, x_{\sharp(b)}^b)$, where each $x_j^b \in \mathbb{R}^{3(\sharp(C_j)-1)}$ is the internal coordinate of the cluster C_j, describing the configuration of the particles inside C_j. The operator H^b is accordingly decomposed as $H^b = H_1 + \cdots + H_{\sharp(b)}$, and each component H_j is defined in the space $\mathcal{H}_j^b = L^2(\mathbb{R}_{x_j^b}^{3(\sharp(C_j)-1)})$, whose tensor product $\mathcal{H}_1^b \otimes \cdots \otimes \mathcal{H}_{\sharp(b)}^b$ is the internal state space $\mathcal{H}^b = L^2(\mathbb{R}_{x^b}^{3(N-\sharp(b))})$. The free energy T_b is defined in the external space $\mathcal{H}_b = L^2(\mathbb{R}_{x_b}^{3(\sharp(b)-1)})$, and the truncated Hamiltonian $H_b = H^b + T_b = I \otimes H^b + T_b \otimes I$ is defined in the total space $\mathcal{H}_{n\ell} = \mathcal{H}_b \otimes \mathcal{H}^b = L^2(\mathbb{R}_x^{3(N-1)})$. v_b is the velocity operator conjugate to the intercluster coordinates x_b. $P_b = P_{H^b}$ is the eigenprojection associated with the subsystem H^b of H, i.e. the orthogonal projection onto the eigenspace of H^b, defined in $\mathcal{H}^b$ and extended obviously to the total space $\mathcal{H}_{n\ell}$. P_b^M is the M-dimensional partial projection of this eigenprojection P_b. We define for a k-dimensional multi-index $M = (M_1, \cdots, M_k)$, $M_j \geq 1$ and $k = 1, \cdots, N-1$,

$$\hat{P}_k^M = \left(I - \sum_{\sharp(b)=k} P_b^{M_k}\right) \cdots \left(I - \sum_{\sharp(d)=2} P_d^{M_2}\right)(I - P^{M_1}),$$

where note that $P^{M_1} = P_a^{M_1} = P_H^{M_1}$ for $\sharp(a) = 1$ is uniquely determined. We also define for a $\sharp(b)$-dimensional multi-index $M_b = (M_1, \cdots, M_{\sharp(b)-1}, M_{\sharp(b)}) = (\hat{M}_b, M_{\sharp(b)})$

$$\tilde{P}_b^{M_b} = P_b^{M_{\sharp(b)}} \hat{P}_{\sharp(b)-1}^{\hat{M}_b}, \quad 2 \leq \sharp(b) \leq N.$$

It is clear that

$$\sum_{2 \leq \sharp(b) \leq N} \tilde{P}_b^{M_b} = I - P^{M_1},$$

provided that the component M_k of M_b depends only on the number k but not on b. In the following we use such M_b's only. Under these circumstances, the following is known to hold.

Theorem 2.1 ([14]). *Let $N = n + 1 \geq 2$ and let $H_{N-1} = H_{n\ell}$ be the Hamiltonian for a local system $(H_{n\ell}, \mathcal{H}_{n\ell})$. Let suitable conditions on the decay rate for the pair potentials $V_{ij}(x_{ij})$ be satisfied (see, e.g., Assumption 1 in [15]). Let $\||x^a|^2 P_a^M\| < \infty$ be satisfied for any integer $M \geq 1$ and cluster decomposition a with $2 \leq \sharp(a) \leq N - 1$. Let $f \in \mathcal{H}^{N-1}$. Then there is a sequence*

$t_m \to \pm\infty$ (as $m \to \pm\infty$) and a sequence M_b^m of multi-indices whose components all tend to ∞ as $m \to \pm\infty$ such that for all cluster decompositions b, $2 \le \#(b) \le N$, and $\varphi \in C_0^\infty(\mathbb{R}_{x_b}^{3(\#(b)-1)})$

$$\|\{\varphi(x_b/t_m) - \varphi(v_b)\}\tilde{P}_b^{M_b^m} e^{-it_m H_{N-1}} f\| \to 0 \tag{2.3}$$

as $m \to \pm\infty$.

The asymptotic relation (2.3) roughly means that, if we restrict our attention to the part $\tilde{P}_b^{M_b^m}$ of the evolution $e^{-itH_{N-1}}f$, in which the particles inside any cluster of b are bounded while any two different clusters of b are scattered, then the quantum-mechanical velocity $v_b = m_b^{-1} p_b$, where m_b is some diagonal mass matrix, is approximated by a classical value $v_b^{(c)} = \lim_{m \to \pm\infty}(v_b \tilde{P}_b^{M_b^m} e^{-it_m H_{N-1}} f, \tilde{P}_b^{M_b^m} e^{-it_m H_{N-1}} f)$ asymptotically as $m \to \pm\infty$ and the local time t of the N body system $H_{N-1} = H_{n\ell}$ is asymptotically equal to the quotient of the configuration by the velocity of the scattered particles (or clusters, exactly speaking):

$$\frac{|x_b|}{\left|v_b^{(c)}\right|}. \tag{2.4}$$

This means by $v_b = m_b^{-1} p_b$ that the local time t is asymptotically and approximately measured if the values of the configurations and momenta for the scattered particles of the local system $(H_{N-1}, \mathcal{H}_{N-1}) = (H_{n\ell}, \mathcal{H}_{n\ell})$ are given.

We note that the time measured by (2.4) is independent of the choice of cluster decomposition b according to Theorem 2.1. This means that t can be taken as a common parameter of motion inside the local system, and can be called *time* of the local system in accordance with the notion of 'common time' in Newton's sense: "relative, apparent, and common time, is some sensible and external (whether accurate or inaccurate) measure of duration by the means of motion, $\cdots$" ([16], p. 6). Once we take t as our notion of time for the system $(H_{n\ell}, \mathcal{H}_{n\ell})$, t recovers the usual meaning of time, by the identity for $e^{-itH_{n\ell}} f$ known as the Schrödinger equation:

$$\left(\frac{1}{i}\frac{d}{dt} + H_{n\ell}\right) e^{-itH_{n\ell}} f = 0.$$

Time $t = t_{(H_{n\ell},\mathcal{H}_{n\ell})}$ is a notion defined only in relation with the local system $(H_{n\ell}, \mathcal{H}_{n\ell})$. To other local system $(H_{mk}, \mathcal{H}_{mk})$, there is associated other local time $t_{(H_{mk},\mathcal{H}_{mk})}$, and between $t = t_{(H_{n\ell},\mathcal{H}_{n\ell})}$ and $t_{(H_{mk},\mathcal{H}_{mk})}$, there is no relation, and they are completely independent notions. In other words, $\mathcal{H}_{n\ell}$ and $\mathcal{H}_{mk}$ are different spaces unless $n = m$ and $\ell = k$. And even when the two local systems $(H_{n\ell}, \mathcal{H}_{n\ell})$ and $(H_{mk}, \mathcal{H}_{mk})$ have a non-vanishing common part: $F_{n+1}^\ell \cap F_{m+1}^k \ne \emptyset$, the common part constitutes its own local system $(H_{pj}, \mathcal{H}_{pj})$, and its local time cannot be compared with those of the two bigger systems $(H_{n\ell}, \mathcal{H}_{n\ell})$ and $(H_{mk}, \mathcal{H}_{mk})$, because these three systems have different base spaces, Hamiltonians, and clocks. More concretely speaking, the times are measured through the quotients (2.4) for each system. But the L^2-representations of the base Hilbert spaces $\mathcal{H}_{n\ell}, \mathcal{H}_{mk}, \mathcal{H}_{pj}$ for those systems are different unless they are identical with each other, and the quotient (2.4) has incommensurable meaning among these representations.

In this sense, local systems are independent mutually. Also they cannot be decomposed into pieces in the sense that the decomposed pieces constitute different local systems.

2.2. Relativity

We note that the center of mass of a local system $(H_{n\ell}, \mathcal{H}_{n\ell})$ is always at the origin of the space coordinate system $x_{(H_{n\ell}, \mathcal{H}_{n\ell})} \in \mathbb{R}^3$ for the local system by the requirement: $\sum_{j \in F_{n+1}} m_j X_j = 0$ in Axiom 2, and that the space coordinate system describes just the relative motions inside a local system by our formulation. The center of mass of a local system, therefore, cannot be identified from the local system itself, except that it is at the origin of the coordinates.

Moreover, just as we have seen in the previous section, we see that, not only the time coordinates $t_{(H_{n\ell}, \mathcal{H}_{n\ell})}$ and $t_{(H_{mk}, \mathcal{H}_{mk})}$, but also the space coordinates $x_{(H_{n\ell}, \mathcal{H}_{n\ell})} \in \mathbb{R}^3$ and $x_{(H_{mk}, \mathcal{H}_{mk})} \in \mathbb{R}^3$ of these two local systems are independent mutually. Thus the space-time coordinates $(t_{(H_{n\ell}, \mathcal{H}_{n\ell})}, x_{(H_{n\ell}, \mathcal{H}_{n\ell})})$ and $(t_{(H_{mk}, \mathcal{H}_{mk})}, x_{(H_{mk}, \mathcal{H}_{mk})})$ are independent between two different local systems $(H_{n\ell}, \mathcal{H}_{n\ell})$ and $(H_{mk}, \mathcal{H}_{mk})$. In particular, insofar as the systems are considered as quantum-mechanical ones, there is no relation between their centers of mass. In other words, the center of mass of any local system cannot be identified by other local systems quantum-mechanically.

Summing these two considerations, we conclude:

1. The center of mass of a local system $(H_{n\ell}, \mathcal{H}_{n\ell})$ cannot be identified *quantum-mechanically* by any local system $(H_{mk}, \mathcal{H}_{mk})$ including the case $(H_{mk}, \mathcal{H}_{mk}) = (H_{n\ell}, \mathcal{H}_{n\ell})$.

2. There is no *quantum-mechanical* relation between any two local coordinates $(t_{(H_{n\ell}, \mathcal{H}_{n\ell})}, x_{(H_{n\ell}, \mathcal{H}_{n\ell})})$ and $(t_{(H_{mk}, \mathcal{H}_{mk})}, x_{(H_{mk}, \mathcal{H}_{mk})})$ of two different local systems $(H_{n\ell}, \mathcal{H}_{n\ell})$ and $(H_{mk}, \mathcal{H}_{mk})$.

Utilizing these properties of the centers of mass and the coordinates of local systems, we may make any postulates concerning

1. the motions of the *centers of mass* of various local systems, and

2. the relation between two local coordinates of any two local systems.

In particular, we may impose *classical* postulates on them as far as the postulates are consistent in themselves.

Thus we assume an arbitrary but fixed transformation:

$$y_2 = f_{21}(y_1) \qquad (2.5)$$

between the coordinate systems $y_j = (y_j^\mu) = (ct_j, x_j)$ for $j = 1, 2$, where c is the speed of light in vacuum and (t_j, x_j) is the space-time coordinates of the local system $L_j = (H_{n_j \ell_j}, \mathcal{H}_{n_j \ell_j})$. We regard these coordinates $y_j = (ct_j, x_j)$ as *classical* coordinates, when we consider the motions of centers of mass and the relations of coordinates of various local systems. We now postulate the general principle of relativity on the physics of the centers of mass:

Axiom 4. *The laws of physics which control the relative motions of the centers of mass of local systems are covariant under the change of the coordinates from* $(ct_{(H_{mk}, \mathcal{H}_{mk})}, x_{(H_{mk}, \mathcal{H}_{mk})})$ *to* $(ct_{(H_{n\ell}, \mathcal{H}_{n\ell})}, x_{(H_{n\ell}, \mathcal{H}_{n\ell})})$ *of the reference frame local systems for any pair* $(H_{mk}, \mathcal{H}_{mk})$ *and* $(H_{n\ell}, \mathcal{H}_{n\ell})$ *of local systems.*

We note that this axiom is consistent with the Euclidean metric adopted for the quantum-mechanical coordinates inside a local system, because Axiom 4 is concerned with classical motions of the centers of mass *outside* local systems, and we are dealing here with a different aspect of nature from the quantum-mechanical one *inside* a local system.

Axiom 4 implies the invariance of the distance under the change of coordinates between two local systems. Thus the metric tensor $g_{\mu\nu}(ct,x)$ which appears here satisfies the transformation rule:

$$g^1_{\mu\nu}(y_1) = g^2_{\alpha\beta}(f_{21}(y_1)) \frac{\partial f^\alpha_{21}}{\partial y^\mu_1}(y_1) \frac{\partial f^\beta_{21}}{\partial y^\nu_1}(y_1), \qquad (2.6)$$

where $y_1 = (ct_1, x_1)$; $y_2 = f_{21}(y_1)$ is the transformation (2.5) in the above from $y_1 = (ct_1, x_1)$ to $y_2 = (ct_2, x_2)$; and $g^j_{\mu\nu}(y_j)$ is the metric tensor expressed in the classical coordinates $y_j = (ct_j, x_j)$ for $j = 1, 2$.

The second postulate is the principle of equivalence, which asserts that the classical coordinate system $(ct_{(H_{n\ell}, \mathcal{H}_{n\ell})}, x_{(H_{n\ell}, \mathcal{H}_{n\ell})})$ is a local Lorentz system of coordinates, insofar as it is concerned with the classical behavior of the center of mass of the local system $(H_{n\ell}, \mathcal{H}_{n\ell})$:

Axiom 5. *The metric or the gravitational tensor $g_{\mu\nu}$ for the center of mass of a local system $(H_{n\ell}, \mathcal{H}_{n\ell})$ in the coordinates $(ct_{(H_{n\ell}, \mathcal{H}_{n\ell})}, x_{(H_{n\ell}, \mathcal{H}_{n\ell})})$ of itself are equal to $\eta_{\mu\nu}$, where $\eta_{\mu\nu} = 0$ for $\mu \neq \nu$, $= 1$ for $\mu = \nu = 1, 2, 3$, and $= -1$ for $\mu = \nu = 0$.*

Since, at the center of mass, the classical space coordinates $x = 0$, Axiom 5 together with the transformation rule (2.6) in the above yields

$$g^1_{\mu\nu}(f_{21}^{-1}(ct_2, 0)) = \eta_{\alpha\beta} \frac{\partial f^\alpha_{21}}{\partial y^\mu_1}(f_{21}^{-1}(ct_2, 0)) \frac{\partial f^\beta_{21}}{\partial y^\nu_1}(f_{21}^{-1}(ct_2, 0)). \qquad (2.7)$$

Also by the same reason: $x = 0$ at the center of mass, the relativistic proper time $d\tau = \sqrt{-g_{\mu\nu}(ct, 0) dy^\mu dy^\nu} = \sqrt{-\eta_{\mu\nu} dy^\mu dy^\nu}$ at the origin of a local system is equal to c times the quantum-mechanical proper time dt of the system.

By the fact that the classical Axioms 4 and 5 of physics are imposed on the centers of mass which are uncontrollable quantum-mechanically, and on the relation between the coordinates of different, therefore quantum-mechanically non-related local systems, the consistency of classical relativistic Axioms 4 and 5 with quantum-mechanical Axioms 1–3 is clear:

Theorem 2.2. *Axioms 1 to 5 are consistent.*

2.3. Observation

Thus far, we did not mention any about the physics which is actually observed. We have just given two aspects of nature which are mutually independent. We will introduce a procedure to yield what we observe when we see nature. This procedure will not be contradictory with the two aspects of nature which we have discussed, as the procedure is concerned solely with "*how nature looks, at the observer,*" i.e. it is solely concerned with "*at the place of the observer, how nature looks,*" with some abuse of the word "place." The validity of the procedure should be judged merely through the comparison between the observation and the prediction given by our procedure.

We note that we can observe only a finite number of disjoint systems, say $L_1, \cdots, L_k$ with $k \geq 1$ a finite integer. We cannot grasp an infinite number of systems at a time. Further each system L_j must have only a finite number of elements by the same reason. Thus these systems $L_1, \cdots, L_k$ may be identified with local systems in the sense of Section 2.1.

Local systems are quantum-mechanical systems, and their coordinates are confined to their insides insofar as we appeal to Axioms 1–3. However we postulated Axioms 4 and 5 on the classical aspects of those coordinates, which make the local coordinates of a local system a classical reference frame for the centers of mass of other local systems. This leaves us the room to define observation as the *classical* observation of the centers of mass of local systems $L_1,\cdots,L_k$. We call this an observation of $L = (L_1,\cdots,L_k)$ inquiring into sub-systems $L_1,\cdots,L_k$, where L is a local system consisting of the particles which belong to one of the local systems $L_1,\cdots,L_k$.

When we observe the sub-local systems $L_1,\cdots,L_k$ of L, we observe the relations or motions among these sub-systems. Internally the local system L behaves following the Hamiltonian H_L associated to the local system L. However the actual observation differs from what the pure quantum-mechanical calculation gives for the system L. For example, when an electron is scattered by a nucleus with relative velocity close to that of light, the observation is different from the pure quantum-mechanical prediction.

In the usual explanation of this phenomenon, one introduces Dirac equation, and calculates differential cross section. However, the calculation only applies to that experiment or to the case which can be described by the Dirac equation, and no gravity is included.

We propose below a procedure which explains gravity as well as quantum-mechanical forces in one framework.

The quantum-mechanical process inside the local system L is described by the evolution

$$\exp(-it_L H_L)f,$$

when the initial state f of the system and the local time t_L of the system are given. The Hamiltonian H_L is decomposed as follows in virtue of the local Hamiltonians $H_1,\cdots,H_k$, which correspond to the sub-local systems $L_1,\cdots,L_k$:

$$H_L = H^b + T + I, \quad H^b = H_1 + \cdots + H_k.$$

Here $b = (C_1,\cdots,C_k)$ is the cluster decomposition corresponding to the decomposition $L = (L_1,\cdots,L_k)$ of L; $H^b = H_1 + \cdots + H_k$ is the sum of the internal energies H_j inside L_j, and is defined in the internal state space $\mathcal{H}^b = \mathcal{H}_1^b \otimes \cdots \otimes \mathcal{H}_k^b$; $T = T_b$ denotes the intercluster free energy among the clusters $C_1,\cdots,C_k$ defined in the external state space $\mathcal{H}_b$; and $I = I_b = I_b(x) = I_b(x_b, x^b)$ is the sum of the intercluster interactions between various two different clusters in the cluster decomposition b (cf. the explanation after Definition 2.3 in Section 2.1).

The main concern in this process would be the case that the clusters $C_1,\cdots,C_k$ form asymptotically bound states as $t_L \to \infty$, since other cases are hard to be observed along the process if the observer's concern is upon the final state of the sub-systems $L_1,\cdots,L_k$.

The evolution $\exp(-it_L H_L)f$ behaves asymptotically as $t_L \to \infty$ as follows for some bound states $g_1,\cdots,g_k$ ($g_j \in \mathcal{H}_j^b$) of local Hamiltonians $H_1,\cdots,H_k$ and for some g_0 belonging to the external state space $\mathcal{H}_b$:

$$\exp(-it_L H_L)f \sim \exp(-it_L h_b)g_0 \otimes \exp(-it_L H_1)g_1 \otimes \cdots \otimes \exp(-it_L H_k)g_k, \quad k \geq 1, \quad (2.8)$$

where $h_b = T_b + I_b(x_b, 0)$. It is easy to see that $g = g_0 \otimes g_1 \otimes \cdots \otimes g_k$ is given by

$$g = g_0 \otimes g_1 \otimes \cdots \otimes g_k = \Omega_b^{+*}f = P_b\Omega_b^{+*}f,$$

provided that the decomposition of the evolution $\exp(-it_L H_L)f$ is of the simple form as in (2.8). Here Ω_b^{+*} is the adjoint of a canonical wave operator [17] corresponding to the cluster decomposition b:

$$\Omega_b^+ = s\text{-}\lim_{t\to\infty}\exp(itH_L)\cdot\exp(-ith_b)\otimes\exp(-itH_1)\otimes\cdots\otimes\exp(-itH_k)P_b,$$

where P_b is the eigenprojection onto the eigenspace of the Hamiltonian $H^b = H_1 + \cdots + H_k$. The process (2.8) just describes the quantum-mechanical process inside the local system L, and does not specify any meaning related with observation up to the present stage.

To see what we observe at actual observations, let us reflect a process of observation of scattering phenomena. We note that the observation of scattering phenomena is concerned with their initial and final stages by what the scattering itself means. At the final stage of observation of scattering processes, the quantities observed are firstly the points hit by the scattered particles on the screen stood against them. If the circumstances are properly set up, one can further indicate the momentum of the scattered particles at the final stage to the extent that the uncertainty principle allows. Consider, e.g., a scattering process of an electron by a nucleus. Given the magnitude of initial momentum of an electron relative to the nucleus, one can infer the magnitude of momentum of the electron at the final stage as being equal to the initial one by the law of conservation of energy, since the electron and the nucleus are far away at the initial and final stages so that the potential energy between them can be neglected compared to the relative kinetic energy. The direction of momentum at the final stage can also be indicated, up to the error due to the uncertainty principle, by setting a sequence of slits toward the desired direction at each point on the screen so that the observer can detect only the electrons scattered to that direction. The magnitude of momentum at initial stage can be selected in advance by applying a uniform magnetic field to the electrons, perpendicularly to their momenta, so that they circulate around circles with the radius proportional to the magnitude of momentum, and then by setting a sequence of slits midst the stream of those electrons. The selection of magnitude of initial momentum makes the direction of momentum ambiguous due to the uncertainty principle, since the sequence of slits lets the position of electrons accurate to some extent. To sum up, the sequences of slits at the initial and final stages necessarily require to take into account the uncertainty principle so that some ambiguity remains in the observation.

However, in the actual observation of a *single* particle, we *have to decide* at which point on the screen the particle hits and which momentum the particle has, using the prepared apparatus like the sequence of slits located at each point on the screen. Even if we impose an interval for the observed values, we *have to assume* that the boundaries of the interval are sharply designated. These are the assumption which we always impose on "observations" implicitly, i.e., we idealize the situation in any observation or in any measurement of a single particle so that the observed values for each particle are sharp for both of the configuration and momentum. In this sense, the values observed actually for each particle must be classical. We have then necessary and sufficient conditions to make predictions about the differential cross section, as we will see in Subsection 1.

Summarizing, we observe just the classical quantities for each particle at the final stage of all observations. In other words, we have to *presuppose* that the values observed for each particle have sharp values, even if we cannot know the values actually. We can apply to this fact the remark stated in the third paragraph of this section about the possibility of defining observation as that of the *classical* centers of mass of local systems, and may assume that the actually observed values follow the classical Axioms 4 and 5. Those sharp values actually observed for each particle give, when summed over the large number of particles, the probabilistic nature of physical phenomena, i.e. that of scattering phenomena.

Theoretically, the quantum-mechanical, probabilistic nature of scattering processes is described by differential cross section, defined as the square of the absolute value of the scattering amplitude obtained from scattering operators $S_{bd} = W_b^{+*} W_d^{-}$, where $W_b^{\pm}$ are usual wave operators. Given the magnitude of the initial momentum of the incoming particle and the scattering angle, the differential cross section gives a prediction about the probability at which point and to which direction on the screen each particle hits on the average. However, as we have remarked, the idealized point on the screen hit by each particle and the scattering angle given as an idealized difference between the directions of the initial and final momenta of each particle have sharp values, and the observation at the final stage is *classical*. We are then required to supplement these classical observations with taking into account the classical relativistic effects on those classical quantities, e.g., on the configuration and the momentum of each particle.

1. As the first step of the relativistic modification of the scattering process, we consider the scattering amplitude $\mathcal{S}(E, \theta)$, where E denotes the energy level of the scattering process and θ is a parameter describing the direction of the scattered particles. Following our remark made in the previous paragraph, we make the following postulate on the scattering amplitude observed in actual experiment:

Axiom 6. *When one observes the final stage of scattering phenomena, the total energy E of the scattering process should be regarded as a classical quantity and is replaced by a relativistic quantity, which obeys the relativistic change of coordinates from the scattering system to the observer's system.*

Since it is not known much about $\mathcal{S}(E, \theta)$ in the many body case, we consider an example of the two body case: Consider a scattering phenomenon of an electron by a Coulomb potential Ze^2/r, where Z is a real number, $r = |x|$, and x is the position vector of the electron relative to the scatterer. We assume that the mass of the scatterer is large enough compared to that of the electron and that $|Z|/137 \ll 1$. Then quantum mechanics gives the differential cross section in a Born approximation:

$$\frac{d\sigma}{d\Omega} = |\mathcal{S}(E, \theta)|^2 \approx \frac{Z^2 e^4}{16 E^2 \sin^4(\theta/2)},$$

where θ is the scattering angle and E is the total energy of the system of the electron and the scatterer. We assume that the observer is stationary with respect to the center of mass of this system of an electron and the scatterer. Then, since the electron is far away from the scatterer after the scattering and the mass of the scatterer is much larger than that of the electron, we may suppose that the energy E in the formula in the above can be replaced by the *classical* kinetic energy of the electron by Axiom 6. Then, assuming that the speed v of the electron relative to the observer is small compared to the speed c of light in vacuum and denoting the rest mass of the electron by m, we have by Axiom 6 that E is observed to have the following relativistic value:

$$E' = c\sqrt{p^2 + m^2 c^2} - mc^2 = \frac{mc^2}{\sqrt{1 - (v/c)^2}} - mc^2 \approx \frac{mv^2}{2\sqrt{1 - (v/c)^2}},$$

where $p = mv/\sqrt{1 - (v/c)^2}$ is the relativistic momentum of the electron. Thus the differential cross section should be observed approximately equal to

$$\frac{d\sigma}{d\Omega} \approx \frac{Z^2 e^4}{4m^2 v^4 \sin^4(\theta/2)}(1-(v/c)^2). \quad (2.9)$$

This coincides with the usual relativistic prediction obtained from the Klein–Gordon equation by a Born approximation. See [11], p. 297 for a case which involves the spin of the electron.

Before proceeding to the inclusion of gravity in the general k cluster case, we review this two body case. We note that the two body case corresponds to the case $k = 2$, where L_1 and L_2 consist of single particle, therefore the corresponding Hamiltonians H_1 and H_2 are zero operators on $\mathcal{H}^0 = \mathbb{C} =$ the complex numbers. The scattering amplitude $\mathcal{S}(E, \theta)$ in this case is an integral kernel of the scattering matrix $\hat{S} = \mathcal{F} S \mathcal{F}^{-1}$, where $S = W^{+*} W^-$ is a scattering operator; $W^\pm = s\text{-lim}_{t \to \pm \infty} \exp(itH_L) \exp(-itT)$ are wave operators (T is negative Laplacian for short-range potentials under an appropriate unit system, while it has to be modified when long-range potentials are included); and $\mathcal{F}$ is Fourier transformation so that $\mathcal{F} T \mathcal{F}^{-1}$ is a multiplication operator by $|\xi|^2$ in the momentum representation $L^2(\mathbb{R}^3_\xi)$. By definition, S commutes with T. This makes $\hat{S}$ decomposable with respect to $|\xi|^2 = \mathcal{F} T \mathcal{F}^{-1}$: For a.e. $E > 0$, there is a unitary operator $\mathcal{S}(E)$ on $L^2(S^2)$, S^2 being two dimensional sphere with radius one, such that for a.e. $E > 0$ and $\omega \in S^2$

$$(\hat{S}h)(\sqrt{E}\omega) = \left(\mathcal{S}(E)h(\sqrt{E}\cdot)\right)(\omega), \quad h \in L^2(\mathbb{R}^3_\xi) = L^2((0,\infty), L^2(S^2_\omega), |\xi|^2 d|\xi|).$$

Thus $\hat{S}$ can be written as $\hat{S} = \{\mathcal{S}(E)\}_{E>0}$. It is known [18] that $\mathcal{S}(E)$ can be expressed as

$$(\mathcal{S}(E)\varphi)(\theta) = \varphi(\theta) - 2\pi i \sqrt{E} \int_{S^2} \mathcal{S}(E, \theta, \omega) \varphi(\omega) d\omega$$

for $\varphi \in L^2(S^2)$. The integral kernel $\mathcal{S}(E, \theta, \omega)$ with ω being the direction of initial wave, is the scattering amplitude $\mathcal{S}(E, \theta)$ stated in the above and $|\mathcal{S}(E, \theta, \omega)|^2$ is called differential cross section. These are the most important quantities in physics in the sense that they are the *only* quantities which can be observed in actual physical observation.

The energy level E in the previous example thus corresponds to the energy shell $T = E$, and the replacement of E by E' in the above means that T is replaced by a *classical relativistic* quantity $E' = c\sqrt{p^2 + m^2 c^2} - mc^2$. We have then seen that the calculation in the above gives a correct relativistic result, which explains the actual observation.

Axiom 6 is concerned with the observation of the final stage of scattering phenomena. To include the gravity into our consideration, we extend Axiom 6 to the intermediate process of quantum-mechanical evolution. The intermediate process cannot be an object of any *actual* observation, because the intermediate observation would change the process itself, consequently the result observed at the final stage would be altered. Our next Axiom 7 is an extension of Axiom 6 from the *actual* observation to the *ideal* observation in the sense that Axiom 7 is concerned with such invisible intermediate processes and modifies the *ideal* intermediate classical quantities by relativistic change of coordinates. The spirit of the treatment developed below is to trace the quantum-mechanical paths by ideal observations so that the quantities will be transformed into classical quantities at each step, but the quantum-mechanical paths will not be altered owing to the *ideality* of the observations. The classical Hamiltonian obtained at the last step will be "requantized" to recapture the quantum-mechanical nature of the process, therefore the ideality of the intermediate observations will be realized in the final expression of the propagator of the observed system.

2. With these remarks in mind, we return to the general k cluster case, and consider a way to include gravity in our framework.

In the scattering process into $k \geq 1$ clusters, what we observe are the centers of mass of those k clusters $C_1, \cdots, C_k$, and of the combined system $L = (L_1, \cdots, L_k)$. In the example of the two body case of the previous subsection, only the combined system $L = (L_1, L_2)$ appears due to $H_1 = H_2 = 0$, therefore the replacement of T by E' is concerned with the free energy between two clusters C_1 and C_2 of the combined system $L = (L_1, L_2)$.

Following this treatment of T in Subsection 1, we replace $T = T_b$ in the exponent of $\exp(-it_L h_b) = \exp(-it_L(T_b + I_b(x_b, 0)))$ on the right hand side of the asymptotic relation (2.8) by the relativistic kinetic energy T'_b among the clusters $C_1, \cdots, C_k$ around the center of mass of $L = (L_1, \cdots, L_k)$, defined by

$$T'_b = \sum_{j=1}^{k} \left(c\sqrt{p_j^2 + m_j^2 c^2} - m_j c^2 \right). \tag{2.10}$$

Here $m_j > 0$ is the rest mass of the cluster C_j, which involves all the internal energies like the kinetic energies inside C_j and the rest masses of the particles inside C_j, and p_j is the relativistic momentum of the center of mass of C_j inside L around the center of mass of L. For simplicity, we assume that the center of mass of L is stationary relative to the observer. Then we can set in the exponent of $\exp(-it_L(T'_b + I_b(x_b, 0)))$

$$t_L = t_O, \tag{2.11}$$

where t_O is the observer's time.

For the factors $\exp(-it_L H_j)$ on the right hand side of (2.8), the object of the (*ideal*) observation is the centers of mass of the k number of clusters $C_1, \cdots, C_k$. These are the ones which now require the relativistic treatment. Since we identify the clusters $C_1, \cdots, C_k$ as their centers of mass moving in a classical fashion, t_L in the exponent of $\exp(-it_L H_j)$ should be replaced by c^{-1} times the classical relativistic proper time at the origin of the local system L_j, which is equal to the quantum-mechanical local time t_j of the sub-local system L_j. By the same reason and by the fact that H_j is the internal energy of the cluster C_j relative to its center of mass, it would be justified to replace the Hamiltonian H_j in the exponent of $\exp(-it_j H_j)$ by the classical relativistic energy *inside* the cluster C_j around its center of mass

$$H'_j = m_j c^2, \tag{2.12}$$

where $m_j > 0$ is the same as in the above.

Summing up, we arrive at the following postulate, which has the same spirit as in Axiom 6 and includes Axiom 6 as a special case concerned with actual observation:

Axiom 7. *In either actual or ideal observation, the space–time coordinates* (ct_L, x_L) *and the four momentum* $p = (p^\mu) = (E_L/c, p_L)$ *of the observed system L should be replaced by classical relativistic quantities, which are transformed into the classical quantities* (ct_O, x_O) *and* $p = (E_O/c, p_O)$ *in the observer's system L_O according to the relativistic change of coordinates specified in Axioms 4 and 5. Here t_L is the local time of the system L and x_L is the internal space coordinates inside the system L; and E_L is the internal energy of the system L and p_L is the momentum of the center of mass of the system L.*

In the case of the present scattering process into k clusters, the system L in this axiom is each of the local systems L_j ($j = 1, 2, \cdots, k$) and L.

We continue to consider the k centers of mass of the clusters $C_1,\cdots,C_k$. At the final stage of the scattering process, the velocities of the centers of mass of the clusters $C_1,\cdots,C_k$ would be steady, say $v_1,\cdots,v_k$, relative to the observer's system. Thus, according to Axiom 7, the local times t_j ($j=1,2,\cdots,k$) in the exponent of $\exp(-it_j H'_j)$, which are equal to c^{-1} times the relativistic proper times at the origins $x_j = 0$ of the local systems L_j, are expressed in the observer's time coordinate t_O by

$$t_j = t_O\sqrt{1-(v_j/c)^2} \approx t_O\left(1 - v_j^2/(2c^2)\right), \quad j=1,2,\cdots,k, \tag{2.13}$$

where we have assumed $|v_j/c| \ll 1$ and used Axioms 4 and 5 to deduce the Lorentz transformation:

$$t_j = \frac{t_O - (v_j/c^2)x_O}{\sqrt{1-(v_j/c)^2}}, \quad x_j = \frac{x_O - v_j t_O}{\sqrt{1-(v_j/c)^2}}.$$

(For simplicity, we wrote the Lorentz transformation for the case of 2-dimensional space–time.)

Inserting (2.10)–(2.12) and (2.11)–(2.13) into the right-hand side of (2.8), we obtain a classical approximation of the evolution:

$$\exp\left(-it_O[(T'_b + I_b(x_b,0) + H'_1 + \cdots + H'_k) - (m_1 v_1^2/2 + \cdots + m_k v_k^2/2)]\right) \tag{2.14}$$

under the assumption that $|v_j/c| \ll 1$ for all $j=1,2,\cdots,k$.

What we want to clarify is the final stage of the scattering process. Thus as we have mentioned, we may assume that all clusters $C_1,\cdots,C_k$ are far away from any of the other clusters and moving almost in steady velocities $v_1,\cdots,v_k$ relative to the observer. We denote by r_{ij} the distance between two centers of mass of the clusters C_i and C_j for $1 \le i < j \le k$. Then, according to our spirit that we are observing the behavior of the centers of mass of the clusters $C_1,\cdots,C_k$ in *classical* fashion following Axioms 4 and 5, the clusters $C_1,\cdots,C_k$ can be regarded to have gravitation among them. This gravitation can be calculated if we assume Einstein's field equation, $|v_j/c| \ll 1$, and certain conditions that the gravitation is weak (see [19], section 17.4), in addition to our Axioms 4 and 5. As an approximation of the first order, we obtain the gravitational potential of Newtonian type for, e.g., the pair of the clusters C_1 and $U_1 = \bigcup_{i=2}^k C_i$:

$$-G\sum_{i=2}^k m_1 m_i/r_{1i},$$

where G is Newton's gravitational constant.

Considering the k body classical problem for the k clusters $C_1,\cdots,C_k$ moving in the sum of these gravitational fields, we see that the sum of the kinetic energies of $C_1,\cdots,C_k$ and the gravitational potentials among them is constant by the classical law of conservation of energy:

$$m_1 v_1^2/2 + \cdots + m_k v_k^2/2 - G \sum_{1 \le i < j \le k} m_i m_j/r_{ij} = \text{constant}.$$

Assuming that $v_j \to v_{j\infty}$ as time tends to infinity, we have $\text{constant} = m_1 v_{1\infty}^2/2 + \cdots + m_k v_{k\infty}^2/2$. Inserting this relation into (2.14) in the above, we obtain the following as a classical approximation of the evolution (2.8):

$$\exp\left(-it_O\left[T'_b + I_b(x_b,0) + \sum_{j=1}^k (m_j c^2 - m_j v_{j\infty}^2/2) - G \sum_{1 \le i < j \le k} m_i m_j/r_{ij}\right]\right). \tag{2.15}$$

What we do at this stage are *ideal* observations, and these observations should not give any sharp classical values. Thus we have to consider (2.15) as a *quantum-mechanical evolution* and we have to recapture the quantum-mechanical feature of the process. To do so we replace p_j in T'_b in (2.15) by a quantum-mechanical momentum D_j, where D_j is a differential operator $-i\frac{\partial}{\partial x_j} = -i\left(\frac{\partial}{\partial x_{j1}}, \frac{\partial}{\partial x_{j2}}, \frac{\partial}{\partial x_{j3}}\right)$ with respect to the 3-dimensional coordinates x_j of the center of mass of the cluster C_j. Thus the actual process should be described by (2.15) with T'_b replaced by a quantum-mechanical Hamiltonian

$$\tilde{T}_b = \sum_{j=1}^{k}\left(c\sqrt{D_j^2 + m_j^2 c^2} - m_j c^2\right).$$

This procedure may be called "requantization," and is summarized as the following axiom concerning the ideal observation.

Axiom 8. *In the expression describing the classical process at the time of the ideal observation, the intercluster momentum $p_j = (p_{j1}, p_{j2}, p_{j3})$ should be replaced by a quantum-mechanical momentum $D_j = -i\left(\frac{\partial}{\partial x_{j1}}, \frac{\partial}{\partial x_{j2}}, \frac{\partial}{\partial x_{j3}}\right)$. Then this gives the evolution describing the intermediate quantum-mechanical process.*

We thus arrive at an approximation for a quantum-mechanical Hamiltonian including gravitational effect up to a constant term, which depends on the system L and its decomposition into $L_1, \cdots, L_k$, but not affecting the quantum-mechanical evolution, therefore can be eliminated:

$$\tilde{H}_L = \tilde{T}_b + I_b(x_b, 0) - G \sum_{1 \leq i < j \leq k} m_i m_j / r_{ij}$$

$$= \sum_{j=1}^{k}\left(c\sqrt{D_j^2 + m_j^2 c^2} - m_j c^2\right) + I_b(x_b, 0) - G \sum_{1 \leq i < j \leq k} m_i m_j / r_{ij}. \quad (2.16)$$

We remark that the gravitational terms here come from the substitution of local times t_j to the time t_L in the factors $\exp(-it_L H_j)$ on the right-hand side of (2.8). This form of Hamiltonian in (2.16) is actually used in [2] with $I_b = 0$ to explain the stability and instability of cold stars of large mass, showing the effectiveness of the Hamiltonian.

Summarizing these arguments from (2.8) to (2.16), we have obtained the following *interpretation* of the observation of the quantum-mechanical evolution: To get our prediction for the observation of local systems $L_1, \cdots, L_k$, the quantum-mechanical evolution of the combined local system $L = (L_1, \cdots, L_k)$

$$\exp(-it_L H_L)f$$

should be replaced by the following evolution, in the approximation of the first order under the assumption that $|v_j/c| \ll 1$ $(j = 1, 2, \cdots, k)$ and the gravitation is weak,

$$(\exp(-it_0 \tilde{H}_L) \otimes \underbrace{I \otimes \cdots \otimes I}_{k \text{ factors}}) P_b \Omega_b^{+*} f, \quad (2.17)$$

provided that the original evolution $\exp(-it_L H_L)f$ decomposes into k number of clusters $C_1, \cdots, C_k$ as $t_L \to \infty$ in the sense of (2.8). Here b is the cluster decomposition $b = (C_1, \cdots, C_k)$

that corresponds to the decomposition $L = (L_1, \cdots, L_k)$ of L; t_O is the observer's time; and

$$\tilde{H}_L = \tilde{T}_b + I_b(x_b, 0) - G \sum_{1 \leq i < j \leq k} m_i m_j / r_{ij} \qquad (2.18)$$

is the relativistic Hamiltonian inside L given by (2.16), which describes the motion of the centers of mass of the clusters $C_1, \cdots, C_k$.

We remark that (2.17) may produce a bound state combining $C_1, \cdots, C_k$ as $t_O \to \infty$ therefore for all t_O, due to the gravitational potentials in the exponent. Note that this is not prohibited by our assumption that $\exp(-it_L H_L)f$ has to decompose into k clusters $C_1, \cdots, C_k$, because the assumption is concerned with the original Hamiltonian H_L but not with the resultant Hamiltonian $\tilde{H}_L$.

Extending our primitive assumption Axiom 6, which was valid for an example stated in Subsection 1, we have arrived at a relativistic Hamiltonian $\tilde{H}_L$, which would describe approximately the intermediate process, under the assumption that the gravitation is weak and the velocities of the particles are small compared to c, by using the Lorentz transformation. We note that, since we started our argument from the asymptotic relation (2.8), which is concerned with the final stage of scattering processes, we could assume that the velocities of particles are almost steady relative to the observer in the correspondent classical expressions of the processes, therefore we could appeal to the Lorentz transformations when performing the change of coordinates in the relevant arguments.

The final values of scattering amplitude should be calculated by using the Hamiltonian $\tilde{H}_L$. Then they would explain actual observations. This is our prediction for the observation of relativistic quantum-mechanical phenomena including the effects by gravity and quantum-mechanical forces.

In the example discussed in Subsection 1, this approach gives the same result as (2.9) in the approximation of the first order, showing the consistency of our spirit. This can be seen by a representation formula of the scattering matrix similar to (3.7) in [18] for the present Hamiltonian $\tilde{H}_L = \tilde{T} + V - GmM/r$ with $V = Ze^2/r$ and M being the mass of the scatterer: Define $\mu = (E + mc^2)^2/c^2 - m^2c^2 = c^{-2}E(E + 2mc^2)$ for $E > 0$, and set

$$(\mathcal{F}_0(E)f)(\omega) = c^{-1}\sqrt{2(E + mc^2)}\mu^{1/4}(\mathcal{F}f)(\sqrt{\mu}\omega), \quad f \in C_0^\infty(\mathbb{R}^3),$$

where $\mathcal{F}$ is Fourier transformation, so that $\mathcal{F}_0(E)$ decomposes $\tilde{T} = c\sqrt{D^2 + m^2c^2} - mc^2$

$$\mathcal{F}_0(E)\tilde{T}f = E\mathcal{F}_0(E)f,$$

and satisfies

$$\int_0^\infty \|\mathcal{F}_0(E)f\|_{L^2(S^2)}^2 dE = \|f\|_{L^2(\mathbb{R}^3)}^2.$$

The scattering matrix $\mathcal{S}(E)$ is then given as follows, in Born approximation of the first order under the assumption that $|Z|/137 \ll 1$, as in (3.7) of [18]:

$$\mathcal{S}(E) \approx I - 2\pi i \mathcal{F}_0(E) \tilde{V} \mathcal{F}_0(E)^*,$$

where $\tilde{V}$ is a modified Coulomb potential obtained from $V = Ze^2/r$ in accordance with the long-range tail of V, and we omitted the gravitational potential in $\tilde{H}_L$, since it is small compared to V. Then, calculating by using oscillatory integrals, we obtain the differential cross

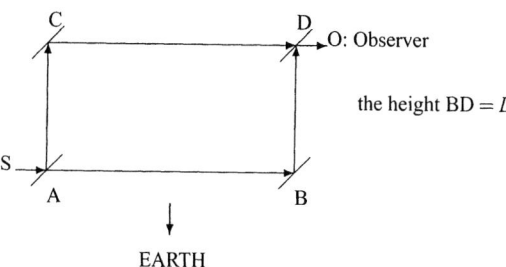

Figure 1.

section $|S(E,\theta)|^2$ equal to (2.9), if we replace the quantum-mechanical quantity $\tilde{T} = E$ by the corresponding classical quantity $E' = c\sqrt{p^2 + m^2c^2} - mc^2 \approx mv^2/\{2\sqrt{1-(v/c)^2}\}$, assuming that the speed v of the electron is small compared to c.

We remark that our stand does not require the resultant Hamiltonian $\tilde{H}_L$ to satisfy the Lorentz invariance or other kinds of invariance under transformations among coordinate systems, unlike the usual attempts require in constructing relativistic quantum theories. We have just given a procedure to predict what we actually observe, but did not propose a physical law. Usual attempts identify physics with observation, and require such kind of invariance of observation. We separate observation from physics, allowing asymmetry to observation, but with preserving two mutually incompatible invariances for physics: Galilei invariance for internal quantum mechanics and general relativistic invariance for external classical physics. This becomes possible by our position that relativity is concerned with the external world outside local systems, but not with the internal physics, which is ruled by quantum mechanics. In fact, we postulated relativity as concerned with the centers of mass of local systems in Axioms 4 and 5, and in Axioms 6 and 7 we clarified the role the relativity plays when observing the centers of mass. We refer the reader to [20] for further philosophical position of ours.

3. EXAMPLES

In this Part 3 we consider two examples of human size and of cosmological size following the spirit of the previous Part, both of which involve the quantum-mechanical aspects and relativistic aspects simultaneously.

3.1. Scattering of One Neutron in A Uniform Gravitational Field

Consider the experiment done by Collela et al. [21] of measuring the interference of one neutron. This experiment is described in some simplification as in Figure 1.

A neutron beam emitted at S is split into two beams by an interferometer at A, and the two beams are recombined at point D by other interferometers or mirrors B and C. The height L of the line BD on the earth can be varied. The dependence on L of the relative phase difference is given as follows, according to the experiment of [21], up to the error of about 1 %:

$$\hbar^{-1} mgLT,$$

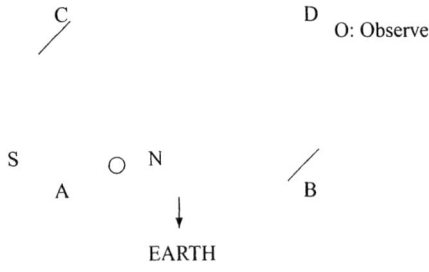

Figure 2.

where m is the mass of the neutron, g is the acceleration by gravity, and T is the (observed) time that the beams travel from C to D or A to B. This experiment shows that quantum mechanics and gravity play important roles *simultaneously* in the size of desktop environment. In fact, the lengths of the lines AB and BD are less than 10 cm in [21].

This experiment can be explained in our context, if we see it as a 3-body scattering process of a neutron N by two mirrors B and C as in Figure 2.

We denote the local system of the three bodies N, B, and C by L. Let the masses of mirrors B and C be M, the neutron mass be m, and assume $0 < m \ll M$. Let x, X_B and X_C denote the 3-dimensional coordinates of N, B and C. Then the Hamiltonian of this system L with $\hbar = 1$ is

$$H = \frac{D_x^2}{2m} + \frac{D_B^2}{2M} + \frac{D_C^2}{2M},$$

where D_x, D_B and D_C are the momentum operators $\frac{1}{i}\frac{\partial}{\partial x} = \frac{1}{i}\left(\frac{\partial}{\partial x_1}, \frac{\partial}{\partial x_2}, \frac{\partial}{\partial x_3}\right)$, $\frac{1}{i}\frac{\partial}{\partial X_B}$, and $\frac{1}{i}\frac{\partial}{\partial X_C}$ for N, B and C. To separate the center of mass, we introduce the two sets of Jacobi coordinates:

$$\begin{cases} x^{(1)} = x - X_C, \\ y^{(1)} = X_B - \frac{mx + MX_C}{m+M}, \end{cases} \quad (3.1)$$

and

$$\begin{cases} x^{(2)} = x - X_B, \\ y^{(2)} = X_C - \frac{mx + MX_B}{m+M}. \end{cases} \quad (3.2)$$

These choices of Jacobi coordinates $(x^{(j)}, y^{(j)})$ ($j = 1, 2$) correspond to two cluster decompositions $b^{(j)} = (C_1^{(j)}, C_2^{(j)})$ of L such that $C_1^{(1)} = \{N, C\}$ and $C_2^{(1)} = \{B\}$, or $C_1^{(2)} = \{N, B\}$ and $C_2^{(2)} = \{C\}$. In either case, $x^{(j)}$ is the internal coordinate inside the cluster $C_1^{(j)}$, and $y^{(j)}$ is the intercluster coordinate between two clusters $C_1^{(j)}$ and $C_2^{(j)}$.

Using these coordinates, we remove the center of mass of the system L. Then we obtain the Hamiltonian H which has the same form for both coordinates:

$$H = H^{(1)} + T^{(1)} = H^{(2)} + T^{(2)},$$

$$H^{(j)} = \frac{(D_x^{(j)})^2}{2\mu}, \quad T^{(j)} = \frac{(D_y^{(j)})^2}{2\nu}.$$

Here $D_x^{(j)} = \frac{1}{i}\frac{\partial}{\partial x^{(j)}}$ and $D_y^{(j)} = \frac{1}{i}\frac{\partial}{\partial y^{(j)}}$ are momentum operators conjugate to $x^{(j)}$ and $y^{(j)}$ ($j = 1, 2$), and μ, ν are the reduced masses:

$$\mu^{-1} = m^{-1} + M^{-1}, \quad \nu^{-1} = M^{-1} + (m+M)^{-1}.$$

Note that the operators $D_x^{(j)}$ and $D_y^{(j)}$ are mutually independent, therefore $H^{(j)}$ commutes with $T^{(j)}$. Thus the propagation of the 3-body system L is given by

$$\exp(-itH)f = \exp(-itT^{(j)})\exp(-itH^{(j)})f, \quad (3.3)$$

where $f = f(x^{(j)}, y^{(j)})$ is the initial wave function at time $t = 0$, just after the neutron has been split into two beams by the interferometer A. Here the *time t* is the local time determined by the Hamiltonian H or the correspondent local system L.

$x^{(j)}$ is the distance vector between N and C, or between N and B, and $y^{(j)}$ is the distance vector between B and the center of mass of the system N+C, or between C and the center of mass of the system N+B. Therefore, as seen from the formula for $y^{(j)}$ in (3.1) or (3.2), we may regard it as

$$y^{(1)} = X_B - X_C \quad \text{or} \quad y^{(2)} = X_C - X_B,$$

for M is larger enough than m. We can thus regard $y^{(j)}$ as constant during the scattering process, hence $f(x^{(j)}, y^{(j)})$ can be regarded as a function of $x^{(j)}$ only. (Exactly speaking, $f(x^{(j)}, y^{(j)})$ can be written as $f(x^{(j)})F(y^{(j)})$ with $F(y^{(j)})$ close to the delta function $\delta_{\pm BC}$ having support at $y^{(j)} = X_B - X_C$ or $y^{(j)} = X_C - X_B$. But as we will see, the factor $F(y^{(j)})$ does not play any essential role in our argument, and we can simply omit it from $f(x^{(j)}, y^{(j)})$.)

Namely $f(x^{(j)}, y^{(j)})$ can be regarded as the wave function of the neutron N, and is split into two wave packets $f_1(x^{(j)}), f_2(x^{(j)})$ at time $t = 0$ by the interferometer A:

$$f = f_1 + f_2.$$

f_1 is the wave packet moving to the direction from A to C, and f_2 is the one from A to B. (3.3) can then be rewritten as follows:

$$\exp(-itH)f = \exp(-itT^{(1)})\exp(-itH^{(1)})f_1 + \exp(-itT^{(2)})\exp(-itH^{(2)})f_2.$$

As remarked in the above, we can regard $y^{(1)} = X_B - X_C$ or $y^{(2)} = X_C - X_B$, therefore we may set $T^{(1)} = T^{(2)} = T$. We thus have

$$\exp(-itH)f = \exp(-itT)\{\exp(-itH^{(1)})f_1 + \exp(-itH^{(2)})f_2\}. \quad (3.4)$$

The description up to here is by the local time t determined by the local system L.

$H^{(1)}$ is the Hamiltonian of the local system consisting of N and C, and the center of mass of N and C is regarded, by $m \ll M$, as located at C with the same height as the observer. Hence, corresponding to (2.11) in Subsection 2, we can set in the first term $\exp(-itH^{(1)})f_1$ of (3.4):

$$t = t_O.$$

$H^{(2)}$ is the Hamiltonian consisting of N and B, and its center of mass is regarded as located at B by $m \ll M$. Therefore that local system has a lower gravitational potential in amount gL compared to the observer O, hence the local time t of the local system $H^{(2)}$ is related with the observer's time t_O as in (2.13) of Subsection 2:

$$t = t_O\sqrt{1 - (2gL)/c^2} \approx t_O(1 - (gL)/c^2).$$

Therefore

$$\exp(-itH^{(2)})f_2 \approx \exp(-it_O \cdot H^{(2)})\exp(it_O \cdot (gL/c^2)H^{(2)})f_2.$$

We note that we can regard $H^{(1)} = H^{(2)}$ by $H = H^{(1)} + T^{(1)} = H^{(2)} + T^{(2)}$ and $T^{(1)} = T^{(2)} = T$. As in (2.12) of Subsection 2, the internal energy $H^{(1)} = H^{(2)}$ of the system N+C or N+B is then approximated by a classical quantity $\mu c^2 = m\left(1 - \frac{m}{m+M}\right) \approx mc^2$.

For the first factor $\exp(-itT)$ on the right hand side of (3.4), the time t in the exponent is the local time of the local system L as in (2.11) of Subsection 2, because $T = T^{(j)}$ is the total intercluster free energy $T_{b^{(j)}}$ corresponding to the cluster decomposition $b^{(j)}$ of the local system L. Since the center of mass of the local system L is at the middle height between B and C, the time t in $\exp(-itT)$ is thus related with t_O as follows:

$$t = t_O\sqrt{1 - (gL)/c^2}.$$

From these, we have the following decomposition of the observed wave function for this 3-body system:

$$\exp(-itH)f \approx \exp(-it_O\sqrt{1 - (gL)/c^2}\,T)\exp(-it_O mc^2)\{f_1 + \exp(it_O \cdot gLm)f_2\}.$$

Setting

$$h_k(t_O) = \exp(-it_O\sqrt{1 - (gL)/c^2}\,T)\exp(-it_O mc^2)f_k, \quad (k = 1, 2)$$

we then have

$$\exp(-itH)f \approx h_1(t_O) + \exp(it_O \cdot gLm)h_2(t_O).$$

Therefore at the time t_O of observation, there remains the desired phase difference, which explains the interference observed in [21]. Note that $T = T^{(j)} = \frac{(D_y^{(j)})^2}{2\nu}$ does not play any essential role in this argument, therefore we did not replace it by a classical quantity.

3.2. Hubble's Law

Hubble's law is a phenomenon that appears when one observes the light emitted from stars and galaxies far away from the earth. The emission of light itself is a quantum-mechanical phenomenon that could be explained by the nonrelativistic quantum field theory as in [11], Section 11-(2). The observation or reception of this emission of light on the earth is explained as a classical observation according to our postulate Axiom 7, by introducing Robertson–Walker metric.

Robertson–Walker metric is the metric derived from the assumptions of *homogeneity* and *isotropy* of the large scale structure of the universe. We refer the reader to [19], Chap. 27 for the details, and we here only outline the argument.

Under the hypotheses of homogeneity and isotropy, the metric is given in general as follows:

$$ds^2 = -(dx^0)^2 + d\sigma^2 = -(dx^0)^2 + a(x^0)^2 \gamma_{ij}(x^k) dx^i dx^j,$$

where x^0 is the time parameter that 'slices' the space–time by means of a one parameter family of some spacelike surfaces, and (x^1, x^2, x^3) is the 'comoving, synchronous space coordinate system' for the universe, in the sense of [19], sections 27.3–27.4. $a(x^0)$ is the so-called "expansion factor" that describes the ratio of expansion of the universe in the usual context of general theory of relativity. A consideration by the use of homogeneity and isotropy yields ([19], section 27.6) that for some functions $f(r)$ $(r = |(x^1, x^2, x^3)|)$ and $h(x^0)$

$$ds^2 = -(dx^0)^2 + e^{f(r)} e^{h(x^0)} \{(dx^1)^2 + (dx^2)^2 + (dx^3)^2\}.$$

Assuming Einstein field equation $G^\mu_\nu - \lambda \delta^\mu_\nu = \kappa T^\mu_\nu$ and calculating, we get with replacing $e^{h(x^0)}$ by a constant times $e^{h(x^0)}$

$$ds^2 = -(dx^0)^2 + e^{h(x^0)} \left(1 + k\frac{r^2}{4r_0^2}\right)^{-2} \{(dx^1)^2 + (dx^2)^2 + (dx^3)^2\},$$

where $k = -1, 0,$ or $+1$. This is called Robertson–Walker metric. Using the polar coordinates (r, θ, φ) and setting $t = x^0$ and

$$\frac{r}{r_0} = u, \quad R(t) = r_0 e^{h(t)/2},$$

one can rewrite ds^2 as follows:

$$ds^2 = -(dt)^2 + R(t)^2 \left(1 + \frac{k}{4}u^2\right)^{-2} [du^2 + u^2\{(d\theta)^2 + (\sin\theta d\varphi)^2\}].$$

Suppose $k = +1$, and consider a 3-dimensional sphere of radius A in a 4-dimensional Euclidean space

$$A^2 = (y^4)^2 + \sum_{k=1}^{3}(y^k)^2.$$

The metric on this sphere is

$$d\sigma^2 = \sum_{k=1}^{3}(dy^k)^2 + (dy^4)^2.$$

This is rewritten by using the equation of the sphere in the above as follows:

$$d\sigma^2 = \sum_{k=1}^{3}(dy^k)^2 + \left\{A^2 - \sum_{k=1}^{3}(y^k)^2\right\}^{-1} \left(\sum_{\ell=1}^{3} y^\ell dy^\ell\right)^2.$$

Set $\rho^2 = \sum_{k=1}^{3}(y^k)^2$, and define v by

$$\rho = A\left(1 + \frac{v^2}{4}\right)^{-1} v.$$

Using polar coordinates (ρ, θ, φ) instead of (y^1, y^2, y^3), and rewriting ρ by the use of v, we have

$$d\sigma^2 = A^2 \left(1 + \frac{v^2}{4}\right)^{-2} [(dv)^2 + v^2 \{(d\theta)^2 + (\sin\theta d\varphi)^2\}].$$

If we set $A = R(t)$, and identify v as u, this formula coincides with the space part $d\sigma^2$ of the aforementioned Robertson–Walker metric ds^2.

In this sense, the space part slice $t = $ constant of the space–time can be regarded as a 3-dimensional sphere of radius $R(t)$ in a 4-dimensional Euclidean space, hence $R(t) = r_0 e^{h(t)/2}$ can be regarded as the radius of the universe and may expand as t grows. The cosmological redshift observed by Hubble [22] gives in this context that $R(t)$ is growing at present (see section 29.2 of [19]), and this is interpreted as a proof of 'expansion' of the universe. However, as we have seen, the 'expansion' is a consequence of the identification of $R(t)$ in the Robertson–Walker metric ds^2 with the radius of a sphere in a virtual 4-dimensional Euclidean space. In this sense, the growth of $R(t)$ in ds^2 does not imply the expansion of the universe in any other senses than it is an 'interpretation.'

The 'expansion' of this type does not contradict the stationary universe ϕ in quantum-mechanical sense specified in Axiom 1. The 'expansion' is an interpretation of the observation *with one observer's coordinate system fixed*. The quantum-mechanical stationary universe ϕ is the inner structure of its own and is independent of the observer's coordinate system. In this sense, the 'expansion' is an 'appearance,' which the universe takes under the 'interference' of the observer to try to reveal its morphology. More philosophically stating, the past and the future do not exist unless one fixes a time coordinate. The 'Big Bang' is an imagination under the *presumption* that the time coordinate exists *a priori*. Unless it is observed with assuming the existence of a time coordinate, the universe can be a stationary state.

Our theory is a reflection and a clarification of this supposition of the existence of time coordinate, adopted *implicitly* in almost all physical theories today.

Example of the previous section is an experiment of human size, and the one in this section is an observation of cosmological size. These two examples together with the one in Section 2 would indicate a unified treatment of physical phenomena from the microscopic size to the cosmological size.

4. OPEN PROBLEMS

In this Part 4 we state some open problems related with our formulation of physics. Some of them are known problems, but do not seem to have been given solutions. We conclude with stating a final goal of our formulation.

4.1. Stability of Matter

As we have seen in Part 2, the relativistic Hamiltonian considered by E. H. Lieb and others (see [2] and the references therein) has reasonable grounds under the assumption that gravitation is weak. It has been thought that the non-invariance with respect to Lorentz transformation is its fault. However, according to our formulation, the non-invariance is not a fault but has natural foundations as a Hamiltonian which describes observational facts.

It is therefore meaningful to research the related spectral and scattering theory for the relativistic Hamiltonian

$$\tilde{H}_L = \sum_{j=1}^{k} \left(c\sqrt{D_j^2 + m_j^2 c^2} - m_j c^2 \right) + I_b(x_b, 0) - G \sum_{1 \leq i < j \leq k} m_i m_j / r_{ij}, \qquad (4.1)$$

which includes the electric potentials and gravitational potentials simultaneously.

The first problem to be treated in this field is the problem of the stability of matter. It is known certain facts about this problem unless the electric potentials and gravitational ones are present simultaneously: If gravitation is absent and I_b is of the form

$$I_b = -e^2 \sum_{j=1}^{k} z|x_j - R|^{-1} + e^2 \sum_{1 \leq i < j \leq k} |x_i - x_j|^{-1},$$

where R is the position of the nucleus with z number of protons, it is known that the stability of the first kind is equivalent to the stability of the second kind, and that atoms are stable when $z \leq 87$. If electric potentials are absent, it is shown [23] that Thomas–Fermi theory is asymptotically exact for fermions. However, nothing seems known for the case which includes both of the electric potentials and gravitational ones as Lieb [2] writes. The research to include both of electric and gravitational potentials would lead us to a deeper understanding of the nature of matter, since any matter includes both kinds of internal forces, e.g., the stability and instability of stars with large number of particles would be understood in a more satisfactory manner than in the present understanding.

4.2. Scattering Theory

The second problem to be considered would be the scattering theory for the Hamiltonian in the formula (4.1). This Hamiltonian has the potentials which are of Coulomb type, therefore, of critical singularity with respect to the free part:

$$\tilde{H}_{L0} = \sum_{j=1}^{k} \left(c\sqrt{D_j^2 + m_j^2 c^2} - m_j c^2 \right).$$

Hence the problem of self-adjointness arises in the first place, and this is closely related with the problem of stability proposed in the previous section. The point is to what extent the free part $\tilde{H}_{L0}$ and the positive part of the sum of the potentials suppress the bad behavior of the negative parts. The gravitational potentials are quite small compared to the electric part and is negligible in the usual human size. But in the size of stars they cannot be neglected, and we have to develop some method which is able to treat the electric and gravitational parts at a time. This is the first problem which we should research in the scattering theory for (4.1).

If some conditions are established for the self-adjointness of (4.1), we should go on to the scattering phenomena governed by the Hamiltonian (4.1). This would give us an image about the phenomena which would be observed when the gravity and electrical forces are present simultaneously.

At first glance, it looks as if there were a problem in our formulation in the point that the approximate relativistic Hamiltonian (4.1) would lose the self-adjointness and stability when the number of particles becomes large. This should, however, be taken as an evidence of the success of our formulation to include gravity. The fact that the universe does not seem to be subject to Boltzmann's heat death, but it, which is in the usual physical context supposed to have been in an equilibrium originally, could develop hot stars, is owing to the instability of gravitation. Our Hamiltonian (4.1) explains this fact so that our inclusion of gravity into (4.1) would be a reasonable one to that extent.

The final problem in this direction is to find a full general relativistic Hamiltonian, which explains the observation without assuming that the gravitation is weak. This seems difficult seeing the present stage of the theory, but we hope that this end would be accomplished in the future.

REFERENCES

1. M. Born, "Zur Quantenmechanik der Stossvorgänge," *Zeitschrift für Physik*, **37**, 863–867 (1926).
2. E. H. Lieb, "The stability of matter: From atoms to stars," *Bull. Amer. Math. Soc.*, **22**, 1–49 (1990).
3. P. A. M. Dirac, *Proc. Roy. Soc.*, **A117**, 610 (1928).
4. F. Streater, "Why should anyone want to axiomatize quantum field theory?" in: *Philosophical Foundations of Quantum Field Theory, Ed. by H. R. Brown and R. Harré*, Clarendon, Oxford University Press, 137–148 (1990).
5. T. Kinoshita and W. B. Lindquist, "Eighth-order magnetic moment of the electron," *Phys. Rev.*, **D27**, 866 (1983).
6. F. J. Dyson, "Divergence of perturbation theory in quantum electrodynamics," *Phys. Rev.*, **75**, 486 (1953).
7. J. Fröhlich, "On the triviality of $\lambda \Phi_d^4$ theories and the approach to the critical point in $d \leq 4$ dimensions," *Nucl. Phys.*, **B 200 [FS4]**, 281–296 (1982).
8. C. J. Isham, "Canonical quantum gravity and the problem of time," in: *Proceedings of the NATO Advanced Study Institute, Salamanca, June 1992*, Kluwer Academic Publishers, (1993).
9. A. Einstein, "The foundation of the general theory of relativity," in: *Translated by W. Perrett and G. B. Jeffery, The Principle of Relativity*, Dover, 111–64 (1923).
10. E. Prugovečki, *Quantum Geometry, A Framework for Quantum General Relativity*, Kluwer Academic Publishers, Dordrecht–Boston–London, (1992).
11. H. Kitada, "Theory of local times," *Il Nuovo Cimento*, **109 B, N. 3**, 281–02 (1994).
12. R. Abraham and J. E. Marsden, *Foundations of Mechanics*, The Benjamin/Cummings Publishing Company, 2nd ed., London–Amsterdam–Don Mills, Ontario–Sydney–Tokyo, (1978).
13. H. L. Cycon et al., *Schrödinger Operators*, Springer-Verlag, (1987).
14. V. Enss, "Introduction to asymptotic observables for multiparticle quantum scattering," in: *Schrödinger Operators, Aarhus 1985, Ed. by E. Balslev, Lect. Note in Math.*, **1218**, Springer-Verlag, 61–92 (1986).
15. H. Kitada, "Asymptotic completeness of N-body wave operators II. A new proof for the short-range case and the asymptotic clustering for long-range systems," in: *Functional Analysis and Related Topics, 1991, Ed. by H. Komatsu, Lect. Note in Math.*, **1540**, Springer-Verlag, 149–189 (1993).
16. I. Newton, "Sir Isaac Newton Principia, Vol. I The Motion of Bodies, Motte's translation Revised by Cajori," in: *Tr. Andrew Motte ed. Florian Cajori*, Univ. of California Press, Berkeley, Los Angeles, London, (1962).
17. J. Dereziński, "Asymptotic completeness of long-range N-body quantum systems," *Annals of Math.*, **138**, 427–476 (1993).
18. H. Isozaki and H. Kitada, "Scattering matrices for two-body Schrödinger operators," *Scientific Papers of the College of Arts and Sciences, The University of Tokyo*, **35**, 81–107 (1986).
19. C. W. Misner, K. S. Thorne, and J. A. Wheeler, *Gravitation*, W. H. Freeman and Company, New York, (1973).
20. H. Kitada and L. Fletcher, "Local time and the unification of physics, Part I: Local time," *Apeiron*, **3**, 38–45 (1996).
21. R. Collela, A. W. Overhauser and S. A. Werner, "Observation of gravitationally induced quantum mechanics," *Phys. Rev. Lett.*, **34**, 1472–1474 (1975).
22. E. P. Hubble, "A relation between distance and radial velocity among extragalactic nebulae," *Proc. Nat. Acad. Sci. U.S.*, **15**, 169–173 (1929).
23. E. H. Lieb and H-T. Yau, "The Chandrasekhar theory of stellar collapse as the limit of quantum mechanics," *Commun. Math. Phys.*, **112**, 147–74 (1987).

ON EMBEDDED EIGENVALUES OF PERTURBED PERIODIC SCHRÖDINGER OPERATORS

Peter Kuchment[1] and Boris Vainberg[2]

[1]Department of Mathematics and Statistics
Wichita State University
Wichita, KS
E-mail: kuchment@twsuvm.uc.twsu.edu
http://www.math.twsu.edu/Faculty/Kuchment/
[2]Department of Mathematics
University of North Carolina at Charlotte
Charlotte, NC
E-mail: brvainbe@uncc.edu

ABSTRACT

The problem of non-existence of eigenvalues imbedded into the continuous spectrum is considered for Schrödinger operators with periodic potentials perturbed by a sufficiently fast decaying "impurity" potentials. Absence of embedded eigenvalues is shown in dimensions two and three if the periodic potential satisfies some additional condition on the corresponding Fermi surface. It is conjectured that generic periodic potentials satisfy this condition. It is stated that separable periodic potentials satisfy it, and hence in dimensions two and three a Schrödinger operator with a separable periodic potential perturbed by a sufficiently fast decaying "impurity" potential has no embedded eigenvalues. The proofs are only sketched. The complete proofs will be provided elsewhere.

1. INTRODUCTION

We consider the Schrödinger operator

$$H_0 = -\Delta + q(x)$$

in $L^2(\mathbb{R}^n)$ with a real valued potential $q \in L^\infty(\mathbb{R}^n)$ periodic with respect to the integer lattice $\mathbb{Z}^n$:

$$q(x+l) = q(x) \text{ for all } l \in \mathbb{Z}^n, x \in \mathbb{R}^n.$$

Let
$$H = H_0 + v(x) = -\Delta + q(x) + v(x) \tag{1.1}$$
be a perturbed operator where potential $v(x)$ decays sufficiently fast at infinity.

It is well known (see [7, 9, 15, 21, 23]) that the spectrum of the unperturbed operator H_0 is absolutely continuous and has band-gap structure:
$$\sigma(H_0) = \bigcup_{i \geq 1} [a_i, b_i],$$
where $a_i < b_i$, and $\lim_{i \to \infty} a_i = \infty$. If v decays fast enough, then the absolutely continuous spectrum of H is the same: $\sigma_{ac}(H) = \sigma(H_0)$, and only some additional point spectrum can arise (see [4], Section 18 in [11], and references therein). A general understanding is that the point spectrum should arise only in the gaps of the continuous one. In other words, there are no eigenvalues of the operator H embedded into its continuous spectrum. However, to prove the absence of embedded eigenvalues rigorously is a very delicate task. There are known examples of embedded eigenvalues (see [8] and Section XIII.13 in [23]). In these examples the "impurity" potential v decays not very fast. In particular, in the first example suggested by J. von Neumann and E. Wigner $q \equiv 0$ and $v = O\left(|x|^{-1}\right)$ as $|x| \to \infty$.

The problem of absence of embedded eigenvalues has been intensively studied. There are many known results on non-existence of embedded eigenvalues for the case of zero underlying potential $q(x)$ (see, for instance, books [8, 13] Section 14.7, and [23] Section XIII.13). The case of periodic $q(x)$ is much less studied. There are papers on behavior of the point spectrum in gaps of the continuous one (see, for instance, [1–6, 10, 12, 17–19, 22, 24, 25, 27]). However, results on absence or existence of embedded eigenvalues for perturbations of a periodic potential are known only in the one-dimensional case of the perturbed Hill operator
$$H = -\frac{d^2}{dx^2} + q(x) + v(x).$$

In [24] a theorem on absence of eigenvalues is proved for the Hill operator. In [8] an example of a Hill operator with a slowly decaying perturbation $v(x)$ and with an embedded eigenvalue is provided.

The purpose of this paper is to study the multi-dimensional case. We introduce a condition ("condition A") on a periodic potential $q(x)$, which guarantees in dimension less than four the absence of embedded eigenvalues for the perturbed operator (1.1) with any sufficiently fast decaying potential $v(x)$. We conjecture that generic periodic potentials satisfy this condition A. We do not have proof of this conjecture, but we show that condition A is satisfied for separable periodic potentials. This leads to a result on absence of embedded eigenvalues when $q(x)$ is separable and $v(x)$ is sufficiently fast decaying. The statement on absence of embedded eigenvalues under the condition A is proved using an approach that was suggested a long time ago by the second author for the case of perturbations of operators with constant coefficients [29]. Some recent developments (in particular, the results of [15] and [20]) made it possible to adjust this method to the periodic case.

The paper is organized as follows. In Section 2 we provide a brief account of the approach used in [29]. Section 3 contains some necessary notions, auxiliary information, and the "condition A." In Section 4 we give a sketch of the proof of the absence of embedded eigenvalues when periodic potential satisfies condition A. In Section 5 we describe the result on separable potentials. Complete proofs will be provided elsewhere.

2. ZERO UNDERLYING POTENTIAL

Let $q(x) \equiv 0$, that is the operator (1.1) has the form

$$H = -\Delta + v(x), \quad x \in \mathbb{R}^n. \tag{2.1}$$

The spectrum of $-\Delta$ in $L^2(\mathbb{R}^n)$ is absolutely continuous and coincides with the real nonnegative semi-axis $\lambda \geq 0$. Adding a fast decaying potential may produce additionally some eigenvalues. The well known Kato's theorem [14] states that the eigenvalues cannot be positive if $v(x) = o(1/|x|)$ as $|x| \to \infty$ (see [26] and references there for some generalizations). Unfortunately, the Kato's proof does not work for perturbations of periodic potentials. The approach used in [29] gives weaker results for the Schrödinger operator (2.1), but on the other hand it is very simple and can be applied to more general equations. The proposition below and its proof can be found in [29] for general elliptic systems.*

Proposition 2.1. *Let $v(x)$ have a compact support. Then operator (2.1) does not have positive eigenvalues.*

Proof: Let

$$(-\Delta + v(x) - \lambda)u = 0, \quad x \in \mathbb{R}^n, \ u \in L^2(\mathbb{R}^n), \ \lambda > 0. \tag{2.2}$$

We have to prove that $u \equiv 0$. We rewrite the equation in the form $(-\Delta - \lambda)u = f$ with $f = -vu$ and apply the Fourier transform. Then we get

$$\left(|\sigma|^2 - \lambda\right)\widetilde{u} = \widetilde{f}(\sigma), \tag{2.3}$$

and therefore

$$\widetilde{u}(\sigma) = \frac{\widetilde{f}(\sigma)}{|\sigma|^2 - \lambda} \quad \text{when } \sigma^2 \neq \lambda. \tag{2.4}$$

We notice that according to the Plancherel's theorem $\widetilde{u}$ belongs to $L^2(\mathbb{R}^n)$. Since $f = -vu$ has a compact support, function $\widetilde{f}(z)$ is an entire function of the first order. Function $\widetilde{f}(\sigma)$ vanishes on the sphere $|\sigma|^2 = \lambda$, since otherwise the right-hand side in (2.4) would have a singularity that would prevent its square integrability. Let us note some properties of the polynomial $p(z) = \sum z_i^2 - \lambda$. First of all, it is irreducible. Secondly, when $\lambda > 0$ it has a "big" set of real zeros. Namely, this set is the sphere of radius $\sqrt{\lambda}$. Since $\widetilde{f}(z) = 0$ at real zeros of the polynomial $p(z) = \sum z_i^2 - \lambda$, we can use the properties of $p(z)$ that we have just mentioned to conclude by analytic continuation type arguments that $\widetilde{f}(z) = 0$ at all complex zeros of this polynomial. Hence, the ratio $\widetilde{f}(z)/p(z)$ is an entire function of the first order (the statement about the order requires some simple estimates). Thus (see (2.4)) $\widetilde{u}(z)$ is an entire function of the first order. We conclude that according to the Paley–Wiener theorem $u(x)$ has a compact support. Now the uniqueness theorem for solutions of the Cauchy problem for equation (2.2) (see [16]) implies that $u \equiv 0$, which proves the proposition.

Let us point out the main ingredients of this proof: Fourier transform that produces an algebraic equation (2.3); Plancherel and Paley–Wiener theorems for the Fourier transform;

---

*We would like to warn the reader of [29] that the term "finite function" in [29] was a literal translation from Russian, and it means "compactly supported function."

irreducibility of the polynomial $p(z)$ and large dimension of its real zeros; an analytic divisibility theorem for entire functions; an uniqueness theorem for solutions of the original PDE. We will show in the following sections that all these elements have some analogs in the case of a periodic underlying potential $q(x)$.

3. PERIODIC CASE AND CONDITION A

We move now to a study of the operator (1.1) with a periodic potential $q(x)$. An analog of Fourier transform for this case is a transform that we will call the Floquet–Bloch transform. If $u(x) \in L^2_{com}(\mathbb{R}^n)$, then the Floquet–Bloch transform of u is a function $\tilde{u}$ of a pair of variables: $x \in \mathbb{R}^n$ and (quasimomentum) $k \in \mathbb{C}^n$. This function $\tilde{u}$ is defined as follows:

$$\tilde{u}(k,x) = \sum_{l \in \mathbb{Z}^n} u(x-l) e^{-ikl}.$$

It is periodic with respect to k with the lattice of periods $2\pi\mathbb{Z}^n \subset \mathbb{C}^n$ and satisfies the Floquet–Bloch cyclic condition with respect to x:

$$\tilde{u}(k,x+l) = \tilde{u}(k,x) e^{-ikl}.$$

Let us note that we use the same notation $\tilde{u}$ for the Floquet–Bloch transform as for the Fourier transform. We hope that this will not lead to any misunderstanding, since we will not use the Fourier transform anymore. A Parseval's equality allows to define the Floquet–Bloch transform for any $u \in L^2(\mathbb{R}^n)$, but in this case only the real values of the quasimomentum k are allowed.

The Floquet–Bloch transform commutes with the differentiation and multiplication by periodic functions. Thus if $u \in L^2(\mathbb{R}^n)$ satisfies the unperturbed equation

$$(-\Delta + q(x) - \lambda) u = f \in L^2(\mathbb{R}^n), \tag{3.1}$$

where q is periodic, then $\tilde{u}$ is a solution of the following problem:

$$\begin{cases} (-\Delta + q(x) - \lambda)\tilde{u} = \tilde{f} \\ \tilde{u}(k,x+l) = \tilde{u}(k,x) e^{-ikl} \end{cases}, \tag{3.2}$$

Sometimes it is convenient to introduce function

$$\hat{u}(k,x) = e^{ikx}\tilde{u}(k,x) = \sum_{l \in \mathbb{Z}^n} u(x-l) e^{ik(x-l)}.$$

Then $\hat{u}$ is periodic with respect to the lattice $\mathbb{Z}^n$ in $\mathbb{R}^n_x$ and is a solution of the problem

$$-(\nabla - ik)^2 \hat{u} + (q(x) - \lambda) \hat{u} = \hat{f}. \tag{3.3}$$

Here

$$u = u(k,x) \in L^2_{loc}(\mathbb{R}^n, L^2(T^n)),$$

where T^n is the torus $\mathbb{R}^n/\mathbb{Z}^n$.

We denote by $A(k)$ the following operator in $L^2(T^n)$:

$$A(k)v = -(\nabla - ik)^2 v + q(x)v$$

for $v \in H^2(T^n)$. Here we denote by $H^2(T^n)$ the standard Sobolev space of functions on T^n. The operator family $A(k) - \lambda$ will be an analog of the polynomial $p(z)$ used in Section 2. The spectrum of $A(k)$ is discrete and depends on k. Now we introduce an analog of the set of zeros of the polynomial $p(z)$. This will be the so called Fermi variety.

Definition 3.1. For given $\lambda \in \mathbb{C}$ the Fermi variety $F_\lambda(q)$ for equation (3.1) is the set of quasimomenta $k \in \mathbb{C}^n$ such that $\lambda \in \sigma(A(k))$. The real Fermi variety is

$$F_{\lambda,\mathbb{R}}(q) = F_\lambda(q) \cap \mathbb{R}^n,$$

that is $F_{\lambda,\mathbb{R}}(q)$ is the set of real quasimomenta k for which $\lambda \in \sigma(A(k))$.

Let us define a cube $Q \subset \mathbb{R}^n$ (the Brillouin zone) by inequalities $0 \leq k_i \leq 2\pi$ for $1 \leq i \leq n$. Let us also denote by $\lambda_j(k)$, $k \in Q$, the eigenvalues of operator $A(k)$ counted in the increasing order. Thus we have a sequence of continuous functions $\lambda_j(k)$ on Q. They are called the band functions. The set of all values of a function $\lambda_j(k)$ for a fixed j defines the jth band $[a_j, b_j]$ of the spectrum $\sigma(H_0)$:

$$\sigma(H_0) = \cup [a_j, b_j]. \tag{3.4}$$

The following fact is very important for the proof of our main result.

Lemma 3.1 (see [15]). *Operator* $(A(k) - \lambda)^{-1}$ *has the form*

$$(A(k) - \lambda)^{-1} = \frac{B(k)}{\varsigma(k)}$$

where $B(k): L^2(T^n) \to H^2(T^n)$ *is an operator valued function,* $\varsigma(k)$ *is a scalar function, and both functions are entire functions of order n in $\mathbb{C}^n$.*

In particular, it follows from Lemma 3.1 that the Fermi variety is an analytic set (which is given by the equation $\varsigma(k) = 0$).

Definition 3.2. An analytic set $A \subset \mathbb{C}^n$ is called irreducible if it cannot be represented as the union of two proper analytic subsets.

Now we are almost in a position to introduce our main condition. Let us note that it follows from (3.4) that a point λ belongs to the spectrum $\sigma(H_0)$ of the operator $H_0 = -\Delta + q(x)$ if and only if the real Fermi variety $F_{\lambda,\mathbb{R}}(q)$ is non-empty. The natural guess is that λ is in the interior of a spectral band when the real Fermi variety is massive, or to be more exact, when it has dimension $n-1$.

Condition A. We say that a periodic potential $q(x)$ satisfies the condition A, if for any λ that belongs to the interior of a spectral band of the corresponding Schrödinger operator

$$H_0 = -\Delta + q(x)$$

any irreducible component of the Fermi variety $F_\lambda(q)$ intersects the real space $\mathbb{R}^n$ by a subset of dimension $n-1$ (that is by a subset that contains a piece of a smooth hypersurface).

4. A THEOREM ON ABSENCE OF EMBEDDED EIGENVALUES

This section is devoted to a sketch of the proof of the following theorem.

Theorem 4.1. *Let $n = 2$ or 3, $q(x) \in L^\infty(\mathbb{R}^n)$ be periodic and satisfy condition A, and*

$$|v(x)| < Ce^{-|x|^r} \text{ with } r > 4/3. \tag{4.1}$$

Then the spectrum of H does not contain embedded eigenvalues. In other words

$$\{\lambda_j\} \cap \bigcup_{i \geq 1}(a_i, b_i) = \emptyset,$$

where $\{\lambda_j\}$ is the set of eigenvalues of H, and

$$\bigcup_{i \geq 1}[a_i, b_i] = \sigma(H_0)$$

is the band structure of the essential spectrum of H.

Sketch of the proof: Let u be an eigenfunction of H with an eigenvalue $\lambda \in (a_i, b_i)$ for some i. Then

$$(-\Delta + q - \lambda)u = -vu, \quad u \in L^2(\mathbb{R}^n),$$

and therefore

$$[-(\nabla - ik)^2 + q(x) - \lambda]\hat{u} = -\hat{f}, \quad x \in T^n, \tag{4.2}$$

where $\hat{u}$ is defined for all real k and

$$\hat{u}, \hat{f} \in L^2_{loc}(\mathbb{R}^n, L^2(T^n)).$$

Standard elliptic estimates give

$$\hat{u} \in L^2_{loc}(\mathbb{R}^n, H^2(T^n)).$$

It follows from Lemma 3.1 and (4.2) that

$$\hat{u} = \frac{B(k)}{s(k)}\hat{f}(k).$$

Estimate (4.1) allows to prove that $\hat{f}(z) = -\widehat{vu}(z)$ is an entire $L^2(T^n)$-valued function of order $r/(r-1)$. Thus,

$$\hat{u} = \frac{g(z)}{s(z)}, \tag{4.3}$$

where $g(z) = B(z)\hat{f}(z)$ is an entire $H^2(T^n)$-valued function of order $\max(n, r/(r-1))$ and $s(z)$ is a scalar entire function of order n. Arguments similar to ones used in the proof of the Proposition 2.1 enable one to prove that under the condition A function $\hat{u}$ in (4.3) is an entire function of order $\max(n, r/(r-1))$. Since $r > 4/3$ and $n \leq 3$, then $\max(n, r/(r-1)) < 4$. This leads to an the estimate (see [15] for details):

$$|u(k,x)| \leq Ce^{-|k|^p} \text{ for some } p > 4/3. \tag{4.4}$$

This allows to apply the Meshkov's uniqueness theorem [20], which guarantees that an equation $(-\Delta + p(x))u = 0$ with a bounded $p(x)$ has only zero solution in the class of functions satisfying (4.4). Thus $u \equiv 0$, which completes the proof.

5. SEPARABLE UNDERLYING POTENTIALS

We believe that a generic periodic potential satisfies the condition A. However, we have no proof of this statement. The following theorem provides a class of potentials which satisfy condition A.

Theorem 5.1. *If a periodic potential $q(x) \subset L^\infty(\mathbb{R}^n)$ is separable, that is*

$$q(x) = \sum q_i(x_i), \tag{5.1}$$

then it satisfies condition A. Moreover, in this case $F_\lambda(q)$ is irreducible modulo the natural periodicity with respect to the lattice $2\pi\mathbb{Z}^n$ and modulo reflections $(k_1,...k_n) \to (\pm k_1,... \pm k_n)$.

As immediate corollary we have

Theorem 5.2. *If $n < 4$, the underlying periodic potential $q(x) \subset L^\infty(\mathbb{R}^n)$ is separable, and $v(x)$ satisfies the estimate (4.1), then the operator (1.1) has no eigenvalues embedded into the continuous spectrum.*

The proof of Theorem 5.1 is too technical and will be provided elsewhere.

6. COMMENTS

1. It is probable that condition A (and hence the result on absence of embedded eigenvalues) can be proven to hold for a generic periodic potential. The authors have not been able to do this so far.

2. We have only proven absence of eigenvalues embedded into the interior of a spectral band. It is likely that eigenvalues cannot occur at the ends of the bands either (maybe except the bottom of the spectrum), if the perturbation potential decays fast enough. This was shown in the one-dimensional case in [24] under the condition

$$\int (1 + |x|) |v(x)| dx < \infty$$

on the perturbation potential. On the other hand, if $v(x)$ only belongs to $L^1(\mathbb{R})$, then eigenvalues at the endpoints of spectral bands can occur [25].

3. Most of the proof of the conditional Theorem 5.1 does not require that the unperturbed operator H_0 is a Schrödinger operator. One can treat general self-adjoint periodic elliptic operators as well. The only obstacle occurs at the last step, when one needs to conclude the absence of fast decaying solutions to the equation. Here we applied results of Meshkov's paper [20], which are applicable only to operators of Schrödinger type. Carrying over Meshkov's results to more general operators would automatically generalize Theorem 5.1. The restriction that the dimension n is less than four also comes from the allowed rate of decay stated in the Meshkov's result.

ACKNOWLEDGMENT

The authors express their gratitude to Professors A. Figotin, S. Molchanov, and V. Papanicolaou for helpful discussions. The work of P. Kuchment was partly supported by an NSF EPSCoR grant and by the NSF Grant DMS 9610444. P. Kuchment expresses his gratitude to NSF and to the State of Kansas for this support. The work of B. Vainberg was partly supported by the NSF Grant DMS-9623727. B. Vainberg expresses his gratitude to NSF for this support.

REFERENCES

1. S. Alama, M. Avellaneda, P. A. Deift, and R. Hempel, On the existence of eigenvalues of a divergence form operator $A + \lambda B$ in a gap of $\sigma(A)$, Asymptotic Anal. 8 (1994), no. 4, 311–314.
2. S. Alama, P. A. Deift, and R. Hempel, Eigenvalue branches of the Schrödinger operator $H - \lambda W$ in a gap of $\sigma(H)$, Commun. Math. Phys. 121 (1989), 291–321.
3. M. S. Birman, The discrete spectrum of the periodic Schrödinger operator perturbed by a decreasing potential, Algebra i Analiz 8 (1996), no. 1, 3–20.
4. M. S. Birman, On the spectrum of singular boundary value problems, Mat. Sbornik 55 (1961), no. 2, 125–173.
5. M. S. Birman, On the discrete spectrum in the gaps of a perturbed periodic second order operator, Funct. Anal. Appl. 25 (1991), 158–161.
6. P. A. Deift and R. Hempel, On the existence of eigenvalues of the Schrödinger operator $H - \lambda W$ in a gap of $\sigma(H)$, Commun. Math. Phys. 103 (1986), 461–490.
7. M. S. P. Eastham, The Spectral Theory of Periodic Differential Equations, Scottish Acad. Press, Edinburgh - London 1973.
8. M. S. P. Eastham and H. Kalf, Schrödinger-type Operators with Continuous Spectra, Pitman. Boston 1982.
9. I. M. Gelfand, Expansion in eigenfunctions of an equation with periodic coefficients, Dokl. Akad. Nauk SSSR 73 (1950), 1117–1120.
10. F. Gesztesy and B. Simon, On a theorem of Deift and Hempel, Commun. Math. Phys. 116 (1988), 503–505.
11. I. M. Glazman, Direct Methods of Qualitative Spectral Analysis of Singular Differential Operators, I.P.S.T., Jerusalem 1965.
12. R. Hempel, Eigenvalue branches of the Schrödinger operator $H \pm \lambda W$ in a spectral gap of H, J. Reine Angew. Math. 399 (1989), 38–59.
13. L. Hörmander, The Analysis of Linear Partial Differential Operators, v. 2, Springer Verlag, Berlin 1983.
14. T. Kato, Growth properties of solutions of the reduced wave equation with a variable coefficient, Comm. Pure Appl. Math. 12 (1959), 403–425.
15. P. Kuchment, Floquet Theory for Partial Differential Equations, Birkhäuser, Basel 1993.
16. E. M. Landis, On some properties of solutions of elliptic equations, Dokl. Akad. Nauk SSSR, 107 (1956), 640–643.
17. S. Z. Levendorskii, Asymptotic formulas with remainder estimates for eigenvalue branches of the Schrödinger operator $H - \lambda W$ in a gap of H. To appear in Transactions of American Mathematical Society.
18. S. Z. Levendorskii and S. I. Boyarchenko, An asymptotic formula for the number of eigenvalue branches of a divergence form operator $A + \lambda B$ in a spectral gap of A. To appear in Communications in Part. Differ. Equat.
19. S. Z. Levendorskii, Lower bounds for the number of eigenvalue branches for the Schrödinger operator $H - \lambda W$ in a gap of H: the case of indefinite W, Comm. partial Diff. Equat. 20 (1995), no. 5–6, 827–854.
20. V. Meshkov, On the possible rate of decay at infinity of solutions of second order partial differential equations, Mat. Sbornik, 182 (1991), no. 3, 364–383. English translation in Math. USSR Sbornik 72 (1992), no. 2, 343–351.
21. F. Odeh, J. B. Keller, Partial differential equations with periodic coefficients and Bloch waves in crystals, J. Math. Phys. 5 (1964), 1499–1504.
22. G. D. Raikov, Eigenvalue asymptotics for the Schrödinger operator with perturbed periodic potential, Invent. Math. 110 (1992), 75–93.
23. M. Reed and B. Simon, Methods of Modern Mathematical Physics v. 4, Acad. Press, NY 1978.
24. F. S. Rofe-Beketov, A test for the finiteness of the number of discrete levels introduced into the gaps of a continuous spectrum by perturbations of a periodic potential, Soviet Math. Dokl. 5 (1964), 689–692.
25. F. S. Rofe-Beketov, Spectrum perturbations, the Knezer-type constants and the effective mass of zones-type potentials, in "Constructive Theory of Functions'84," Sofia 1984, p.757–766.
26. G. Roach, B. Zhang, A transmission problem for the reduced wave equation in inhomogeneous media with an infinite interface, Proc. R. Soc. London, A, 436 (1992), 121–140
27. A. V. Sobolev, Weyl asymptotics for the discrete spectrum of the perturbed Hill operator, Adv. Sov. Math. 7 (1991), 159–178.
28. E. C. Titchmarsh, Eigenfunction Expansions Associated with Second-Order Differential Equations, Part II, Claredon Press, Oxford 1958.

29. B. Vainberg, Principles of radiation, limiting absorption and limiting amplitude in the general theory of partial differential equations, Russian Math. Surveys 21 (1966), no. 3, 115–193.

6

ON PRINCIPAL EIGENVALUES FOR INDEFINITE-WEIGHT ELLIPTIC PROBLEMS

Yehuda Pinchover

Department of Mathematics
Technion-Israel Institute of Technology
32000 Haifa, Israel
E-mail: pincho@tx.technion.ac.il

1. INTRODUCTION

Consider the quantum mechanical system $H_\mu = -\Delta - \mu V$ in $\mathbb{R}^d$, where $\mu \in \mathbb{R}$ is a spectral parameter and $V \in C_0^\infty(\mathbb{R}^d)$. It is well known that for $d \geq 3$, the Schrödinger operator H_μ has no bound states provided that $|\mu|$ is sufficiently small. On the other hand, for $d = 1, 2$, B. Simon proved the following delicate result (see [35,37], and also the discussion in the Notes of [35]).

Theorem 1.1. *Suppose that $d = 1, 2$. Then H_μ has a negative eigenvalue for all negative μ if and only if*

$$\int_{\mathbb{R}^d} V(x)dx \leq 0.$$

The explanation of the above phenomena is closely related to the following notions.

Let P be a linear, second order, elliptic operator defined in a subdomain Ω of a non-compact, connected, Riemannian manifold X of dimension d. Here P is an elliptic operator with real Hölder continuous coefficients which in any coordinate system $(U; x_1, \ldots, x_d)$ has the form

$$P(x, \partial_x) = -\sum_{i,j=1}^d a_{ij}(x)\partial_i\partial_j + \sum_{i=1}^d b_i(x)\partial_i + c(x), \tag{1.1}$$

where $\partial_i = \partial/\partial x_i$. We assume that for every $x \in \Omega$ the real quadratic form

$$\sum_{i,j=1}^d a_{ij}(x)\xi_i\xi_j, \quad \xi = (\xi_1, \ldots, \xi_d) \in \mathbb{R}^d \tag{1.2}$$

Spectral and Scattering Theory, edited by Ramm,
Plenum Press, New York, 1998

is positive definite.

We denote the cone of all positive (classical) solutions of the elliptic equation $Pu = 0$ in Ω by $\mathcal{C}_P(\Omega)$. We denote by P^* the formal adjoint of P.

Let $\{\Omega_k\}_{k=1}^{\infty}$ be *an exhaustion of* Ω, i.e. a sequence of smooth, relatively compact domains such that $\overline{\Omega}_k \subset \Omega_{k+1}$ and $\cup_{k=1}^{\infty} \Omega_k = \Omega$. Assume that $\mathcal{C}_P(\Omega) \neq \emptyset$. Then for every $k \geq 1$ the Dirichlet Green function $G_P^{\Omega_k}(x,y)$ exists and is positive. By the generalized maximum principle, $\{G_P^{\Omega_k}(x,y)\}_{k=1}^{\infty}$ is an increasing sequence which, by the elliptic Harnack inequality, converges uniformly in any compact subdomain of Ω either to $G_P^{\Omega}(x,y)$, the positive *minimal Green function* of P in Ω and P is said to be *subcritical operator* in Ω, or to infinity and in this case P is *critical* in Ω. The operator P is said to be *supercritical* if $\mathcal{C}_P(\Omega) = \emptyset$ [23, 26, 32].

These notions of criticality and subcriticality which were first introduced by B. Simon for Schrödinger operators [38] have been proved to be valuable tools for proving many results concerning positive solutions of linear (and sometimes nonlinear) elliptic and parabolic equations (see for example, [16, 19, 20, 23, 28, 30–32, 38, 40] and the references therein). The criticality theory approach has the advantage that it applies to general critical or subcritical operators in a general domain.

It follows that P is critical (resp. subcritical) in Ω if and only if P^* is critical (resp. subcritical) in Ω. Furthermore, if P is critical in Ω then $\mathcal{C}_P(\Omega)$ is a one-dimensional cone and any positive supersolution of the equation $Pu = 0$ in Ω is a solution. In this case $\phi \in \mathcal{C}_P(\Omega)$ is called a *ground state of P in Ω* [1].

Remark. We would like to point out that the criticality theory is also valid for the class of weak solutions of elliptic equations in divergence form and also for the class of strong solutions of strongly elliptic equations with locally bounded coefficients. Nevertheless, for the sake of clarity we prefer to present our results for the class of classical solutions.

Subcriticality is a stable property in the following sense. If P is subcritical in Ω and $V \in C_0^{\alpha}(\Omega)$ then there exists $\epsilon > 0$ such that $P - \mu V$ is subcritical for all $|\mu| < \epsilon$ [23, 26]. On the other hand, if P is critical in Ω and $V \in C^{\alpha}(\Omega)$ is a nonzero, nonnegative function then for any $\epsilon > 0$ the operator $P + \epsilon V$ is subcritical and $P - \epsilon V$ is supercritical in Ω.

Note that $P = -\Delta$ is critical in $\mathbb{R}^d$ if $d = 1, 2$, and subcritical for $d \geq 3$. Moreover, the Schrödinger operator H_μ considered in Theorem 1.1 has a negative eigenvalue if and only if H_μ is supercritical.

We turn now to the critical case. The following result demonstrates again the instability of criticality. The theorem which is the main result of [28] extends Theorem 1.1 to the case of a general (non-selfadjoint), linear, second order elliptic operator which is critical in a domain Ω.

Theorem 1.2. *Let P be a critical elliptic operator on Ω and let ϕ_0 (resp. $\tilde{\phi}_0$) be the ground state of the operator P (resp. P^*) in Ω. Assume that $W \in C_0^{\alpha}(\Omega)$. Then there exists $\mu < 0$ such that $P - \mu W(x)$ is subcritical in Ω if and only if*

$$\int_{\Omega} W(x) \phi_0(x) \tilde{\phi}_0(x) dx > 0. \quad (1.3)$$

This type of result appears first in a theorem of Picone for the following Neumann indefinite eigenvalue problem

$$-u'' = tW(x)u \quad x \in (a,b)$$
$$u'(a) = u'(b) = 0$$

([25], see also [8]). The indefinite character of the problem is due to the changing of sign of the function W.

In the last two decades theorems of this kind were proved for linear and nonlinear problems by many authors (see [2,4,5,7,9,10,13–15,18,21,28,36,37,39] and the references therein). Most of these results deal with either symmetric operators or problems in smooth, bounded domains. So, one can either use variational characterization of the principal eigenvalue or perturbation spectral theory of compact operators and the Krein–Rutman theorem.

The study of such problems has been motivated in part by the wish to understand related linear and nonlinear indefinite boundary value problems such as

$$Pu = \mu W(x) f(u(x)) \quad \text{for } x \in \Omega, \quad u = 0 \quad \text{for } x \in \partial\Omega. \tag{1.4}$$

Problems of this kind appear in mathematical physics and in many reaction diffusion problems like population genetics, ecological problems and superconductivity.

The aim of this paper is to extend Theorem 1.2 to a wider class of perturbations which we call *weak perturbations* (see Theorem 4.1). It turns out that small and semismall perturbations which were studied in [24, 27] are weak perturbations. Moreover, most of the results of the type of Theorem 1.2 which were mentioned above are special cases of Theorem 4.1.

The outline of this paper is as follows. In Section 2, we give some basic definitions, collect some results and fix notations. In Section 3, we define the notion of weak perturbation of subcritical and critical operators, study some of the properties and give some examples of such perturbations. Finally, our main result (Theorem 4.1) is proved in Section 4.

2. PRELIMINARIES

In this section we collect some terminology and results which we need in this paper.

Consider a noncompact, connected, C^3-smooth Riemannian manifold X of dimension d. We associate to any subdomain $\Omega \subseteq X$ an exhaustion $\{\Omega_n\}_{n=1}^{\infty}$ and a sequence $\{\chi_n(x)\}_{n=1}^{\infty}$ of smooth cutoff functions in Ω such that $\chi_n(x) \equiv 1$ in Ω_n, $\chi_n(x) \equiv 0$ in $\Omega \setminus \Omega_{n+1}$, and $0 \leq \chi_n(x) \leq 1$ in Ω. Let $W \in C^\alpha(\Omega)$, $0 < \alpha \leq 1$. We denote $W_n(x) = \chi_n(x) W(x)$ and $W_n^*(x) = W(x) - W_n(x)$. We say that a function $f \in C(\Omega)$ *vanishes at infinity of* Ω if for every $\epsilon > 0$ there exists $n \in \mathbb{N}$ such that $|f(x)| < \epsilon$ in $\Omega \setminus \Omega_n$.

The next notion was introduced by S. Agmon in [1].

Definition 2.1. Let P be an elliptic operator defined in a domain $\Omega \subseteq X$. A function $u \in C(\Omega \setminus \Omega_n)$ is said to be a *positive solution of the operator P of minimal growth in a neighborhood of infinity in* Ω if u satisfies the following two conditions:

(i) There exists $n \in \mathbb{N}$ such that u is a positive solution of the equation $Pu = 0$ in $\Omega \setminus \overline{\Omega_n}$;

(ii) If v is a continuous function on $\Omega \setminus \Omega_k$ for some $k \geq n$ which is a positive solution of the equation $Pu = 0$ in $\Omega \setminus \overline{\Omega_k}$, and $u \leq v$ on $\partial \Omega_k$, then $u \leq v$ on $\Omega \setminus \Omega_k$.

Remark. If P is subcritical in Ω then $G_P^\Omega(x, x_0)$ is a positive solution of the equation $Pu = 0$ of minimal growth in a neighborhood of infinity in Ω. On the other hand, if P is critical in Ω then the ground state is a positive solution of the equation $Pu = 0$ in Ω which has minimal growth in a neighborhood of infinity in Ω.

Definition 2.2. Let P_i, $i = 1, 2$ be two subcritical operators in $\Omega \subseteq X$. We say that the Green functions $G_{P_1}^\Omega(x,y)$ and $G_{P_2}^\Omega(x,y)$ are *equivalent* (resp. *semi-equivalent*) if there exists $C > 0$ such that

$$C^{-1} G_{P_2}^\Omega(x,y) \leq G_{P_1}^\Omega(x,y) \leq C G_{P_2}^\Omega(x,y) \tag{2.1}$$

for all $x, y \in \Omega$, $x \neq y$ (resp. for some $x \in \Omega$ and all $y \in \Omega \setminus \{x\}$).

Many papers deal with sufficient conditions, in terms of proximity near infinity in Ω between two given subcritical operators, which imply that the Green functions are equivalent (see, [6,23,24,26,27,29, and the references therein]). The following notion is closely related to this problem.

Definition 2.3. Let P be a subcritical operator in $\Omega \subseteq X$ and let $W \in C^\alpha(\Omega)$. We say that W is a *small perturbation* of P in Ω if

$$\lim_{k \to \infty} \left\{ \sup_{x,y \in \Omega \setminus \Omega_k} \int_{\Omega \setminus \Omega_k} \frac{G_P^\Omega(x,z) |W(z)| G_P^\Omega(z,y)}{G_P^\Omega(x,y)} \, dz \right\} = 0. \tag{2.2}$$

The following refinement of Definition 2.3 was recently introduced by M. Murata [24].

Definition 2.4. Let P be a subcritical operator in Ω and let $W \in C^\alpha(\Omega)$. Fix a reference point $x_0 \in \Omega_1$. We say that W is a *semismall perturbation* of P in Ω if

$$\lim_{k \to \infty} \left\{ \sup_{y \in \Omega \setminus \Omega_k} \int_{\Omega \setminus \Omega_k} \frac{G_P^\Omega(x_0,z) |W(z)| G_P^\Omega(z,y)}{G_P^\Omega(x_0,y)} \, dz \right\} = 0. \tag{2.3}$$

A small perturbation is semismall [24]. It is known [24, 27] that if the operators P and $P + W$ are subcritical in Ω and W is a small (resp. semismall) perturbation of P in Ω then $G_P^\Omega(x,y)$ and $G_{P+W}^\Omega(x,y)$ are equivalent (resp. semi-equivalent). This follows from pointwise estimates of the iterated Green kernels corresponding to the Neumann series expansion of the Green function $G_{P+\varepsilon W}^\Omega(x,y)$ in terms of $G_P^\Omega(x,y)$ and W. Moreover, the corresponding Martin boundaries are homeomorphic.

On the other hand, suppose that W is a nonzero function such that the Green functions $G_P^\Omega(x,y)$ and $G_{P+|W|}^\Omega(x,y)$ are (resp. semi-) equivalent and let $K \in C^\alpha(\Omega)$ be an arbitrary function which vanishes at infinity of Ω. Then the function $K(x)W(x)$ is a (resp. semi-) small perturbation of the operator P in Ω. Moreover, using the resolvent equation it follows that if the (best) equivalence constant of $G_P^\Omega(x,y)$ and $G_{P+|W_n^*|}^\Omega(x,y)$ tends to 1 as n tends to infinity then W is a (resp. semi-) small perturbation.

Remark. There are several examples where the semi-equivalence of the Green functions implies that the perturbation is small.

Another class of perturbations was recently studied by A. Ancona [6]. In this paper, Ancona investigates the question of the equivalence of the Green functions of two uniformly elliptic, weakly coercive operators with uniformly bounded coefficients. The perturbation is allowed to be not only in the zero order term but also in the higher order terms. It is proved that the Green functions and the Martin boundaries of such two operators are equivalent provided that the "distance near infinity" between the operators is sufficiently small. Moreover, the equivalence constant tends to 1 if the supports of the perturbations are concentrated at infinity in Ω. Therefore, a zero order perturbation of Ancona's type is a small perturbation. Note that unlike the notion of small and semismall perturbations the definition of Ancona's distance does not depend explicitly on the behavior of the Green functions at infinity but only on the difference between the coefficients of the operators.

The discussion above and in particular Remark 2 implies that the results in [6, 24, 27] concerning the equivalence of the Green functions and the Martin boundaries seems to be almost optimal. Nevertheless, it turns out that Theorem 1.2 can be extended to a wider class of perturbations, namely, *weak perturbations*.

Throughout this paper we use the following key lemma many times (see [28, Theorem 3.1]). For the completeness, we present here an outline of the proof.

Lemma 2.1. *Let P be an elliptic operator defined on a domain $\Omega \subseteq X$ and assume that P is of the form (1.1). Consider the one parameter family of operators $P_t = P + tW + (1-t)V$, where $V, W \in C^\alpha(\Omega)$ and $0 \leq t \leq 1$. Suppose that $\mathcal{C}_{P_i}(\Omega) \neq \emptyset, i = 0, 1$. Then $\mathcal{C}_{P_t}(\Omega) \neq \emptyset$ for all $0 < t < 1$. Moreover, P_t is subcritical for all $0 < t < 1$ unless $P_0 = P_1$ and P_0 is critical in Ω.*

Proof: Let v_i be positive supersolutions of the equations $P_i u = 0$ in Ω, $i = 0, 1$. One can check easily that if $0 < t < 1$ then

$$v_t(x) = (v_0(x))^{1-t}(v_1(x))^t$$

is a positive supersolution of the equation $P_t u = 0$ in Ω. Moreover, v_t is a solution of $P_t u = 0$ in Ω if and only if $P_0 = P_1$ and $v_0 = Cv_1 \in \mathcal{C}_{P_0}(\Omega)$. These properties easily imply the statements of the Lemma. □

3. WEAK PERTURBATION

In this section we introduce the notion of weak perturbation and study some properties of such kind of perturbations.

Definition 3.1. *Let P be a subcritical operator in $\Omega \subseteq X$. A function $W \in C^\alpha(\Omega)$ is said to be a weak perturbation of the operator P in Ω if the following condition holds true.*

(∗) *For every $\eta \in \mathbb{R}$ there exists $N \in \mathbb{N}$ such that the operator $P - \eta W_n^*(x)$ is subcritical in Ω for any $n \geq N$.*

A function $W \in C^\alpha(\Omega)$ is said to be *a weak perturbation* of a critical operator P in Ω if there exists a nonzero, nonnegative function $V \in C_0^\alpha(\Omega)$ such that the function W is a weak perturbation of the subcritical operator $P + V$ in Ω.

Remarks.

1. Using Lemma 2.1 it can be easily shown that $W \in C^\alpha(\Omega)$ is a weak perturbation of a critical operator P in Ω if and only if for *any* nonzero, nonnegative function $V \in C_0^\alpha(\Omega)$ the function W is a weak perturbation of the subcritical operator $P + V$ in Ω.

2. It follows immediately from Definition 3.1 that if W is a weak perturbation of P in Ω then it is also a weak perturbation of $P + V$ for any nonnegative function $V \in C^\alpha(\Omega)$. Also, the set of all functions W which are weak perturbations of P in Ω is a vector space.

3. If $|W|$ is a weak perturbation of a subcritical operator P in Ω then P and W satisfy $(*)$ with respect to any exhaustion of Ω and any associated sequence of cutoff functions.

4. Using the results in [24, 27], it follows that if W is a small or even semismall perturbation of an operator P in Ω then $|W|$ is a weak perturbation of P in Ω. In particular, if one deals only with zero order perturbations then the perturbations considered recently by A. Ancona [6] are also weak. For example, if W is a perturbation of a weakly coercive operator P such that the operators P and $P+W$ satisfy the assumptions of Remark 1.2 in [6] then $|W|$ is a weak perturbations of P in X.

5. One can use known integral estimates on the number of the negative eigenvalues of subcritical operators to show that a perturbation is weak. For example, let $d \geq 3$. By the Cwikel–Lieb–Rozenblum bound (see, [35]), if $W \in L^{d/2}(\mathbb{R}^d)$ then $|W|$ is a weak perturbation of $-\Delta$ in $\mathbb{R}^d$. Moreover, such a function W is a weak perturbation of a symmetric subcritical operator P with periodic coefficients in $\mathbb{R}^d$ [11]. For further results in this direction see also [12, 17]. Note that $W(x) = (1+|x|)^{-2}$ is not a weak perturbation of $-\Delta$ in $\mathbb{R}^d, d \geq 3$. On the other hand, for any $\epsilon > 0$ the function $W(x) = (1+|x|)^{-(2+\epsilon)}$ is a small perturbation of $-\Delta$ in $\mathbb{R}^d, d \geq 3$ [23, 26, 29, 33, 34]. For some other examples of weak perturbations, see [23, 24, 26, 27, 29, and the references therein].

The following example shows that a weak perturbation is a weaker notion than semismall perturbation.

Example 3.1. Let $P = -\Delta$ in $\mathbb{R}^d$, $d \geq 3$, and consider a nonnegative function $W \in L^{d/2}(\mathbb{R}^d)$ such that

$$\int_{\mathbb{R}^d} W(y)(1+|y|)^{2-d} dy = \infty.$$

By the Cwikel–Lieb–Rozenblum inequality W is a weak perturbation. Suppose that W is a semismall perturbation, then the ground state ϕ_0 of the operator $-\Delta - \lambda_0 W$ behaves near infinity like the Green function of the Laplacian and therefore,

$$\phi_0(x) = \lambda_0 \int_{\mathbb{R}^d} G_{-\Delta}^{\mathbb{R}^d}(x,y) W(y) \phi_0(y) dy$$

$$\geq C \int_{\mathbb{R}^d} |x-y|^{2-d} W(y)(1+|y|)^{2-d} dy$$

$$\geq C \int_{|x-y|<2|x|} |x|^{2-d} W(y)(1+|y|)^{2-d} dy$$

$$\geq C|x|^{2-d} \int_{|y|<|x|} W(y)(1+|y|)^{2-d} dy \geq C\phi_0(x) F(|x|),$$

where $F(s)$ is a real function which tends to infinity as s tends to infinity. This is a contradiction.

Lemma 3.2. *(i) If W is a weak perturbation of a subcritical operator P in Ω then there exists $\epsilon > 0$ such that the operator $P - \mu W(x)$ is subcritical in Ω for any $|\mu| < \epsilon$. Moreover, if the operator $P - \mu W$ is subcritical in Ω then there exists $\delta > 0$ such that $P - (\mu+t)W$ is subcritical in Ω for all $|t| < \delta$.*

(ii) If W is a weak perturbation of a critical or subcritical operator P in Ω and for some $\mu \in \mathbb{R}$ the operator $P - \mu W(x)$ is subcritical or critical in Ω then W is also a weak perturbation of the operator $P - \mu W(x)$ in Ω.

Proof: (i). There exists $N \in \mathbb{N}$ such that $P \pm W_N^*$ is subcritical in Ω. Since W_N has compact support and P is subcritical there exists $\delta > 0$ such that $P \pm \delta W_N$ is subcritical in Ω. Set $\epsilon = \frac{\delta}{1+\delta}$. It follows from Lemma 2.1 that $P \pm \epsilon W_N^* \pm (1-\epsilon)\delta W_N = P \pm \epsilon W$ is subcritical in Ω. Thus, by Lemma 2.1, $P - \mu W$ is subcritical in Ω for any $|\mu| \leq \epsilon$.

Suppose now that $P - \mu W$ is subcritical in Ω for some $\mu > 0$ (the case $\mu < 0$ can be treated similarly). It is enough to show that $P - \mu_1 W$ is subcritical in Ω for some $\mu_1 > \mu$. There exists $N \in \mathbb{N}$ such that $P - 2\mu W_N^*$ is subcritical in Ω. On the other hand, there exists $\epsilon > 0$ such that $P - \mu W - \epsilon W_N$ is subcritical in Ω. Set $\beta = \frac{2\mu}{2\mu+\epsilon}$. Then $P - \beta(\mu W + \epsilon W_N) - (1-\beta)2\mu W_N^* = P - \mu_1 W$ is subcritical in Ω, where $\mu_1 = \frac{2\mu(\mu+\epsilon)}{2\mu+\epsilon}$. Since $\mu_1 > \mu$ the first part of the lemma is proved.

(ii). Without loss of generality, we may assume that $\mu > 0$. Suppose first that P and $P - \mu W$ are subcritical in Ω. By the first part of the lemma, there exists $\delta = \delta(\mu) > 0$ such that $P - (\mu + \delta)W$ is subcritical. On the other hand, for every $\eta \in \mathbb{R}$ there exists $N \in \mathbb{N}$ such that $P - \eta W_n^*$ is subcritical in Ω for all $n \geq N$. Set $\beta = \frac{\mu}{\mu+\delta}$. Then $P - \beta(\mu + \delta)W - (1-\beta)\eta W_N^* = P - \mu W - \tilde{\eta} W_n^*$ is subcritical in Ω for all $n \geq N$, where $\tilde{\eta} = \frac{\delta\eta}{\mu+\delta}$. Hence, we can make $|\tilde{\eta}|$ arbitrarily large provided we choose $|\eta|$ large enough.

Suppose now that P is subcritical or critical and $P - \mu W$ is critical in Ω. Then for any nonzero, nonnegative $V \in C_0^\alpha(\Omega)$, the operators $P + V$ and $P + V - \mu W$ are subcritical. Moreover, By Remark 3, W is a weak perturbation of $P + V$ in Ω. Therefore, we are again in the situation of the the previous case.

Finally, assume that P is critical and $P - \mu W$ is subcritical in Ω. Take a nonzero, nonnegative $V \in C_0^\alpha(\Omega)$. There exists $\epsilon = \epsilon(\mu) > 0$ such that $P - \mu W - \epsilon V$ is subcritical in Ω. On the other hand, W is a weak perturbation of $P + V$ in Ω and by the previous part, W is also a weak perturbation of the subcritical operator $P + V - \mu W$. Consequently, for every $\eta \in \mathbb{R}$ there exists $N \in \mathbb{N}$ such that $P + V - \mu W - \eta W_n^*$ is subcritical for all $n \geq N$. Take $\beta = \frac{\epsilon}{1+\epsilon}$. It follows that $P + \beta(V - \mu W - \eta W_n^*) + (1-\beta)(-\mu W - \epsilon V) = P - \mu W - \beta\eta W_n^*$ is subcritical in Ω for all $n \geq N$ and this completes the proof. □

Remark. It follows from Lemma 3.2 that if $|W|$ is a weak perturbation of $-\Delta$ in $\mathbb{R}^d, d \geq 3$ and $-\Delta + W$ is subcritical then $|W|$ is strongly subcritical in the sense of [16]. Therefore, the results in [16] apply to such perturbations.

Let $V \in C^\alpha(\Omega)$ be a nonzero, nonnegative function and $W \in C^\alpha(\Omega)$. Define

$$\mu_+(\lambda, V) = \sup\{\mu \in \mathbb{R} | \mathcal{C}_{P-\lambda V - \mu W}(\Omega) \neq \emptyset\} \qquad (3.1)$$

$$\mu_-(\lambda, V) = \inf\{\mu \in \mathbb{R} | \mathcal{C}_{P-\lambda V - \mu W}(\Omega) \neq \emptyset\} \qquad (3.2)$$

and denote

$$S_+(\lambda, V) = \{\mu \in \mathbb{R} | P - \lambda V - \mu W \text{ is subcritical in } \Omega\}.$$

Note that if P is critical in Ω then either $\mu_-(0, V) = 0$ or $\mu_+(0, V) = 0$. Moreover, for weak perturbations it follows from Lemma 2.1 and Lemma 3.2 that

Lemma 3.3. *Let P be a subcritical or critical operator in Ω and $V \in C^\alpha(\Omega)$ be a nonzero, nonnegative function. Assume that W is a weak perturbation of P in Ω. If $|\mu_\pm(0, V)| < \infty$ then the operator $P - \lambda V - \mu_\pm(\lambda, V)W$ is critical in Ω for all $\lambda \leq 0$. In other words, we have $S_+(\lambda, V) = (\mu_-(\lambda, V), \mu_+(\lambda, V))$. Moreover, for every $\lambda < 0$, $\mu_-(\lambda, V) < 0 < \mu_+(\lambda, V)$ and if P is subcritical then we also have $\mu_-(0, V) < 0 < \mu_+(0, V)$.*

4. PROOF OF THE MAIN THEOREM

In this section we state and prove our main result which extends Theorem 1.2 (where the perturbation W is a function with compact support) to the case of a weak perturbation.

Theorem 4.1. *Let P be a critical operator in Ω and $W \in C^\alpha(\Omega)$ a weak perturbation of the operator P in Ω, where $0 < \alpha \leq 1$. Let $x_0 \in \Omega_1$ be a fixed reference point.*

Denote by ϕ_0 (resp. $\tilde{\phi}_0$) the ground state of the operator P (resp. P^*) in Ω such that $\phi_0(x_0) = 1$ (resp. $\tilde{\phi}_0(x_0) = 1$). Assume also that

$$\int_\Omega |W(x)|\phi_0(x)\tilde{\phi}_0(x)dx < \infty. \tag{4.1}$$

(i) If there exists $\mu < 0$ such that $P - \mu W(x)$ is subcritical in Ω then

$$\int_\Omega W(x)\phi_0(x)\tilde{\phi}_0(x)dx > 0. \tag{4.2}$$

(ii) Assume that for some nonnegative, nonzero function $V \in C_0^\alpha(\Omega)$ there exist $\mu_0 < 0$ and a positive constant C such that

$$G^\Omega_{P+V-\mu W}(x,x_0) \leq C\phi_0(x) \quad \text{and} \quad G^\Omega_{P+V-\mu W}(x_0,x) \leq C\tilde{\phi}_0(x) \tag{4.3}$$

for all $x \in \Omega \setminus \Omega_1$ and $\mu_0 \leq \mu < 0$. If the integral condition (4.2) holds true then there exists $\mu < 0$ such that $P - \mu W(x)$ is subcritical in Ω.

Remarks.

1. If $W \in C^\alpha(\Omega)$ is a small perturbation of a subcritical operator P in Ω then the ground states $\phi_\pm(x)$ of the operator $P - \mu_\pm(0,V)W(x)$ are comparable with $G^\Omega_{P-\mu W}(x,x_0)$ for all μ such that $\mu_- < \mu < \mu_+$. ([27], see also [24], where this result is proved for a semismall perturbation of P^*). In particular, this assertion holds true for a perturbation W of the type considered by Ancona in [6] (see also [3, 14, 15, 20, 22, 23, 29, 36, 40]). Thus, small perturbations and even perturbations which are semismall of both P and P^* all satisfy assumption (4.3).

2. Recently, T. Weidl [39] proved an abstract result on the number of bound states appearing below the spectrum of a semi-bounded symmetric operator in Hilbert space in the case of a "weak" non-sign-defined perturbation. As an application he considered a perturbation of a critical Schrödinger operator $H = -\Delta - \beta_0 V$ by a function W in $\mathbb{R}^d$. It is assumed that there exist $\beta > \beta_0 > 0$ and $\alpha > 0$ such that the negative spectra of the operators $H = -\Delta - \beta V$ and $H = -\Delta - \beta_0 V - \alpha|W|$ are finite. Under some additional assumptions it is shown ([39, Theorem 8.1]) that the statement of Theorem 1.2 holds true. Note that in Theorem 4.1 we have no assumption on the critical operator P but only on the perturbation W. On the other hand, if W is a weak perturbation of a subcritical or critical Schrödinger operator H then the negative spectrum of $H - \alpha W$ is finite for all $\alpha \in \mathbb{R}$.

3. We are not aware of any example of a weak perturbation W of a critical operator P in Ω such that the integrability condition (4.1) is not satisfied.

Proof of Theorem 4.1: (i) Suppose first that $P - \mu W$ is subcritical for some $\mu < 0$. By Lemma 3.2 (ii), W is also a weak perturbation of the operator $P - \mu W$ in Ω. Consequently, there exists N_0 such that for every $n \geq N_0$ the operator $P - \mu W + \mu W_n^* = P - \mu W_n$ is subcritical in Ω. Since the function W_n has compact support, Theorem 1.2 implies that for all $n \geq N$

$$\int_\Omega W_n(x) \phi_0(x) \tilde{\phi}_0(x) dx > 0. \tag{4.4}$$

By our assumption, the function $W\phi_0\tilde{\phi}_0$ is integrable in Ω. Thus, using the Lebesgue dominated convergence theorem we obtain

$$\int_\Omega W(x) \phi_0(x) \tilde{\phi}_0(x) dx \geq 0. \tag{4.5}$$

In order to show that we have strict inequality in (4.5) we proceed as in the proof of Theorem 1.4 in [28]. Suppose that we have equality in (4.5). Let $V \in C_0^\infty(\Omega)$ be a nonnegative, nonzero function. Since the operator $P - \mu W$ is subcritical in Ω there exists $\epsilon > 0$ such that the operator $P - \mu W - \epsilon V$ is subcritical in Ω. Therefore, we can use (4.5) with the function $W(x) + \frac{\epsilon}{\mu}V(x)$ which is a weak perturbation of P in Ω. Thus,

$$\int_\Omega (W(x) + \frac{\epsilon}{\mu} V(x)) \phi_0(x) \tilde{\phi}_0(x) dx \geq 0. \tag{4.6}$$

Since $\mu < 0$ and

$$\int_\Omega V(x) \phi_0(x) \tilde{\phi}_0(x) dx > 0,$$

we see that (4.6) contradicts our assumption of an equality in (4.5).

(ii) Assume now that

$$\int_\Omega W(x) \phi_0(x) \tilde{\phi}_0(x) dx > 0 \tag{4.7}$$

and suppose that $\mu_-(0, V) = 0$, where $V \in C_0^\alpha(\Omega)$ is a nonnegative, nonzero function. In particular, W changes sign in Ω and therefore, $\mu_-(\lambda, V) > -\infty$ for any $\lambda \leq 0$. Lemma 3.3 implies that if $\lambda < 0$ then $\mu_-(\lambda, V) < 0$ and furthermore, the operator $P - \lambda V - \mu_-(\lambda, V)W$ is critical in Ω. Denote by $\phi_{(\lambda,-)}(x)$ (resp. $\tilde{\phi}_{(\lambda,-)}(x)$) the ground state of the operator $P - \lambda V - \mu_-(\lambda, V)W$ (resp. $P^* - \lambda V - \mu_-(\lambda, V)W$) such that $\phi_{(\lambda,-)}(x_0) = 1$ (resp. $\tilde{\phi}_{(\lambda,-)}(x_0) = 1$).

Note that, $\lim_{\lambda \to 0_-} \mu_-(\lambda, V) = \mu_-(0, V) = 0$. It follows from remark 2 that if $\lambda < 0$ and $|\lambda|$ is sufficiently small, then both $\phi_{(\lambda,-)}(x)$ and $G^\Omega_{P+V-\mu_-(\lambda,V)W}(x, x_0)$ are positive solutions of minimal growth in a neighborhood of infinity in Ω of the equation $(P - \mu_-(\lambda, V)W(x))u = 0$. Using our assumption (4.3), the Harnack inequality and the generalized maximum principle we infer that there exist $\lambda_1 < 0$ and a constant $C_1 > 0$ such that

$$\phi_{(\lambda,-)}(x) \leq C_1 \phi_0(x) \quad \text{and} \quad \tilde{\phi}_{(\lambda,-)}(x) \leq C_1 \tilde{\phi}_0(x) \tag{4.8}$$

for every $\lambda_1 \leq \lambda < 0$ and all $x \in \Omega$. In particular, $W\phi_{(\lambda,-)}\tilde{\phi}_{(\lambda,-)} \in L^1(\Omega)$. Recall that for every $\lambda < 0$ the operator $P - \lambda V$ is subcritical while $P - \lambda V - \mu_-(\lambda, V)W$ is critical in Ω. Moreover, W is a weak perturbation of the operator $P - \lambda V - \mu_-(\lambda, V)W$. Consequently, by first part of the theorem for all $\lambda_1 \leq \lambda < 0$ we have

$$\int_\Omega W(x) \phi_{(\lambda,-)}(x) \tilde{\phi}_{(\lambda,-)}(x) dx < 0. \tag{4.9}$$

Recall that, $\lim_{\lambda \to 0_-} \mu_-(\lambda, V) = \mu_-(0, V) = 0$. Moreover, ϕ_0 (resp. $\tilde{\phi}_0$) is the unique positive normalized solution of the equation $Pu = 0$ (resp. $P^*u = 0$) in Ω. Using the Harnack inequality and elliptic Schauder estimates it follows that as $\lambda \nearrow 0$ the functions $\phi_{(\lambda,-)}$ (resp. $\tilde{\phi}_{(\lambda,-)}$) converge uniformly in any compact subset of Ω to ϕ_0 (resp. $\tilde{\phi}_0$).

Consequently, inequalities (4.8), (4.9) and the Lebesgue dominated convergence theorem imply that

$$\int_\Omega W(x)\phi_0(x)\tilde{\phi}_0(x)dx \leq 0$$

which contradicts our assumption (4.7). □

Remark. Consider a critical operator P defined on Ω and a class of functions W such that the second part of Theorem 4.1 holds true. It follows that for any nonzero, nonnegative function V with compact support there exists $\epsilon > 0$ such that $P + V - \mu W$ is subcritical in Ω for all $|\mu| < \epsilon$. Thus, for the second part of the theorem one should assume that the perturbation is in some sense "small."

ACKNOWLEDGMENT

The author wishes to acknowledge stimulating conversations with Professor M. Murata on the subject of this paper. This research was supported by the Fund for the Promotion of Research at the Technion.

REFERENCES

1. S. Agmon, On positivity and decay of second order elliptic equations on Riemannian manifolds, *in:* "Methods of Functional Analysis and Theory of Elliptic Equations," D. Greco, ed., pp.19–52, Liguori, Naples, 1982.
2. S. Alama and G. Tarantello, On semilinear elliptic equations with indefinite nonlinearities, *Cal. Var.* **1** (1993), 439–475.
3. W. Allegretto, Principal eigenvalues for indefinite-weight elliptic problems in $\mathbb{R}^n$, *Proc. Amer. Math. Soc.* **116**, (1992), 701–706.
4. W. Allegretto and A. B. Mingarelli, On the non-existence of positive solutions for a Schrödinger equation with an indefinite weight-function, *C. R. Math. Acad. Sci. Canada* **8** (1986), 69–73.
5. W. Allegretto and A. B. Mingarelli, Boundary problems of the second order with an indefinite weight-function, *J. Reine Angew. Math.* **398** (1989), 1–24.
6. A. Ancona, Comparison of Green's functions for elliptic operators on manifolds or domains, *J. Analyse. Math.*, to appear.
7. W. Arendt and C. K. J. Batty, The spectral function and principal eigenvalues for Schrödinger operators, *Potential Analysis* **5** (1996), 207–230.
8. M. Bôcher, The smallest characteristic number in a certain exceptional case, *Bull. Amer. Math. Soc.* **21** (1914), 6–9.
9. H. Berestycki, I. Capuzzo-Dolcetta and L. Nirenberg, Superlinear indefinite elliptic problems and nonlinear Liouville theorems, *Topological Methods in Nonlinear Analysis* **4** (1994), 59–78.
10. H. Berestycki, I. Capuzzo-Dolcetta and L. Nirenberg, Variational methods for indefinite superlinear homogeneous elliptic problems, *NoDEA* **2** (1995), 553–572.
11. M. Sh. Birman and M. Z. Solomyak, Estimates for the number of negative eigenvalues of the Schrödinger operator and its generalizations, *in:* "Estimates and Asymptotics for Discrete Spectra of Integral and Differential Equations," M. Sh. Birman ed., Adv. Sov. Math. 7, pp. 1–55, Amer. Math. Soc., Providence, 1991.
12. M. Sh. Birman and M. Z. Solomyak, Schrödinger operator. Estimates for number of the bound states as function-theoretical problem, *in:* "Spectral Theory of Operators," S. G. Gindikin ed., Amer. Math. Soc. Transl. (2) **150**, pp. 1–54, Amer. Math. Soc., Providence, 1992.

13. J. Bochenek, On positive eigenvectors of a linear eigenvalue problem, *Ann. Polonici Math.* **51** (1990), 77–87.
14. K. J. Brown, C. Cosner and J. Fleckinger, Principal eigenvalues for problems with indefinite weight function on $\mathbb{R}^N$, *Proc. Amer. Math. Soc.* **109** (1990), 147–155.
15. K. J. Brown and A. Tertikas, The existence of principal eigenvalues for problems with indefinite weight function on $\mathbb{R}^k$, *Proc. Royal. Soc. Edinburgh* **123A** (1993), 561–569.
16. E. B. Davies and B. Simon, L^p norms of non-critical Schrödinger semigroups, *J. Funct. Anal.* **102** (1991), 95–115.
17. Yu. V. Egorov and V. A. Kondrat'ev, Estimates of the negative spectrum of an elliptic operator, *in:* "Spectral Theory of Operators," S. G. Gindikin ed., Amer. Math. Soc. Transl. (2) **150**, pp. 111–140, Amer. Math. Soc., Providence, 1992.
18. J. Fleckinger and A. B. Mingarelli, On the eigenvalues of non-definite elliptic operators, *in:* "Proceedings International Conference on Differential Equations," I. W. Knowles and R. T. Lewis, eds., North-Holland Mathematics Studies, No. 92, pp. 219–227, 1984.
19. F. Gesztesy and Z. Zhao, On critical and subcritical Sturm–Liouville operators, *J. Funct. Anal.* **98** (1991), 311–345.
20. F. Gesztesy and Z. Zhao, On positive solutions of critical Schrödinger operators in two dimensions, *J. Funct. Anal.* **127** (1995), 235–256.
21. P. Hess and T. Kato, On some linear and nonlinear eigenvalue problems with an indefinite weight function, *Comm. Partial Differential Equations* **5** (1980), 999–1030.
22. J. López-Gómez, The maximum principle and the existence of principal eigenvalues for some linear weighted boundary value problems, *J. Differential Equations* **127** (1996), 263–294.
23. M. Murata, Structure of positive solutions to $(-\Delta + V)u = 0$ in $\mathbb{R}^n$, *Duke Math. J.* **53** (1986), 869–943.
24. M. Murata, Semismall perturbations in the Martin theory for elliptic equations, *Israel J. Math.*, to appear.
25. M. Picone, Sui valori eccezionali di un parametro, *Ann. Scuola Norm. Sup. Pisa* **11** (1910), 1–143.
26. Y. Pinchover, On positive solutions of second-order elliptic equations, stability results and classification, *Duke Math. J.* **57** (1988), 955–980.
27. Y. Pinchover, Criticality and ground states for second-order elliptic equations, *J. Differential Equations*, **80** (1989), 237–250.
28. Y. Pinchover, On criticality and ground states for second order elliptic equations, II, *J. Differential Equations* **87** (1990), 353–364.
29. Y. Pinchover, On the equivalence of Green functions of second order elliptic equations in $\mathbb{R}^n$, *Differential and Integral Equations* **5** (1992), 481–493.
30. Y. Pinchover, On the localization of binding for Schrödinger operators and its extension to elliptic operators, *J. Analyse Math.* **66** (1995), 57–83.
31. Y. Pinchover, On positivity, criticality and spectral radius of the shuttle operator for elliptic operators, *Duke Math. J.* **85** (1996), 431–445.
32. R. G. Pinsky, "Positive Harmonic Functions and Diffusion," Cambridge Studies in Advanced Mathematics, Vol. 45, Cambridge University Press, Cambridge, 1995.
33. A. G. Ramm, Uniqueness theorem for the solution of the Dirichlet problem, *Siber. Math. J.* **19** (1978), 1003–1005.
34. A. G. Ramm, Sufficient conditions for zero not to be an eigenvalue of the Schrödinger operator, *J. Math. Phys.* **26** (1987), 1341–1343.
35. M. Reed and B. Simon, "Methods of Modern Mathematical Physics. IV. Analysis of Operators," Academic Press, New York, 1978.
36. G. Rozenblioum and M. Solomyak, On principal eigenvalues for indefinite problems in the Euclidean space, preprint.
37. B. Simon, The bound state of weakly coupled Schrödinger operators in one and two dimensions, *Ann. Physics* **97** (1976), 279–288.
38. B. Simon, Large time behavior of the L^p norm of Schrödinger semigroups, *J. Funct. Anal.* **40** (1981), 66–83.
39. T. Weidl, Remarks on virtual bound states for semi-bounded operators, preprint.
40. Z. Zhao, Subcriticality and gaugeability of the Schrödinger operator, *Trans. Amer. Math. Soc.* **334** (1992), 75–96.

SCATTERING BY OBSTACLES IN ACOUSTIC WAVEGUIDES

A. G. Ramm[1] and G. N. Makrakis[2]

[1] Department of Mathematics
Kansas State University
Manhattan, Kansas
E-mail: ramm@math.ksu.edu

[2] Institute of Applied and Computational Mathematics, FO.R.T.H
P.O. Box 1527, 71 110, Heraklion, Crete, Greece
and Department of Mathematics
University of Crete
E-mail: makrakg@calderon.iacm.forth.gr

ABSTRACT

We study the scattering of acoustic waves by an obstacle embedded in a uniform waveguide with planar boundaries, which is a fundamental model in shallow ocean acoustics. Under certain geometric assumptions on the shape of the obstacle, we prove a Rellich-type uniqueness theorem in the case of soft obstacles. The solvability of the scattering problem is proved via boundary integral equations and the limiting absorption principle is established. An eigenfunction expansion in terms of the scattering solutions reveals the appropriate (partial) scattering amplitudes for the waveguide. We show that these scattering amplitudes for the propagating modes, at a fixed frequency, define uniquely the shape of a soft obstacle.

1. INTRODUCTION

In this paper we deal with some aspects of the scattering of acoustic waves by an obstacle embedded in a uniform waveguide with planar boundaries, which is a fundamental model for shallow-ocean acoustics [1]. This model has been already considered for acoustically soft obstacles (Dirichlet boundary condition) by Gilbert and Xu [6–8, 29] who have also dealt with the case of a horizontally stratified ocean.

It is well known that the Dirichlet Laplacian has purely continuous spectrum for a large class of unbounded domains, including exterior of bounded domains, some classes of generalized (perturbed) conical domains, generalized cylinders and locally perturbed generalized cylinders (see, e.g., Refs. [5, 11, 17, 26]). The relation between the spectral properties of linear operators, in particular, the Schrödinger operators, and the behavior at large times of the solutions to exterior time-dependent problems was studied by Ramm [11], and for the waveguide by Ramm and Werner [25].

If the local perturbation violates certain geometrical restrictions (like the one in the Assumptions (**2.A**) or (**2.B**) in Section 2 below), it is possible to have an unbounded sequence of eigenvalues which are embedded in the essential spectrum (see Section 7 below, and Ref. [26] and the references therein). The same situation arises also in the case of the Neumann Laplacian. It has been shown that in certain two-dimensional water-wave and acoustic problems governed by the Helmholtz equation in strips (see, e.g., Refs. [2, 4]), trapped modes can exist in the case of hard obstacles. Therefore, the conditions which the obstacles must satisfy for the uniqueness of the solution to the scattering problem in waveguides to hold are not well understood. It is an open problem to clarify these assumptions. In particular, uniqueness theorem for scattering or boundary-value problem for a hard obstacle in the waveguide is presently not known and non-uniqueness results are given in Refs. [2, 4].

We present some uniqueness results for soft obstacles in a waveguide in Section 2. The first of our proof uses an idea of Rellich [26] but is based on the new identity (2.4), and the second one is based on the Rellich's identity. A uniqueness theorem similar to our Theorem 2.1 for the case of soft obstacles has been given earlier by Morgenröther and Werner [9], but our proof is different. The second uniqueness Theorem 2.1' is new.

In Section 3 we investigate the solvability of the boundary value problem for a soft obstacle using boundary integral equations. In Section 4 we prove the limiting absorption principle for the problem at hand, following Ramm [17] (see also Ref. [11], Chapt. VII, Sec. 5). An appropriate definition of the scattering solutions is given in Section 5, and an eigenfunction expansion leading to a natural definition of the partial scattering amplitudes for the waveguide problem is also constructed there. In Section 6 the uniqueness of an inverse obstacle scattering problem at fixed frequency (that is, the assertion that the scattering data, namely partial scattering amplitudes for the propagating modes, uniquely define the shape of the obstacle) is proved following the reasoning in Ref. [11], pp. 85–86, and the results and ideas in Ref. [18]. In Section 7 an example of eigenvalues embedded in the continuous spectrum of the Dirichlet Laplacian in a locally perturbed and rotationally symmetric waveguide is given. This example is taken essentially from the paper by Witsch [27].

In Section 8 some recent papers by R. Kleinman, T. Angell and coauthors are discussed. In these papers a uniqueness theorem is claimed for a boundary value problem for hard obstacle in a waveguide. We point out a mistake in their proof. The uniqueness theorem, claimed by the mentioned authors, is wrong, and a counterexample can be found in Refs. [2, 4]. We also point out that the "new conditions" introduced in these recent papers in place of the radiation condition, are void: they are never satisfied. We follow Ref. [24] in Section 8.

We do not want to increase the volume of the paper, and that is why some results similar to the ones in Refs. [12–23] are not stated and discussed. We refer the reader to Refs. [16, 18] for analysis of the scattering problems in the case of Lipschitz boundaries, to Refs. [12–14, 18] for some methods of the proof of the uniqueness of the solution to inverse obstacle scattering problems, in particular for non-smooth obstacles, for instance, obstacles with Lipschitz (and even less regular) boundaries, to Ref. [22] for an analytical example of non-uniqueness of the solution to an inverse geophysical problem, and to Ref. [19] for a study of the dependence of the scattering amplitude on the boundary of the obstacle.

We consider the homogeneous waveguide, but the theory we develop is applicable without essential changes to the horizontally stratified waveguide. This is clear from the presentation (see also the closing paragraph at the end of Section 3).

We will use the following notations. Let $L := \{x = (\widehat{x}, x_3) : \widehat{x} \in \mathbb{R}^2, 0 \le x_3 \le h\}$, where $\widehat{x} = (x_1, x_2)$ and h is a positive constant, and D be a bounded connected subdomain of L with C^2 boundary S having a unit normal N *pointing into* D. We denote the complement of D in L by $D' := L \backslash D$, and let $D_r := \{x : |\widehat{x}| \le r, x \in D'\}$, $\partial D_r = S \cup \Gamma_r \cup \partial L_r$, $\Gamma_r := \{x : |\widehat{x}| = r, x \in D'\}$ and $\partial L_r := \{x : x \in \partial L, |\widehat{x}| \le r\}$. By $r_0 > 0$ we denote such a number, that the obstacle D lies in the cylinder $\{x : x \in L, |\widehat{x}| < r_0\}$. Let $L_0^2(D')$ stand for the set of $L^2(D')$-functions vanishing near infinity (but not necessarily near S), and Ω_0 for the exterior in L of the region D_{r_0}.

Scattering of acoustic waves in the layer L is governed by the Helmholtz equation

$$\Delta u + k^2 u = 0 \quad \text{in} \quad D', \quad k = \text{const} > 0. \tag{1.1}$$

We assume that u satisfies either the boundary conditions

$$u(\widehat{x}, 0) = u(\widehat{x}, h) = 0, \tag{1.2}$$

or

$$u(\widehat{x}, h) = \partial_{x_3} u(\widehat{x}, 0) = 0, \tag{1.3}$$

on the surfaces $x_3 = 0$ and $x_3 = h$ of the layer L. We fully investigate the case $u|_S = 0$ (soft obstacle), and we comment on the case $\partial_N u |_S = 0$ (hard obstacle).

If the obstacle is absent the above problem can be solved by taking the Fourier transform of Equation (1.1) with respect to the variable $\widehat{x}$. This yields the eigenvalue problem

$$\phi''(x_3) + k^2 \phi(x_3) = \lambda^2 \phi(x_3), \tag{1.4}$$

$$\phi(0) = \phi(h) = 0, \tag{1.5}$$

or

$$\phi(h) = \phi'(0) = 0, \tag{1.6}$$

corresponding to the boundary conditions (1.2) or (1.3), respectively. Here

$$\phi := \frac{1}{2\pi} \int_{\mathbb{R}^2} u(\widehat{x}, x_3) \exp(i \boldsymbol{\lambda} \cdot \widehat{x}) d\widehat{x},$$

and $\phi := \phi(x_3, \lambda, k), \boldsymbol{\lambda} := (\lambda_1, \lambda_2), |\boldsymbol{\lambda}| := \lambda$.

The problems (1.4), (1.5) (or (1.4), (1.6)) have an infinite set of simple eigenvalues

$$\lambda \in \mathcal{K} := \left\{ \lambda_m = m \frac{\pi}{h}, \quad m = 1, 2, \dots \right\} \tag{1.7}$$

or

$$\lambda \in \mathcal{K}' := \left\{ \lambda_m = (m + \frac{1}{2}) \frac{\pi}{h}, \quad m = 0, 1, 2, \dots \right\}, \tag{1.8}$$

respectively. The corresponding eigenfunctions $\phi_m(x_3)$ form a complete orthonormal set in $L_2(0, h)$ if we also impose the normalization condition

$$\int_0^h |\phi_m|^2 dx_3 = 1,$$

where

$$\phi_m(x_3) = \sqrt{\frac{2}{h}} \sin(m\pi x_3/h), \qquad m = 1, 2, ..., \tag{1.9}$$

if condition (1.5) holds, or

$$\phi_m(x_3) = \sqrt{\frac{2}{h}} \cos\left(\left(m + \frac{1}{2}\right)\pi x_3/h\right), \qquad m = 0, 1, 2, ..., \tag{1.10}$$

if condition (1.6) holds.

2. THE UNIQUENESS THEOREM

In the sequel we prove a uniqueness theorem for the scattering problem when the obstacle D is soft. We assume in this paper, unless otherwise stated, that the obstacle satisfies the following geometric condition:

Assumption 2.A. There exists a direction e_2 parallel to $\partial L = \{x : x_3 = 0 \text{ or } x_3 = h\}$ and a point $O \in D$, such that if O has the coordinate $x_2 = 0$, then

$$x_2 N_2 \leq 0 \quad \text{on} \quad S, \tag{A}$$

N_2 being the x_2-component of the normal N which points into D.

Theorem 2.1. *Let $k \notin \mathcal{K}$. If u is a solution of the problem*

$$\Delta u + k^2 u = 0 \quad \text{in} \quad D', \tag{2.1}$$

$$u \big|_{\partial D'} = 0, \tag{2.2}$$

$$u \in L^2(D'), \tag{2.3}$$

then $u \equiv 0$ in D'. The same conclusion holds if (2.3) is replaced by the condition

$$\lim_{r \to \infty} \int_{\Gamma_r} [|u|^2 + |u_r|^2] ds = 0 \tag{2.3'}$$

Proof: We start with the identity

$$0 = x_2 u_2 (u_{jj} + k^2 u) = (x_2 u_2 u_j)_j + \left(\frac{k^2 u^2 x_2}{2} - \frac{x_2 (\nabla u)^2}{2}\right)_2 + \frac{(\nabla u)^2 - k^2 u^2}{2} - u_2^2 \tag{2.4}$$

where $u_j := \partial u/\partial x_j$, and over the repeated indices summation is understood. Integrating (2.4) by parts over the region D_r we get

$$0 = \int_{\partial D_r} \left(x_2 u_2 u_j N_j + \frac{k^2 u^2 - (\nabla u)^2}{2} x_2 N_2\right) ds + \int_{D_r} \left(\frac{(\nabla u)^2 - k^2 u^2}{2} - u_2^2\right) dx. \tag{2.5}$$

Scattering by Obstacles in Acoustic Waveguides

One has $N_2 = 0$ on ∂L_r, and since $u = 0$ on $S \cup \partial L_r$, it follows that $\nabla u = N u_N$ and $u_j = N_j u_N$ on $S \cup \partial L_r$. According to Lemma 2.3 (see below) we have

$$\int_{D'} (|\nabla u|^2 - k^2 |u|^2) dx = 0. \qquad (2.6)$$

Therefore, taking $r \to \infty$ in (2.5), we get

$$\int_{D'} u_2^2 dx = \int_S x_2 N_2 \left(u_N^2 - \frac{u_N^2}{2} \right) ds + \lim_{r \to \infty} \int_{\Gamma_r} x_2 \left(u_2 u_N + \frac{k^2 u^2 - |\nabla u|^2}{2} N_2 \right) ds. \qquad (2.7)$$

Furthermore, from Lemma 2.2 below it follows that

$$|u| + |\nabla u| \leq c \exp(-\epsilon |\hat{x}|), \qquad \epsilon > 0. \qquad (2.8)$$

Therefore the second integral in (2.7) tends to zero as $r \to \infty$, and we get

$$0 \leq \int_{D'} u_2^2 dx = \frac{1}{2} \int_S u_N^2 x_2 N_2 ds, \qquad (2.9)$$

which, because of the geometric assumption (2.A), implies $u_N = 0$ on S. From this and (2.2) we have $u = u_N = 0$ on S. By the uniqueness of the solution to the Cauchy problem for elliptic equations, one gets $u = 0$ in D'. $\square$

We prove now Lemmas 2.2 and 2.3.

Lemma 2.2. *If u solves (2.1)–(2.3), or (2.1)–(2.3'), then (2.8) holds.*

Proof: Let us write

$$u(\hat{x}, x_3) = \sum_{m=1}^{\infty} \phi_m(x_3) u_m(\hat{x}), \qquad r := |\hat{x}| > r_0, \qquad (2.10)$$

where

$$u_m(\hat{x}) := \int_0^h u(\hat{x}, x_3) \phi_m(x_3) dx_3. \qquad (2.11)$$

Then (2.1) and (2.10) imply

$$\hat{\Delta} u_m + k_m^2 u_m = 0 \text{ in } \mathbb{R}^2, \qquad k_m^2 := k^2 - \lambda_m^2, \qquad (2.12)$$

where $\hat{\Delta} = \partial_{x_1}^2 + \partial_{x_2}^2$.

By the assumption $k \in \mathcal{K}$, $k_m \neq 0$. Therefore, if (2.3) holds then

$$\sum_{m=1}^{\infty} \int_{\Omega_0} |u_m|^2 d\hat{x} < \infty, \qquad (2.13)$$

where Ω_0 was defined above formula (1.1).

If (2.3') holds, then each u_m satisfies condition (2.3'), because the functions ϕ_m are orthogonal.

Thus, (2.10), (2.12) and (2.13) or (2.13') imply that

$$u_m = 0 \quad \text{if} \quad k^2 > \lambda_m^2, \tag{2.14}$$

and

$$|u_m| + |\nabla u_m| \leq c \exp(-\epsilon|\hat{x}|), \quad \epsilon > 0, \quad \text{if} \quad k^2 < \lambda_m^2. \tag{2.15}$$

The conclusions (2.14) and (2.15) follow from (2.10) since

$$u_m(\hat{x}) = \sum_{l=-\infty}^{\infty} \left(c_l^{(1)} H_{|l|}^{(1)}(k_m r) + c_l^{(2)} H_{|l|}^{(2)}(k_m r) \right) \exp(il\theta), \tag{2.16}$$

if $k_m = (k^2 - \lambda_m^2)^{1/2} > 0$. Here $r = |\hat{x}| > r_0$, $D \subset D_{r_0}$, $H_l^{(j)}(r)$, $j = 1, 2$ are the Hankel functions and $c_l^{(j)}$ are arbitrary complex constants. From the asymptotics of Hankel functions as $r \to \infty$, it follows that (2.13) and (2.16) imply $c_l^{(j)} = 0$ for all l and $j = 1, 2$, and m such that $k^2 > \lambda_m^2$. Thus (2.14) holds.

If $k^2 < \lambda_m^2$, that is $k_m = i|k_m|$, the Hankel function $H_l^{(1)}(i|k_m|r)$ decay exponentially as $r \to +\infty$ if $|k_m| \geq \epsilon > 0$ and $H_l^{(2)}(i|k_m|r)$ grow exponentially as $r \to \infty$. Thus $c_l^{(2)} = 0$ for all l and (2.15) is established with $\epsilon = \min\{|k_m| : k^2 < \lambda_m^2\}$. Lemma 2.2 is proved. □

Let us now consider different boundary conditions on ∂L and S. Assume that (2.2) is replaced by

$$u = 0 \quad \text{at} \quad \partial L_+ := \{x \in L : x_3 = h\} \tag{2.17}$$

$$\partial_{x_3} u = 0 \quad \text{at} \quad \partial L_- := \{x \in L : x_3 = 0\} \tag{2.18}$$

$$u = 0 \quad \text{on} \quad S, \quad \text{or} \quad u_N = 0 \quad \text{on} \quad S. \tag{2.19}$$

Then the arguments above should be changed only in the treatment of the integral over ∂D_r in (2.5). Namely, at L_- one has $u_N = 0$ and on ∂L one has $N_2 = 0$ as before. The integral over Γ_r vanishes, as $r \to \infty$, as before. Therefore only the integral over S should be investigated for the boundary conditions (2.19). The Dirichlet boundary condition (2.19) has been investigated already in the proof of Theorem 2.1. Consider the Neumann boundary condition. Equation (2.5) now yields:

$$\int_{D'} u_2^2 dx = \frac{1}{2} \int_S x_2 N_2 [k^2 u^2 - (\nabla u)^2] ds =: I. \tag{2.20}$$

If $u_N = 0$ on S, then $\nabla u = \nabla_t u$, where $\nabla_t u$ is the gradient in the tangential to S plane orthogonal to N. Thus, the integral I in (2.20) can be written as

$$I = \frac{1}{2} \int_S x_2 N_2 [k^2 u^2 - (\nabla_t u)^2] ds. \tag{2.21}$$

If Assumption (2.A) holds, then one could conclude that $I \leq 0$ and, therefore, could obtain a uniqueness theorem from equation (2.20), provided that

$$k^2 u^2 \geq (\nabla_t u)^2 \quad \text{on} \quad S, \tag{2.22}$$

for the solution to (2.1) satisfying the second boundary condition (2.19).

However, this is not possible since inequality (2.22) *does not hold* in general.

So, it is an open problem to find out when a uniqueness theorem, analogous to Theorem 2.1, holds in the case of acoustically hard obstacle, that is for the Neumann boundary condition on S. In fact, examples of non-uniqueness are discussed in Refs. [2–4].

Let us now prove Lemma 2.3.

Lemma 2.3. *If u satisfies conditions (2.1)–(2.3) then $\nabla u \in L^2(D')$ and (2.15) holds.*

Proof: From (2.1) and Green's formula we get

$$0 = \int_{D_r} \bar{u}(\Delta + k^2)u\,dx = \int_{D_r} (k^2|u|^2 - |\nabla u|^2)\,dx + \int_{\partial D_r} \bar{u}u_N\,ds. \tag{2.23}$$

Thus

$$\int_{D_r} (|\nabla u|^2 - k^2|u|^2)\,dx = \int_{\partial D_r} \bar{u}u_N\,ds := I. \tag{2.24}$$

If boundary conditions (2.2) (or (2.17)–(2.19)) hold, then

$$I = I(r) := \int_{\Gamma_r} \bar{u}u_N\,ds = \int_{\Gamma_r} \bar{u}u_r\,ds, \tag{2.25}$$

where $\Gamma_r := \{x : |\hat{x}| = r\}$.

Lemma 2.2 is proved if we prove that

$$I(r_n) \to 0 \quad \text{as} \quad r_n \to \infty, \tag{2.26}$$

for some sequence r_n.

If (2.3') holds, then (2.26) is obvious. Suppose (2.3) holds and we want to prove (2.26). Let

$$w(r) := \int_{\Gamma_r} |u|^2 r\,d\theta\,dx_3. \tag{2.27}$$

Then

$$w'(r) := \frac{dw}{dr} = 2\operatorname{Re} I(r) + \frac{1}{r}w(r), \tag{2.28}$$

so

$$2\operatorname{Re} I(r) = w'(r) - \frac{1}{r}w(r). \tag{2.29}$$

Since u solves (2.1), (2.2), Green's formula yields

$$0 = \int_{\Gamma_r} (\bar{u}u_r - \bar{u}_r u)\,ds = 2i\operatorname{Im}\int_{\Gamma_r} \bar{u}u_r\,ds. \tag{2.30}$$

Thus $\operatorname{Im} I(r) = 0$ and (2.29) can be written as

$$I(r) = \frac{1}{2}\left(w'(r) - \frac{1}{r}w(r)\right). \tag{2.31}$$

Lemma 2.3 follows from (2.31), and the following Lemma 2.4. □

Lemma 2.4. *Assume $w(r) \geq 0$, $\int_a^\infty w(r)dr < \infty$, $w \in C^2(a,\infty)$, $a = \text{const} \geq 0$. Then there is a sequence $r_n \to \infty$ such that*

$$|w'(r_n)| + |w(r_n)| \to 0. \tag{2.32}$$

Proof: If $\int_a^\infty w(r)dr < \infty$ and $w(r) \geq 0$ decreases monotonically to zero in a neighborhood of infinity, then $w'(r) < 0$ and $w'(r_n) \to 0$ for some sequence $r_n \to \infty$, since $\sup_{r>a} \int_a^r w'(r)dr < \infty$.

If $w(r)$ does not decrease monotonically to zero near infinity, then $w(r)$ has infinitely many nonnegative local maxima and minima. Let $w(r_n)$ are local minima of $w(r)$, $r_n \to \infty$. Then $w'(r_n) = 0$ and one can select a subsequence r'_n out of the sequence r_n, such that $w(r'_n) \to 0$ as $r'_n \to \infty$. Indeed, since $\int_a^\infty w(r)dr < \infty$, if such a subsequence would not exist, then, for some $\epsilon > 0$, one would have $w(r_n) \geq \epsilon > 0$ as $r_n \to \infty$. Then $w(r) \geq \epsilon$ for $r \geq r_1$ and this contradicts the assumptions $\int_a^\infty w(r)dr < \infty$, $w(r) \geq 0$. Lemma 2.4 is proved. □

Remark 2.1. Consider a compact perturbation of the boundary ∂L which is denoted Σ. Assume that $S \subset L$ and that the region L_Σ, which is the layer with the compactly perturbed boundary, satisfies the Assumption (2.A). Then Theorem 2.1 holds in L_Σ with boundary condition $u|_\Sigma = 0$. The proof is the same as above.

Remark 2.2. Similar results hold for an obstacle in a cylindrical waveguide and for a compact perturbation of the boundary of the cylinder provided that the perturbed cylinder lies inside the unperturbed one and Assumption (2.A) holds for the perturbed cylinder.

Recently AGR found another geometrical assumption and a short proof of the uniqueness theorem similar to Theorem 2.1.

Assumption 2.B. There exists a point O inside D such that if O is the origin, and N is the unit normal to S pointing into D, then

$$x \cdot N \leq 0 \qquad \forall x \in S.$$

This assumption holds, in particular, for convex obstacles.

Theorem 2.1'. *The conclusion of Theorem 2.1 holds under the assumption* **(2.B)**.

Proof: The proof is based on the well-known Rellich's identity:

$$\int_S (2x \cdot \nabla u \, u_N - x \cdot N |\nabla u|^2) ds = 2 \int_D x \cdot \nabla u \Delta u \, dx + (2-n) \int_D |\nabla u|^2 dx.$$

Here n is the dimension of the space and one may assume without loss of generality that u is real-valued in the proof of Theorems 2.1 and 2.1'.

If $u = 0$ on S then $\nabla u = N u_N$ on S. Equation (2.1) and the Rellich's identity yield:

$$\int_S x \cdot N u_N^2 ds = -k^2 \int_{D'} [(x_j u^2)_j - n u^2] dx + (2-n) \int_{D'} |\nabla u|^2 dx = 2 \int_{D'} |\nabla u|^2 dx > 0.$$

Here we have used (2.6) and (2.2). Assumption **(2.B)** and the last equation imply $u = 0$ in D'. Theorem 2.1' is proved. □

3. BOUNDARY VALUE PROBLEMS IN D'

In this section we investigate the boundary value problem

$$(\Delta + k^2)u = f \quad \text{in} \quad D', \quad f \in L_0^2(D'), \quad k \notin \mathcal{K} \tag{3.1}$$

$$u|_S = \phi(s), \quad u|_{\partial L} = 0 \quad \text{or} \quad u|_{L^+} = 0 \quad u_N|_{L^-} = 0, \tag{3.2}$$

$$u = \sum_{m=1}^{\infty} \phi_m(x_3) u_m(\hat{x}), \quad \hat{x} = (x_1, x_2), \quad r > r_0, \tag{3.3}$$

where r_0 was defined above formula (1.1), and u_m satisfy the radiation conditions

$$\sqrt{r}(u_{mr} - ik_m u_m) \xrightarrow[r \to \infty]{} 0, \quad r = |\hat{x}|, \quad k_m = (k^2 - \lambda_m^2)^{1/2} > 0, \tag{3.4}$$

$$|u_m| = O(\exp(-\epsilon r)) \quad \text{if} \quad k^2 < \lambda_m^2, \quad \epsilon > 0. \tag{3.5}$$

Theorem 3.1. *For any $f \in L_0^2(D')$ and any $\phi \in H^{3/2}(S)$, the problem (3.1)–(3.5) has a unique solution.*

Proof: Uniqueness follows from Theorem 2.1. Indeed, if u solves homogeneous equations (3.1) and (3.2), then

$$\lim_{r \to \infty} \int_{\Gamma_r} (\bar{u} u_r - \bar{u}_r u) ds = 0,$$

and because the system of functions ϕ_m is orthogonal, it follows that each u_m satisfies the above relation as well. This, and the radiation condition (3.4) imply that each of u_m satisfy condition (2.3').
The equation

$$\Delta u + k_m^2 u_m = 0 \quad \text{in} \quad \Omega_0, k_m > 0$$

and condition (2.3') for u_m imply that $u_m = 0$ in Ω_0, (see Ref. [11], p. 25). If $\lambda_m^2 > k^2$, then condition (3.5) says that u_m decay exponentially. Therefore any solution u to the homogeneous problem (3.1)–(3.5) belongs to $L^2(D')$ and solves (2.1)–(2.3). By Theorem 2.1, this $u = 0$ in D', as claimed.

Existence is a consequence of the uniqueness and of the Fredholm property of the problem (3.1)–(3.5). Let us establish this property.
First, we introduce the function

$$w := u - \int_{D'} g(x, y, k) f(y) dy, \tag{3.6}$$

where g is the Green function of the Helmholtz operator in L. If, for example, (1.5) is assumed, then

$$g(x,y,k) = \sum_{m=1}^{\infty} \frac{2}{h} \sin(m\pi x_3/h) \sin(m\pi y_3/h) \frac{i}{4} H_0^1(k_m|\hat{x}-\hat{y}|).$$

The function w solves the problem

$$(\Delta + k^2)w = 0 \quad \text{in} \quad D', \tag{3.7}$$

$$w|_{\partial L} = 0, \quad w|_S = h := \phi - \int_{D'} g(s,y,k)f(y)dy, \tag{3.8}$$

and it satisfies (3.3)–(3.5).

Therefore, one may assume $f = 0$ in (3.1) without loss of generality but then one has to consider the nonhomogeneous boundary condition (3.8). We look for the solution to (3.7)–(3.8) in the form

$$w(x) = \int_S g(x,t,k)\mu(t)dt, \quad x \in D'. \tag{3.9}$$

The function (3.9) satisfies (3.7) and the first condition (3.8) for any $\mu(t)$. It will also satisfy the second condition (3.8), if

$$Q(k)\mu := \int_S g(x,t,k)\mu(t)dt = h(x), \quad x \in S. \tag{3.10}$$

The operator $Q(k)$ in (3.10) acts from $H^l(S) \to H^{l+1}(S)$ for any positive integer l (Ref. [11], p. 81) provided that S is sufficiently smooth. For our purposes, $l = 0$ is sufficient, and then Lipschitz S is admissible. However, in the argument below we use the space $H^2(S)$, and because of this we assume that S is C^2. For a theory with Lipschitz S see Refs. [16] and [18].

By changing g to g_ρ if necessary, one can always assume that Q or Q_ρ is injective. Here $Q_\rho\mu := \int_S Q_\rho\mu dt$ and g_ρ is the Green function in $L\setminus B_\rho$, B_ρ being a ball of small radius ρ, $B_\rho \subseteq D$ (see Ref. [11], pp. 20, 29). If Q is invertible, then the map $Q : H^l(S) \to H^{l+1}(S)$ is equivalent to a Fredholm map on $H^l(S)$. Indeed, take $k_1 \neq k$ and put $Q_1 := Q(k)$. Let Q_1 be injective. Then $Q = Q - Q_1 + Q_1 = Q_1[I + Q_1^{-1}(Q - Q_1)]$. The operator $Q_1^{-1}(Q - Q_1)$ is compact in $H^l(S)$ since $Q - Q_1 : H^l \to H^{l+2}$ while $Q_1^{-1} : H^{l+1} \to H^l$. Thus $Q_1^{-1}Q$ is a Fredholm operator equivalent to Q (see, e.g. Refs. [11] and [15]). Theorem 2.1 yields injectivity of Q_ρ for a suitable choice of ρ. Thus, the corresponding equation (3.10) is uniquely solvable for Q replaced by Q_ρ. If Q is not injective, the above argument shows that it has a finite dimensional null-space and equation (3.10) is solvable, if h satisfies finitely many orthogonality relations.

However, the boundary value problem (3.1)–(3.5) is uniquely and unconditionally solvable for any f and ψ, the solution given by

$$u(x) = \int_{D'} g(x,y,k)f(y)dy + w(x), \tag{3.11}$$

$$w(x) = \int_S g_\rho(x,t,k)(Q_\rho^{-1}h)(t)dt, \tag{3.12}$$

for a properly chosen small $\rho > 0$, as our argument above shows. □

Other methods can be also used for proving existence of the solution to the boundary value problem (3.1)–(3.5) (see, e.g. Refs. [11] and [16]). One can also treat the case when k^2 is replaced by $k^2 n(x_3)$ (horizontally stratified waveguide). In this case the eigenvalues (1.7) and (1.8) and the eigenfunctions (1.9) and (1.10) will be different, but the scheme of the study remains essentially the same.

4. RESOLVENT KERNEL OF THE LAPLACIAN AND LIMITING ABSORPTION PRINCIPLE

Let us now consider the boundary value problem

$$(\Delta + k^2 + i\epsilon) G_\epsilon(x,y) = -\delta(x-y) \quad \text{in } D', \qquad k \notin \mathcal{K} \tag{4.1}$$

$$G_\epsilon|_S = 0, \quad G_\epsilon|_{\partial L} = 0, \quad \text{or} \quad G_\epsilon|_{\partial L_+} = 0, \quad G_{\epsilon N}|_{\partial L_-} = 0, \tag{4.2}$$

$$G_\epsilon \in L^2(D'). \tag{4.3}$$

This problem is uniquely solvable if $\epsilon > 0$ (Ref. [11], p. 309).

Theorem 4.1. *The limit of $G_\epsilon(x,y)$, as $\epsilon \to 0$, exists in the sense of uniform convergence on compact subsets $\{(x,y,k) : 0 < R_1 \leq |x-y| \leq R_2, a < k < b, (a,b) \cap \mathcal{K} = \emptyset\}$, where R_1, R_2 are some positive constants.*

Proof: To prove this limiting absorption principle we start with the equations

$$G_\epsilon(x,y) = g_\epsilon(x,y) - \int_S g_\epsilon(x,t,k) \mu_\epsilon dt, \quad \mu_\epsilon = \frac{\partial}{\partial N_t} G_\epsilon(t,y), \tag{4.4}$$

where $g_\epsilon(x,y)$ is the Green function for the Helmholtz operator in L and N_t is the normal to S pointing into D'.

Differentiating the integral in (4.4) along the normal to S, and using the standard formulas for the exterior normal derivative of the single layer potentials, we get

$$\mu_\epsilon + A_\epsilon \mu_\epsilon = 2 \frac{\partial g_\epsilon}{\partial N_s}, \quad A_\epsilon \mu = 2 \int_S \frac{\partial g_\epsilon(s,t)}{\partial N_s} \mu(t) dt \tag{4.5}$$

The operator A_ϵ is compact in $L^2(S)$ for any $0 \leq \epsilon \leq \epsilon_0$, ϵ_0 being a positive number. Consequently, the operator $I + A_\epsilon$ is Fredholm and it depends continuously (in the operator norm) on ϵ, $0 \leq \epsilon \leq \epsilon_0$. For each $\epsilon \in [0, \epsilon_0]$ the operator $I + A_\epsilon$ is injective, as we will prove later. Therefore

$$\sup_{0 \leq \epsilon \leq \epsilon_0} \|(I + A_\epsilon)^{-1}\| \leq c < \infty, \tag{4.6}$$

where $c > 0$ is a constant, and

$$\mu_\epsilon \xrightarrow[L^2(S)]{} \mu_0 := (I+A_0)^{-1} 2\frac{\partial g(s,t,k)}{\partial N_s}, \tag{4.7}$$

where $g(x,s,k)$ is the limit of g_ϵ at $\epsilon = 0$.

Thus, as $\epsilon \to 0$, one gets

$$G_\epsilon \to G := g(x,y,k) - \int_S g(x,s,k)\mu_0(s,y,k)ds. \tag{4.8}$$

To complete the proof we need to prove (4.6). The continuity of the inverse operator $(I+A_\epsilon)^{-1}$ with respect to ϵ follows from (4.6) and the identity

$$A^{-1} - B^{-1} = B^{-1}(B-A)A^{-1}$$

There are two claims to prove:

(a) $(I+A_0)^{-1}$ exists, and

(b) the estimate (4.6) holds.

Let us assume (a) and prove (4.6). Suppose the contrary, that is

$$\sup_{0 \leq \epsilon \leq \epsilon_0} \|(I+A_\epsilon)^{-1}\| = \infty$$

Then there is a sequence $\epsilon_n \to \tilde{\epsilon}$, $0 \leq \tilde{\epsilon} \leq \epsilon_0$, such that $\|(I+A_{\epsilon_n})^{-1}\| \to \infty$. Since

$$F_n := (I+A_{\epsilon_n})^{-1} = (I+A_{\tilde{\epsilon}})^{-1} + (I+A_{\tilde{\epsilon}})^{-1}(A_{\tilde{\epsilon}} - A_{\epsilon_n})(I+A_{\epsilon_n})^{-1},$$

and

$$\|A_{\tilde{\epsilon}} - A_{\epsilon_n}\| \to 0, \quad \text{as} \quad n \to \infty,$$

one gets

$$(I+T_n)F_n = (I+A_{\tilde{\epsilon}})^{-1},$$

where

$$T_n := -(I+A_{\tilde{\epsilon}})^{-1}(A_{\tilde{\epsilon}} - A_{\epsilon_n}), \quad \|T_n\| \to 0 \quad \text{as} \quad n \to \infty.$$

Thus

$$\|F_n\| \leq \|(I+T_n)^{-1}\| \|(I+A_{\tilde{\epsilon}})^{-1}\| \leq c$$

where c does not depend on n. This contradiction proves (4.6).

It remains to be proved that $(I+A_\epsilon)^{-1}$ exists for all $\epsilon \in [0,\epsilon_0]$. Since the operators $I+A_\epsilon$ are Fredholm-type, it is sufficient to check that its null-space is trivial. Assume that

$$(I+A_\epsilon)\mu = 0, \quad \mu \neq 0.$$

Then the function $w_\epsilon := \int_S g_\epsilon(x,t)\mu(t)dt$ solves the problem

$$(\nabla^2 + k^2 + i\epsilon)w_\epsilon = 0 \quad \text{in} \quad D \cup D', \qquad w_{\epsilon N}^+ = 0 \quad \text{on} \quad S, \tag{4.9}$$

where $w_{\epsilon N}^+$ is the limiting value on S of the normal derivative from inside of D. Thus $w_\epsilon = 0$ in D, if $\epsilon > 0$ and $w_\epsilon = 0$ on S. On the other hand, the solution w_ϵ of the problem

$$(\nabla^2 + k^2 + i\epsilon)w_\epsilon = 0 \quad \text{in} \quad D', \qquad w_\epsilon = 0 \quad \text{on} \quad S, \quad w_\epsilon \in L^2(D'), \tag{4.10}$$

vanishes in D'. Thus, if $\epsilon > 0$, $w_\epsilon = 0$ in $D \cup D'$, and $\mu = w_{\epsilon N}^+ - w_{\epsilon N}^- = 0$, where $w_{\epsilon N}^-$ is the limiting value of the normal derivative on S from outside of D.

If $\epsilon = 0$, the above argument is valid for $k^2 \notin \sigma(\Delta_N(D))$, that is, k^2 is not a Neumann eigenvalue of the Laplacian in D, and yields the conclusion $w_0 = 0$ in D. To get the conclusion $w_0 = 0$ in D', one notes that w_0 solves the homogeneous problem (3.1)–(3.5) and therefore Theorem 3.1 implies the desired conclusion.

In case $k^2 \in \sigma(\Delta_N(D))$, one should repeat the above argument with g replaced by g_ρ (defined in Section 3, in the second paragraph below equation (3.10)), where $\rho > 0$ is chosen so that $k^2 \notin \sigma(\Delta_N(D \setminus B_\rho))$. Such a choice is always possible (see Ref. [11]). Therefore, the limiting absorption principle for $G(x,y,k)$ is established. □

A proof of this principle, based on the compactness argument, can also be given (see Refs. [10] and [11]).

5. SCATTERING SOLUTIONS

We define the scattering solutions using the approach introduced in Ref. [17] (see also Ref. [11], pp. 46–48). If $G(x,y,k)$ is the resolvent kernel of the Laplacian in D', then we study the limit of G as $|\hat{y}| := r \to \infty$, $\hat{y}/r = -\alpha$, $\alpha \in S^1$, S^1 being the unit circle, $\hat{y} := (y_1, y_2)$. It follows from (4.8) and the symmetry of the Green function $G(x,y,k) = G(y,x,k)$, that, as $r \to \infty$ and $\hat{y}/r = -\alpha$,

$$G(x,y,k) \sim \sum_{m=1}^\infty \gamma_m(r)\phi_m(y_3)\psi_m(x,\alpha,k_m), \tag{5.1}$$

where

$$\gamma_m(r) := \frac{i}{4}\sqrt{\frac{2}{\pi k_m r}} \exp\left(ik_m r - \frac{i\pi}{4}\right), \tag{5.2}$$

and $k_m = (k^2 - \lambda_m^2)^{1/2}$ if $k^2 > \lambda_m^2$, $k_m = i(\lambda_m^2 - k^2)^{1/2}$ if $k^2 < \lambda_m^2$, the square roots are chosen so that $\sqrt{t} > 0$ if $t > 0$.

The **scattering solutions** are the coefficients ψ_m in (5.1), defined by

$$\psi_m := \phi_m(x_3)\exp(ik_m\alpha \cdot \hat{x}) - \int_S \phi_m(t_3)\exp(ik_m\alpha \cdot t)\mu_0(t,x,k)dt := \psi_m^{(0)} + v_m, \tag{5.3}$$

where $\mu_0(t,x,k)$ is defined in equation (4.8), and

$$\psi_m^{(0)}(x,\alpha,k_m) = \phi_m(x_3)u_m^{(0)}(\hat{x},\alpha,k_m),$$
$$u_m^{(0)}(\hat{x},\alpha,k_m) = \exp(ik_m\alpha \cdot \hat{x}) \quad \text{if} \quad k^2 > \lambda_m^2. \tag{5.4}$$

If the obstacle is absent, then, as $r \to \infty, \widehat{y}/r = -\alpha$,

$$g(x,y,k) \sim \sum_{m=1}^{\infty} \gamma_m(r) \phi_m(y_3) \psi_m^{(0)}(\widehat{x}, \alpha, x_3, k_m), \qquad (5.5)$$

where g is the resolvent kernel of the Laplacian in L in the absence of D.

The asymptotics (5.1) and (5.5) are obtained by taking asymptotics termwise in the series for G and g respectively.

The series (5.5) converges for $x_3 \neq y_3$, and also for $x_3 = y_3$ provided that $|\alpha \cdot x| < |x|$. We do not claim that these series converge pointwise for all $\widehat{x} \in \mathbb{R}^2$. Indeed, if $k^2 < \lambda_m^2$, then the corresponding term in the expansion of $g(x,y,k)$ is

$$\phi_m(x_3) \phi_m(y_3) \frac{i}{4} H_0^{(1)}(k_m|\widehat{x}-\widehat{y}|), \qquad k_m = i(\lambda_m^2 - k^2)^{1/2},$$

and its asymptotics as $r = |\widehat{y}| \to \infty, \widehat{y}/r = -\alpha$, is

$$\phi_m(x_3) \phi_m(y_3) \frac{i}{4} e^{-\frac{i\pi}{4}} \sqrt{\frac{2}{\pi k_m r}} \exp(-|k_m|r) \exp(-|k_m|\alpha \cdot \widehat{x}),$$

where $\alpha \cdot \widehat{x} = \alpha_1 x_1 + \alpha_2 x_2$. The last expression increases exponentially as $|\widehat{x}| \to \infty$, if $\alpha \cdot \widehat{x} < 0$. However, if $r > |\widehat{x}|$, then

$$-|k_m|(r + \alpha \cdot \widehat{x}) \leq -|k_m|\epsilon, \qquad \epsilon > 0.$$

Therefore the asymptotic series (5.5) converges uniformly on compact subsets of $\mathbb{R}^2 \times S^1 \times (\mathbb{R}_+ \setminus \mathcal{K})$. Note also that the functions $\psi_m(x)$ and $\psi_m^{(0)}(x)$ are not uniformly bounded in D'. For example, the function $\psi_m^{(0)}(x)$ is exponentially growing as $|\widehat{x}| \to \infty$, $\alpha \cdot \widehat{x} > 0$, $\lambda_m^2 > k^2$. If $\lambda_m^2 < k^2$, then $\psi_m^{(0)}(x) = \phi_m(x_3) \exp(ik_m \alpha \cdot \widehat{x})$ is uniformly bounded in D'.

Denote by M the integer such that $\lambda_m^2 < k^2$ if $m \leq M$, $\lambda_m^2 > k^2$ if $m > M$. In the sequel, we consider as an **incident field** in D' the functions $\psi_m^{(0)}(x, \alpha, k_m)$, $m \leq M$. Fixing $m \leq M$, we formulate the **scattering problem** in D' as follows:

Find the solution ψ_m to the problem

$$(\Delta + k^2)\psi_m = 0 \quad \text{in} \quad D', \qquad (5.6)$$

$$\psi_m |_L = 0 \quad (\text{or } \psi_{mN}|_{L_-} = 0, \qquad \psi_m|_{L_+} = 0), \qquad (5.7)$$

$$\psi_m = 0 \quad \text{on} \quad S, \qquad (5.8)$$

$$\psi_m(x,k) = \psi_m^{(0)}(x, \alpha, k_m) + v_m(x,k), \qquad (5.9)$$

where $v_m(x,k)$ satisfies equation (5.6), the boundary conditions (5.7) on L, $v_m = -\psi_m^{(0)}$ on S, and it admits the representation (analogous to (3.3))

$$v_m(x,k) = \sum_{l=1}^{\infty} v_m^l(\widehat{x}, k) \phi_l(x_3), \qquad r > r_0. \qquad (5.10)$$

Here

$$v_m^l(\hat{x},k) = \int_0^h v_m(x,k)\phi_l(x_3)dx_3, \qquad (5.11)$$

and v_m^l satisfy the radiation conditions

$$\int_{S_r} \left|\frac{\partial v_m^l}{\partial r} - ik_l v_m^l\right|^2 ds \xrightarrow[r\to\infty]{} 0, \quad \text{if} \quad l \le M, \qquad (5.12)$$

and

$$|v_m^l| \xrightarrow[r\to\infty]{} 0, \quad \text{if} \quad l > M. \qquad (5.13)$$

Note that v_m solves a problem analogous to (3.7)–(3.8). Thus, using a formula similar to (3.9), and formulas (5.1)–(5.5), we get the representation

$$v_m(x,k) = -\int_S \phi_m(t_3) u_m^{(0)}(t,\alpha,k_m)\mu(t,x,k)dt, \qquad (5.14)$$

where $\mu := \frac{\partial G(t,x,k)}{\partial N_t}$, N_t being the normal to S at the point t, for the case of Dirichlet boundary conditions on $\partial D'$. Then v_m^l are given by

$$v_m^l(\hat{x},k) = -\int_S \phi_m(t_3) u_m^{(0)}(t,\alpha,k_m) \int_0^h \mu(t,x,k)\phi_l(x_3)dx_3 dt. \qquad (5.15)$$

Note that, the representation (5.14) can be used in the numerical calculation of v_m, after solving the integral equation

$$\mu + A\mu = 2\frac{\partial g}{\partial N}, \qquad A\mu := 2\int_S \frac{\partial g(s,t,k)}{\partial N_s}\mu(t)dt, \qquad (5.16)$$

N_s being the normal to S at s *pointing into* D'. Equation (5.16) is similar to (4.5).

Let us now state and prove the eigenfunction expansion theorem and sketch its proof following Ref. [18] and Ref. [11], p. 48.

Theorem 5.1. *Let* $f \in L^2(D')$. *Denote*

$$\hat{f}_m(\alpha,k_m) := \int_{D'} f(x)\overline{\psi_m(x,\alpha,k_m)}dx, \quad m = 1,2,..., \qquad (5.17)$$

where $\psi_m(x,\alpha,k_m)$ are the scattering solutions (see (5.1), (5.3), and (5.6)–(5.13)).

Then

$$f(x) = \frac{1}{4\pi^2}\sum_{m=1}^\infty \int_{S^1} d\alpha \int_0^\infty d\xi \xi \hat{f}_m(\alpha,\xi)\psi_m(x,\alpha,\xi), \qquad (5.18)$$

Proof: The Dirichlet Laplacian has no spectrum in the interval $(-\infty, \lambda_1)$ (see Eq. (1.7)). Thus (see Ref. [11], p. 48, Eq. (16)) we have

$$f(x) = \frac{1}{i\pi} \int_{\lambda_1}^{\infty} dk\, k \int_{D'} dy f(y) [G(x,y,k) - \overline{G(x,y,k)}], \tag{5.19}$$

where the bar stands for complex conjugate.

Using Green's formula and formula (5.1), we express the $G - \overline{G}$ in terms of the scattering solutions:

$$G(x,y,k) - \overline{G(x,y,k)} = \lim_{r \to \infty} \int_{|\hat{t}|=r} \int_0^h dt_3 d\hat{t} [\overline{G(t,y,k)} G_r(x,t,k)$$
$$- G(x,t,k) \overline{G_r(t,y,k)}] := J. \tag{5.20}$$

From (5.20), (5.1) and (5.2), with $\alpha = -\frac{\hat{t}}{|\hat{t}|}$, we have

$$J = \sum_{m,l=1}^{\infty} \gamma_m(r) \overline{\gamma_l(r)} r (2ik_m) \int_0^h dt_3 \phi_m(t_3) \phi_l(t_3)$$
$$\times \int_{S^1} d\alpha \psi_m(x, \alpha, k_m) \overline{\psi_l(y, \alpha, k_l)}, \tag{5.21}$$

or

$$J = \sum_{m=1}^{\infty} \frac{i}{4\pi} \int_{S^1} \psi_m(x, \alpha, k_m) \overline{\psi_m(y, \alpha, k_m)} d\alpha. \tag{5.22}$$

Here we have used the orthogonality of ϕ_m and the conditions $\text{Im}\, \gamma_m = 0$, $\text{Im}\, \psi_m = 0$ if $k^2 < \lambda_m^2$. These equations imply that, for each $m = 1, 2, 3, \ldots$, the interval $k^2 < \lambda_m^2$ does not contribute to the eigenfunction expansion. Let $\xi^2 := k^2 - \lambda_m^2$, $kdk = \xi d\xi$. From (5.20), (5.21) and (5.22), we get the eigenfunction expansion formula (5.18) where $\hat{f}_m(\alpha, k_m)$ is defined in (5.17). Theorem 5.1 is proved. □

The result in Theorem 5.1 is similar to the main result by Goldstein [5], for cylindrical waveguides, but our proof is quite different, much shorter, and, as mentioned above, it follows the proof in Refs. [17] and [11].

6. INVERSE SCATTERING PROBLEMS

Suppose that two obstacles, D_1 and D_2, generate the same scattering data. We want to derive from this that $D_1 = D_2$. We assume that the boundary condition on D_1 and D_2 is the Dirichlet one. As the scattering data we take the partial scattering amplitudes $A_{ml}(\alpha, \alpha', k_m, k_l)$, $l \leq M$, for all $\alpha, \alpha' \in S^1$. Here m, $1 \leq m \leq M$, is assumed fixed. In other words, we send an incident wave $\psi_m^{(0)}(x, \alpha)$ and we observe the (partial) scattering amplitudes $A_{ml}(\alpha, \alpha', k_m, k_l)$, $1 \leq l \leq M$, defined as the coefficients in the asymptotics

$$v_m^l(\hat{x}, k) \sim A_{ml}(\alpha, \alpha', k_m, k_l) \gamma_l(|\hat{x}|), \quad |\hat{x}| \to \infty, \quad \frac{\hat{x}}{|\hat{x}|} = \alpha', \tag{6.1}$$

where $v_m^l(\hat{x}, k)$ are defined in (5.11). Using the now standard arguments (Ref. [11], pp. 84–88) and the uniqueness Theorem 2.1, we prove the following uniqueness theorem for the inverse problem.

Theorem 6.1. *If* $A_{ml}^{(1)} = A_{ml}^{(2)}$, *for all* $1 \leq l \leq M$, *all* $\alpha, \alpha' \in S^1$, *then* $D_1 = D_2$.

Proof: The basic idea of the proof is exactly the same as in Ref. [11]. Using (6.1) and the assumption of the theorem, the difference of the scattering solutions $w := u^{(1)} - u^{(2)}$ is $o(r^{-1/2})$ outside of $D_{12} := D_1 \cup D_2$, and according to Theorem 2.1, w vanishes in $D'_{12} := L \backslash D_{12}$. Therefore, in any connected component $\tilde{D}$ of $D_{12} \backslash \overline{D^{12}}$, $D^{12} := D_1 \cap D_2$, and the bar denotes the closure, one has $u^{(1)} = u^{(2)}$ in $\tilde{D}$, and, denoting by $v := u^{(1)} = u^{(2)}$, one obtains $v = 0$ on $\partial \tilde{D}$. If $\tilde{D}$ is not empty one gets a contradiction. The Dirichlet Laplacian in a compact region $\tilde{D}$ has discrete spectrum, and in particular, the dimension of the eigenspace for any fixed eigenvalue is finite. On the other hand, choosing various α one can get as large dimension of the eigenspace corresponding to eigenvalue k^2 of the Dirichlet Laplacian in $\tilde{D}$ as one wishes (see, e.g. Ref. [11], p. [29]). Hence $\tilde{D}$ must be empty, D_1 must coincide with D_2, and Theorem 6.1 is proved for the Dirichlet boundary condition on S. □

Remarks.

1. A proof of the uniqueness Theorem 6.1 similar to that in Ref. [13] (see also Ref. [12]) is also valid.

2. If the Neumann condition on S is assumed, then the argument in the proof of Theorem 6.1 is not valid, because the Neumann Laplacian in domains with some non-smooth boundaries (the domain $\tilde{D}$ in the proof of Theorem 6.1, for example) may have non-discrete spectrum. However, in this case (provided that a uniqueness theorem similar to Theorem 2.1 is established) the argument in Ramm [18] comes to rescue.

In Ref. [18] a new idea is introduced for the proof of the uniqueness of the solution to inverse obstacle scattering problem (IOSP). In order to explain the significance of this idea, let us recall that for some years there were assertions (both published and unpublished) that the method of M. Schiffer for proving uniqueness of the solution to IOSP breaks down in the case of Neumann boundary condition because the spectrum of the Neumann Laplacian in $\tilde{D}$ is not necessarily discrete (unlike the spectrum of the Dirichlet Laplacian) even in the case when the boundaries of the domains D_1 and D_2 are infinitely smooth. This observation was considered as one that invalidates the Schiffer's beautiful argument in the case of Neumann boundary condition. One way out of this difficulty is given in Ref. [14]. The other way is given in Ref. [18]. The idea in Ref. [18] is to avoid the reference to the discreteness of the spectrum of the Neumann (or the Dirichlet) Laplacian altogether, and to use instead the fact that the Hilbert space $L^2(\tilde{D})$ is separable, and therefore can have at most countable orthonormal basis. It is proved in Ref. [18] that if D_1 and D_2 are bounded Lipschitz domains which produce the same scattering data for k in an open interval belonging to the positive semiaxis, and for all directions of the scattered wave, and if $\tilde{D}$ contains an open subset, then the Neumann (and the Dirichlet) Laplacian in $\tilde{D}$ has continuum of orthonormal in $L^2(\tilde{D})$ eigenfunctions, which is impossible since this space is separable. Thus, uniqueness of the solution to IOSP is established for Lipschitz obstacles and for both Neumann and Dirichlet boundary conditions, in domains, in which a uniqueness theorem is valid, which allows one to recover the scattered field outside the obstacle from the knowledge of the asymptotics of this field. In particular, this is the case for the domains considered in the present paper.

7. EXISTENCE OF IMBEDDED EIGENVALUES ON THE CONTINUOUS SPECTRUM OF THE DIRICHLET LAPLACIAN IN SOME COMPACTLY PERTURBED WAVEGUIDES

In this section we sketch the proof of the existence of imbedded eigenvalues of the Dirichlet Laplacian in a compactly perturbed waveguide provided that the perturbed waveguide has rotationally symmetric cross-section and the cross-section of the perturbed part of the waveguide is larger than the cross-section of the original waveguide. This result is similar to the one by Witsch [27] and was obtained as a partial answer to the question posed in Ref. [11], p. 395, Problem 6. In Ref. [26] the result is given for a compact perturbation of a half-cylinder, but the idea of our proof is the same.

Let us assume that the perturbed waveguide W has the $z := x_3$-axis as the axis of symmetry, that for $|z| > b$ the cross-section of the waveguide does not depend on z and is a circle of radius 1, while there is an interval $a_1 < z < a_2$ of the z-axis where the radius of the cross-section is larger than 1. Under the above assumptions, we claim that:

Claim. There are infinitely many positive eigenvalues of the Dirichlet Laplacian A in W.

Let us prove this claim. Note that by symmetry an eigenfunction of A in a perturbed cylinder $W_m \subset W$, formed by the surface of W and the planes $\phi = 0$ and $\phi = \pi/m$ (where m is a positive integer which can be chosen as large as one wishes, and ϕ is the polar angle in the cross-section of W), extends to an eigenfunction of A in W with the same eigenvalue by reflections in the planes $\phi = \ell\pi/m, \ell = 0, 1, 2, ..., 2m$, as one can check. By reflection, for example, in the plane $\phi = 0$, we mean the transformation $u(r, \phi, z) = -u(r, -\phi, z)$, which allows one to check that the reflected function satisfies the Helmholtz equation in $\tilde{W}$ if the function u satisfied it in the W_m. Here $\tilde{W}$ is the reflected region which is the union of W_m and its reflection.

Let $W_s \subset W_m$ be a cylinder with height d and the cross-section which coincide with one of the cross-sections of W_m with radius $r > 1$. The Dirichlet Laplacian in W_s has discrete spectrum and its first eigenvalue Λ is arbitrary large if m is sufficiently large. It is easy to see that $\Lambda > O(m^2)$ as m grows. Since $W_s \subset W_m$, the variational principle for the eigenvalues implies that A in W_m has eigenvalues below the lowest point of its essential spectrum which are smaller than those of A in W_s. The eigenvalues of A in W_s one can calculate by the separation of variables. They are equal to $\nu_l + \mu_j$, where ν_l are the eigenvalues of the Laplacian in the cross-section of W_s and μ_j are the eigenvalues of the operator $\frac{d^2}{dz^2}$ on the interval $[0, d]$, (where d is the height of the cylinder W_s), with the zero boundary conditions at the ends. Note that $\nu_l = r^{-2} p_l$, where p_l are the eigenvalues of A in the cross-section W_1 similar to the cross-section of W_s but with radius 1. Let $l = 1$ and we drop the index l in what follows. Since $r > 1$, one can get as many eigenvalues $\nu + \mu_j < p$ as one wishes, by choosing m sufficiently large. All these eigenvalues lie below the essential spectrum of A in the region $W_1 \times (-\infty, +\infty)$, that is, below p. Let $R > r$, W_R be the cross-section similar to W_1 and let $T_R := W_R \times (-\infty, +\infty)$. We assume that $W_m \subset T_R$. By the variational principle the eigenvalues of A below its essential spectrum in W_m, that is, below p, are greater than the lowest point of spectrum of A in W_R, equal to $\frac{p}{R^2}$, since $W_m \subset W_R$. Thus, we have proved that there are eigenvalues of A in W_m on the interval $(\frac{p}{R^2}, p)$, where p can be chosen as large as one wishes by taking m sufficiently large. But, as we explained above, each eigenvalue of A in W_m is an eigenvalue of A in W (with the reflected eigenfunction). Thus, the claim is proved. □

8. REMARK ON RECENT PAPERS BY R. KLEINMAN, T. ANGELL AND COAUTHORS

After this work was finished and presented on June 4, 1997, at the ISAAC Congress, held at the University of Delaware, AGR learned of the two recent papers devoted to the same subject and available on the internet www.udel.edu

In this section comments are given on these papers which are reports of the center for the mathematics of waves at the University of Delaware, CW-1996-4 and CW-1996-11, supported by AFOSR grant F9620-96-1-0039, CNRS-NSF grant INT-9415493, and NATO grant CR6-940999.

These are papers CW-1996-4 (CW4), by Angell, Kleinman, Lesselier and Rozier, Uniqueness and complete families for an acoustic waveguide problem, and CW-1996-11, by Kleinman and Angell, Reciprocity, radiation conditions and uniqueness.

In CW4 an attempt is made to prove uniqueness theorem for the scattering problem for an acoustically hard obstacle placed in two-dimensional waveguide with boundaries which are two parallel straight lines.

Note that in Refs. [2] and [4] one can find counterexamples to such a uniqueness theorem, the proof of which in CW4 is erroneous: in particular, inequality (2.37) is obviously false since K^2 can be taken arbitrary large and $u(x)$ is a fixed function independent of K and not vanishing identically.

Theorem 2.2 in CW4 asserts that "the only solution...which satisfy conditions of Theorem 2.1 is $u \equiv 0$," while in fact $u \equiv 0$ *does not satisfy condition (2.7) of Theorem 2.1* since this condition is stated in CW4 with the strict inequality sign. We discuss below condition (2.7) in the form (8.1).

The main point in CW4 and CW11 is the introduction of the "new set of conditions..." in place of the radiation condition and its variants (see Ref. [11] where the role of the radiation condition is explained and its variants are discussed). The basic "new condition" is condition (2.7) in CW4, (which we give with the non-strict inequality sign by the reason explained in the preceding paragraph):

$$\lim_{r \to \infty} [\text{Im} \int_{|x|=r} \bar{u} u_N ds] = c = \text{const} \geq 0, \qquad (8.1)$$

where the unit normal N is pointing outside the ball $|x| < r$, and condition (8.1) "should hold for a solution $u \not\equiv 0$ to Helmholtz equation and *any of its orthogonal decomposition components, that is, for any $u_j, j = 1, 2$, where $u = u_1 + u_2$ and $\int_{|x|=r} u_1 \bar{u_2} ds = 0$*."

We want to show that the "new condition" (8.1) is void, it is never satisfied: one can always find an orthogonal decomposition of u for which u_1 does not satisfy (8.1). For simplicity, let us assume that the obstacle is placed in the whole space, this case is also included in the considerations of CW4 and CW11.

Choose any function $f(x) = h(x) + ig(x)$, where h and g are real-valued, such that:

a) f does not satisfy condition (1), and, in particular, does not vanish identically on the surface $|x| = r$, and

b) $\int_{|x|=r} f \bar{u} ds \neq 0$.

There are infinitely many such functions.

Define $u_1 = cf, u_2 = u - u_1$, where $c = \text{const}$ is chosen so that $\int_{|x|=r} u_1 \bar{u_2} ds = 0$. This yields the formula for c:

$$|c| = \frac{\int_{|x|=r} f\bar{u}ds}{\int_{|x|=r} |f|^2 ds}, \quad \arg c = -\arg\left[\int_{|x|=r} f\bar{u}ds\right].$$

The constructed u_1 does not satisfy condition (8.1): if f does not satisfy (8.1) then clearly cf does not satisfy it for any constant $c \neq 0$.

We have proved that the "new condition" from CW4 and CW11 is never satisfied.

To construct examples of f which do not satisfy (8.1), one notes that the expression on the left-hand side in (8.1), in which $f = h + ig$ is substituted in place of u, is

$$\lim_{|x|\to\infty} \int_{|x|=r} (hg_N - gh_N)ds = \int_{\mathbb{R}^3} (h\Delta g - g\Delta h)dx,$$

where we assume that $h, g \in C^2$, and the integral on the right-hand side converges.

Take, for instance, $h = 1$ and $\Delta g = G$, where G is an arbitrary $L_0^2(\mathbb{R}^3)$ function such that $\int_{\mathbb{R}^3} h\Delta g dx = \int_{\mathbb{R}^3} Gdx < 0$, then condition (8.1) is not satisfied by $u = f$. There are infinitely many such f, and one can find among them many satisfying the condition $\int_{|x|=r} f\bar{u}ds \neq 0$, and such that condition (8.1) is not satisfied.

A specific example of f in $\mathbb{R}^3$ is $f := F := r^2 - icr^4, c = const > 0$, for which the expression under the sign of limit in (8.1) is negative and tends to $-\infty$ as $r = |x|$ grows. If $\int_{|x|=r} \bar{u}ds \neq 0$, then one can choose $c > 0$ sufficiently large so that $\int_{|x|=r} F\bar{u}ds \neq 0$, and then F is the desired specific example. If $\int_{|x|=r} \bar{u}ds = 0$, then, since $u(x)$ does not vanish identically, there is a real-valued normalized in $L^2(S^2)$ spherical harmonic Y_ℓ such that $\int_{|x|=r} Y_\ell \bar{u}ds \neq 0$. Let $f := Y_\ell F$. Then this f is the desired specific example.

The authors of CW4 write that "uniqueness in a waveguide was established by Sveshnikov by an application of limiting absorption principle" and refer to the paper by Sveshnikov A. G., Doklady Acad. Nauk USSR, 80, (1951), 341–344. In fact, the paper by Sveshnikov which deals with the waveguide whose boundaries are planes, is the paper Sveshnikov A. G., Doklady Acad. Nauk USSR, 73, (1950), 917–920. In this paper it is incorrectly claimed that the limiting absorption principle in the absence of an obstacle holds for the Helmholtz equation in L for all $k > 0$. In fact, this principle does not hold for $k \in \mathcal{K}$, in two- and three-dimensional spaces, as mentioned in Ref. [25], p. 20. One can see this from the explicit formula for the resolvent kernel of the Dirichlet Laplacian in L, for example.

In CW4 reference is made to [28], where, according to the authors of CW4, for the first time the uniqueness theorem for the boundary-value problem for soft obstacle in a waveguide was established. In fact, the proof of the uniqueness theorem in [28] is wrong. The correct proof of such a uniqueness theorem appears earlier in [9]. The author of [28] in his Ph.D. thesis used the correct proof from [9].

REFERENCES

1. D. S. Ahluwalia and J. B. Keller, *Exact and asymptotic representation of the sound field in a stratified ocean*, in Wave Propagation and Underwater Acoustics, eds. J. B. Keller and J. S. Papadakis, Springer-Verlag, pp (1977), 14–85.
2. M. Callan, C. M. Linton and D. V. Evans, *Trapped modes in two-dimensional waveguides*, J. Fluid Mech. **229** 51–64 (1991).
3. D. V. Evans, M. Levitin, D. Vassiliev, *Existence theorems for trapped modes*, J. Fluid Mech. **261** 21–31 (1994).

4. D. V. Evans, C. M. Linton and F. Ursell, *Trapped mode frequencies embedded in the continuous spectrum*, Q. J. Mech. appl. Math. **46(2)** 253–274 (1993).
5. C. I. Goldstein, *Eigenfunction expansion associated with the Laplacian for certain domains with infinite boundaries I*, Trans. Amer. Math. Soc. **135** 1–31 (1969).
6. R. P. Gilbert and Y. Xu, *Starting fields and far fields in ocean acoustics*, Wave Motion **11** 507–54 (1989).
7. R. P. Gilbert and Y. Xu, *Dense sets and the projection theorem for acoustic waves in a homogeneous finite depth ocean*, Math. Meth. in the Appl. Sci. **12** 67–76 (1989).
8. R. P. Gilbert and Y. Xu, *The propagation problem and far-field patterns in a stratified finite-depth ocean*, Math. Meth. in the Appl. Sci. **12** 199–208 (1990).
9. K. Morgenröther and P. Werner, *Resonances and standing waves*, Math. Meth. in the Appl. Sci. **9** 105–126 (1987).
10. K. Morgenröther and P. Werner, *On the principles of limiting absorption and limiting amplitude for a class of locally perturbed waveguides. Part 1: Time-independent theory*, Math. Meth. in the Appl. Sci. **10** 125–144 (1988).
11. A. G. Ramm, *Scattering by obstacles* (Reidel, 1986).
12. A. G. Ramm, *Multidimensional inverse scattering: solved and unsolved problems*, Proc. Intern. Conf. on Dynamical Syst. and Applic., eds. G. Ladde and M. Sabandham (Atlanta, Georgia, 1994), pp.287–296.
13. A. G. Ramm, *Scattering amplitude as a function of the obstacle*, Appl. Math. Lett. **6 (5)** 85–87 (1993).
14. A. G. Ramm, *New method for proving uniqueness theorems for obstacle inverse scattering problem*, Appl. Math. Lett. **6 (6)** 89–92 (1993).
15. A. G. Ramm, *The scattering problem analyzed by means of an integral equation of the first kind*, J. Math. Anal. Appl. **201** 324–327 (1996).
16. A. G. Ramm and A. Ruiz, *Existence and uniqueness of scattering solutions in non-smooth domains*, J. Math. Anal. Appl. **201** 329–338 (1996).
17. A. G. Ramm, *Investigation of the scattering problem in some domains with infinite boundaries, I, II*, Vestnik (1963) 45–66, 67–767 (19).
18. A. G. Ramm, *Uniqueness theorems for inverse obstacle scattering problems in Lipschitz domains*, Applic. Analysis **59** 377–383 (1995).
19. A. G. Ramm, *Continuous dependence of the scattering amplitude on the surface of an obstacle*, Math. Methods in the Appl. Sci. **18** 121–126 (1995).
20. A. G. Ramm, *Stability of the solution to inverse obstacle scattering problem*, J. Inverse and Ill-Posed Problems **2(3)** 269–275 (1994).
21. A. G. Ramm, *Stability estimates for obstacle scattering*, J. Math. Anal. Appl. **188(3)** 743–751 (1994).
22. A. G. Ramm, *Examples of nonuniqueness for an inverse problems of geophysics*, Appl. Math. Lett. **8(4)** 87–90 (1995).
23. A. G. Ramm, *Stability of the solution to 3D inverse scattering problems with fixed-energy data*, Inverse Problems in Mechanics, (Proc. ASME, AMD-Vol. 186 1994), pp. 99–102.
24. A. G. Ramm, *Remark on recent papers by R. Kleinman, T. Angell and coauthors*,
25. A. G. Ramm and P. Werner, *On limit amplitude principle*, Jour. fuer die reine und angew. Math. **360** 19–46 (1985).
26. F. Rellich, *Das Eigenwertproblem von $\Delta u + \lambda u = 0$ in unendlichen Gebieten*, Jber. d. deutsch. Math.-Verein. **53** 47–65 (1943).
27. K. J. Witsch, *Examples of embedded eigenvalues for the Dirichlet–Laplacian in domains with infinite boundaries*, Math. Meth. in the Appl. Sci. **12** 177–182 (1990).
28. Y. Xu, *The propagating solution and far field patterns for acoustic harmonic waves in a finite depth ocean*, Appl. Anal. **35** 129–151 (1990).
29. Y. Xu, *An injective far-field pattern operator and inverse scattering problem in a finite depth ocean*, Proc. Edinburgh Math. Soc. **43** 295–311 (1991).

8

RECOVERY OF COMPACTLY SUPPORTED SPHERICALLY SYMMETRIC POTENTIALS FROM THE PHASE SHIFT OF THE S-WAVE

A. G. Ramm*

Department of Mathematics
Kansas State University
Manhattan, KS
E-mail: ramm@math.ksu.edu

ABSTRACT

It is proved that the phase shift $\delta_0(k)$ known for all $k > 0$, determines a real-valued locally integrable spherically symmetric potential $q(x)$, $|q(x)| \leq C\exp(-c|x|^\gamma)$, $\gamma > 1$, where γ is a fixed number, c and C are positive constants. A procedure for finding the unknown potential from the above phase shift is proposed, an iterative process for finding the potential from the above data is described and its convergence is proved.

1. INTRODUCTION AND THE MAIN RESULT

The usual statement of the inverse scattering problem on the half-axis consists in finding $q(x) \in L_{1,1} := \{q : q = \bar{q}, \int_0^\infty (1+x)|q(x)|dx < \infty\}$ from the scattering data $\{S(k), \lambda_j^2, s_j\}$, where $S(k) = \frac{f(-k)}{f(k)}$ is the S-matrix, $f(k)$ is the Jost function, $-\lambda_j^2$ are the bound states, $\lambda_j > 0$, $1 \leq j \leq n$, n is the number of bound states, and s_j are the normalizing constants [7, 10].

The purpose of this paper is to consider the case when it is known a priori that

$$q(x) = 0 \text{ for } |x| \geq a, \tag{1.0}$$

or, for some fixed number $\gamma > 1$, that

$$|q(x)| \leq C\exp(-|x|^\gamma), \quad \gamma > 1, \quad C, \quad c = \text{const} > 0, \tag{1.0'}$$

---
*The author thanks R. Airapetyan for helpful discussions.

Spectral and Scattering Theory, edited by Ramm,
Plenum Press, New York, 1998

and to prove that in this case the potential $q(x)$ is uniquely determined by the phase shift known as a function of k on an arbitrary fixed subset of the semiaxis $(0, \infty)$.

This is of interest in physics. Indeed, assuming $q(\mathbf{x}), \mathbf{x} \in R^3$, to be spherically symmetric: $q = q(r), r = |\mathbf{x}|$, one arrives at the one-dimensional inverse scattering problem for the equation

$$u'' + k^2 u - q(x)u = 0, \quad x > 0, \quad x := r, \tag{1.1}$$

$$u(0,k) = 0, \tag{1.2}$$

$$u(x,k) = \exp[i\delta(k)]\sin[kx + \delta(k)] + o(1), \quad x \to \infty, \tag{1.3}$$

where $\delta(k)$ is the phase shift for the partial wave with angular momentum $\ell = 0$. This phase shift is related to $S(k)$, the $S-$ matrix, by the formula:

$$S(k) = \exp[2i\delta(k)]. \tag{1.4}$$

Here

$$\delta(+\infty) = 0, \quad \delta(-k) = -\delta(k), \quad k \in \mathbb{R}, \tag{1.5}$$

$$f(k) = |f(k)|\exp[-i\delta(k)]. \tag{1.6}$$

The scattering data in 3D case is the scattering amplitude $A(\alpha', \alpha, k)$ which for $q = q(r)$ takes the form:

$$A(\alpha', \alpha, k) = A(\alpha' \cdot \alpha, k) = \sum_{\ell=0}^{\infty} A_\ell(k) Y_\ell(\alpha') \overline{Y_\ell(\alpha)}, \tag{1.7}$$

where $\ell = (\ell, m), -\ell \leq m \leq \ell$, $Y_{\ell m}$ are the orthonormal in $L^2(S^2)$ spherical harmonics, α' and α are unit vectors in the direction of scattered and incident waves, S^2 is the unit sphere in $\mathbb{R}^3$,

$$S_\ell(k) = 1 + \frac{ikA_\ell}{2\pi} \quad \ell = 0, 1, 2, \ldots \tag{1.8}$$

and $S(k) := S_0(k)$. Therefore the knowledge of $A(\alpha', \alpha, k)$ for $q = q(r)$ is equivalent to the knowledge of the set $\{A_\ell(k)\}_{\ell=0,1,2,\ldots}$, which is equivalent to the knowledge of the set $\{S_\ell(k)\}_{\ell=0,1,2,\ldots}$. This knowledge implies directly the knowledge of $S(k) := S_0(k)$, or, which is equivalent, the knowledge of $\delta(k) := \delta_0(k)$, the phase shift. The knowledge of $\delta(k), 0 < k < \infty$, does not determine, in general, λ_j^2 and s_j, $1 \leq j \leq n$, it determines only $S(k)$ by formula (1.4).

It is therefore of interest to find out for what class of the potentials the phase shift $\delta(k)$, $0 < k < \infty$, determines $q(x)$ (and therefore λ_j^2 and s_j) uniquely. This question and related questions were discussed in several earlier papers [1, 3, 4, 21, 22], where mostly the inverse problem on the whole axis was studied. We note especially paper [3] which uses the uniqueness of recovery of the potential from the corresponding Weyl's function. In [5,6] the phase problem is discussed in connection with inverse scattering. In [9] the case of inverse scattering with the known bound states and phase shift, but unknown normalizing constants is discussed.

The purpose of this paper is to prove that $\delta(k)$ determines $q(x)$ satisfies conditions (1.0) or, more generally, (1.0'). Bargmann potentials show that one cannot weaken the rate of decay of $q(x)$ to exponential uniqueness of the recovery of exponential decaying $q(x)$ from $\delta(k)$ does not hold. We also emphasize the algorithmic aspect of the inversion procedure, and in this direction the idea used in [21, 22] is useful.

Note that the scattering amplitude known at a fixed $k > 0$ for α' and α running through some open subsets of S^2, however small, determines $q(\mathbf{x})$, $\mathbf{x} \in R^3$, uniquely if q is real-valued and compactly supported [11] (or decays faster than any exponential [20]).

Our main results are:
1) uniqueness theorem 1.1,
and
2) reconstruction procedure based on theorems 3.1, 4.1 and 4.2.

In this section we state the uniqueness theorem 1.1, its proof is given in section 2. In section 3 an algorithm for recovery of the potential from the phase shift is described. In section 4 an iterative solution of the inverse problem is justified.
The uniqueness theorem we prove is:

Theorem 1.1. *If $q(x) \in L_{1,1}$ satisfies condition (1.0'), then $\delta(k)$ determines $q(x)$ uniquely.*

2. PROOF OF THE UNIQUENESS THEOREM

The strategy of the proof is the following.

Step 1

First we construct $f(k)$ uniquely from the given $S(k) = \exp[2i\delta(k)]$ by solving the Riemann problem:

$$f(k) = S(-k)f(-k), \quad -\infty < k < \infty, \tag{2.1}$$

$f(k)$ is analytic in $\mathbb{C}_+$ and $f(-k)$ is analytic in $\mathbb{C}_-$, $\mathbb{C}_+ := \{k : \text{Im}\,k > 0\}$ is an upper half-plane of the complex k-plane, and $\mathbb{C}_-$ is in the lower one. Index of a function $F(k)$ is denoted $\text{ind}\,F(k)$ and is defined as follows:

$$\text{ind}\,F(k) := \int_{-\infty}^{\infty} d\log F(k) = \frac{1}{2\pi}\Delta_{-\infty,\infty}\arg F(k),$$

where $\Delta_C \arg F$ is the increment of the argument of $F(k)$ on the contour C.

One has: $\text{ind}\,S(-k) = n \geq 0$ if $f(0) \neq 0$ and $\text{ind}\,S(-k) = \frac{1}{2} + n$ if $f(0) = 0$, so that problem (2.1) is solvable and its general solution bounded at infinity, depends on $n + 1$ arbitrary constants [2].

However, there is a unique solution to (2.1) with the properties

$$f(\infty) = 1 \text{ in } \overline{\mathbb{C}}_+, \quad f(i\lambda_j) = 0, \quad \dot{f}(i\lambda_j) \neq 0, \tag{2.2}$$

where $f(x)$ is analytic in $\overline{\mathbb{C}}_+$,

$$\dot{f}(x,k) = \frac{\partial f}{\partial k}, \quad f'(x,k) := \frac{\partial f(x,k)}{\partial x}. \tag{2.3}$$

Step 2

If $f(k)$ is found then $i\lambda_j$, $1 \leq j \leq n$, are uniquely determined as the only zeros of $f(k)$ in the upper half-plane $\mathbb{C}_+ := \{k : \text{Im}\, k > 0\}$.

Step 3

From the Marchenko equation

$$A(x,y) + \int_x^\infty F(y+t)A(x,t)dt = -F(x+y), \quad y \geq x \geq 0, \tag{2.4}$$

where

$$F(x) := \frac{1}{2\pi}\int_{-\infty}^\infty [1 - S(k)]\exp(ikx)dk + \sum_{j=1}^n s_j \exp(-\lambda_j x), \tag{2.5}$$

it follows that the solutions $A_m(x,y)$, $m = 1,2$, to (2.4), corresponding to the F_m which differ only by s_{jm}, $m = 1,2$, differ by a term which is of the form $\sum_{j=1}^n b_j(y)\exp(-\lambda_j x)$.

Therefore the corresponding potentials

$$q_m(x) = -2\frac{d}{dx}A_m(x,x) \tag{2.6}$$

differ by a term which decays not faster than an exponential.

Thus, for a unique choice of s_j one can get $q(x)$ in (2.6) decaying faster than any exponential function $\exp(-c|x|)$ for any constant $c > 0$.

Let us go through these steps in detail:

Step 1

Assume that

$$h_+(k) := h(k) = S(-k)h(-k) := S(-k)h_-(k), \quad -\infty < k < \infty, \tag{2.7}$$

and

$$h_\pm(\infty) = 1 \text{ in } \overline{\mathbb{C}}_+, \tag{2.7'}$$

$h(k)$ is an entire analytic function with exactly n simple zeros $i\nu_j$ on the imaginary axis in $\mathbb{C}_+$, and $\dot{h}(i\nu_j) \neq 0$. Here and below $h_+ (h_-)$ are analytic in $\mathbb{C}_+ (\mathbb{C}_-)$ functions. Let us first prove that $\nu_j = \lambda_j$, $1 \leq j \leq n$, where $i\lambda_j$ are the poles of $S(k)$ in $\mathbb{C}_+$. Note that if $q(x)$ satisfies (1.0) or (1.0'), then $f(x,k)$, the Jost solution to equation (1.1) (which is the unique solution to (1.1) with asymptotics $f(x,k) = \exp(ikx) + o(1)$ as $x \to +\infty$, $k > 0$) and the Jost function $f(k) := f(0,k)$ are entire functions of k and $S(k) := \frac{f(-k)}{f(k)}$ is a meromorphic in $\mathbb{C}$ function of k.

Let us first prove the following lemma:

Lemma 2.1. *If $f(i\lambda) = 0$, $\lambda > 0$, then $f(-i\lambda) \neq 0$.*

Proof: The Wronskian formula [10, p. 279]

$$f(x,k)f'(x,-k) - f'(x,k)f(x-k) = -2ik. \tag{2.8}$$

can be continued to $\mathbb{C}'_+$ analytically if q satisfies (1.0) or (1.0′) and at $x = 0$ and $k = i\lambda$, $\lambda > 0$, yields, if $f(i\lambda) = 0$:

$$-f'(0, i\lambda)f(0, -i\lambda) = 2\lambda > 0. \tag{2.9}$$

Formula (2.9) shows that $f(0, -i\lambda) := f(-i\lambda) \neq 0$. $\square$

Since $\delta(k)$ corresponds by the assumption to a $q(x)$ (satisfying (1.0) or (1.0′)), there exists $f(k)$ such that

$$S(k) = \frac{f(-k)}{f(k)} = \exp[2i\delta(k)], \tag{2.10}$$

and $f(k)$ satisfies (2.2).

We want to prove that any $h(k)$ which satisfies (2.6), (2.7), and which is an entire function of k, must be equal to $f(k)$ in $\mathbb{C}_+$, and in particular, must have zeros in $\mathbb{C}_+$ at the points $i\lambda_j$, $1 \leq j \leq n$, and only at these points.

Let us write (2.6) as

$$\frac{h_+}{f_+} := \frac{h(k)}{f(k)} = \frac{h(-k)}{f(-k)} := \frac{h_-(k)}{f_-(k)}, \quad -\infty < k < \infty. \tag{2.11}$$

The left-hand side (right-hand side) of (2.11) is analytic in $\mathbb{C}_+$ ($\mathbb{C}_-$) except, possibly, at the zeros of $f(k)$ in $\mathbb{C}_+$, (of $f_-(k)$ in $\mathbb{C}_-$) and tends to 1 as $k \to \infty$ in $\overline{\mathbb{C}}_+$ ($\overline{\mathbb{C}}_-$).

If $f(i\lambda_j) = 0$, $\lambda_j > 0$, then (2.11) shows that

$$h(i\lambda_j) = f(i\lambda_j) \frac{h(-i\lambda_j)}{f(-i\lambda_j)} = 0, \tag{2.12}$$

where we have used Lemma 2.1, which says that $f(-i\lambda_j) \neq 0$. Since $h(k)$ has exactly n zeros in $\mathbb{C}_+$ by the assumption, it follows that $\nu_j = \lambda_j$, $1 \leq j \leq n$, and the function $\frac{h_+(k)}{f_+(k)} := \frac{h(k)}{f(k)}$ is analytic in $\mathbb{C}_+$. Likewise, the function $\frac{h_-(k)}{f_-(k)} := \frac{h(k)}{f(k)}$ is analytic in $\mathbb{C}_-$. They coincide on the real axis. Therefore each is an analytic continuation of the other. Thus $\frac{h(k)}{f(k)}$ is analytic in $\mathbb{C}$ and equals to 1 at infinity. By Liouville theorem it follows that $\frac{h(k)}{f(k)} \equiv 1$, so $h(k) = f(k)$, $k \in \mathbb{C}$, and Step 1 is completed.

Step 2

In Step 1 we have found the unique entire function $f(k)$ which solves the Riemann problem (2.1), has exactly n purely imaginary zeros in $\mathbb{C}^n$ and tends to 1 at infinity.

Since this solution is analytic in $\mathbb{C}_+$, its zeros are uniquely determined. Thus $\delta(k), k \in \mathbb{R}$, determines λ_j, $1 \leq j \leq n$, uniquely if $q(x)$ satisfies (1.0) or (1.0′). Step 2 is completed.

Step 3

Let us choose arbitrary numbers $s_j > 0$, $1 \leq j \leq n$, and consider the Marchenko equation:

$$A(x,y) + \int_x^\infty F(y+t)A(x,t)dt = -F(x+y), \quad y \geq x \geq 0, \qquad (2.13)$$

where $F(x)$ is given by (2.5).

Lemma 2.2. *There is at most one choice of s_j such that the solution $A(x,y)$ to (2.13) has the property that the potential $q(x) = -2\frac{dA(x,x)}{dx}$ decays faster than any exponential as $x \to +\infty$.*

Proof: One solution with the above property does exist since we have assumed that $q(x)$ satisfies (1.0) or (1.0'): for such a potential $A(x,y)$ satisfies (2.13) with the constants s_j, $1 \leq j \leq n$, uniquely determined by $q(x)$, and the corresponding $A(x,y)$ has the property stated in Lemma 2.2.

Let us prove that no other choice of s_j yields this property.

Consider any other choice $s_j' > 0$, $s_j' \neq s_j$. Let $B(x,y)$ be the solution to (2.13) with $F(x)$ replaced by $F(x) + M(x)$, where

$$M(x) = \sum_{j=1}^n (s_j' - s_j)\exp(-\lambda_j x). \qquad (2.14)$$

Let

$$a(x,y) := B(x,y) - A(x,y). \qquad (2.15)$$

Then (2.13) and a similar equation for B, namely,

$$B(x,y) + \int_x^\infty [M(y+t) + F(y+t)]B(x,t)dt = -F(x+y) - M(x+y), \qquad (2.16)$$

imply:

$$a(x,y) + \int_x^\infty F(y+t)a(x,t)dt = -M(x+y) - \int_x^\infty M(y+t)[A(x,t) + a(x,t)]dt. \qquad (2.17)$$

Assume that q satisfies (1.0). Then one can prove (see e.g. [10, p. 279]) that

$$F(y) = 0 \text{ for } y > 2a, \quad A(x,y) = 0 \text{ for } y \geq x \geq a. \qquad (2.18)$$

Therefore (2.17) implies:

$$a(x,x) = -M(2x) - \int_x^\infty M(x+t)a(x,t)dt, \quad x > a. \qquad (2.19)$$

If $s_j' \neq s_j$ for at least one j, $1 \leq j \leq n$, then $M(2x)$ decays, as $x \to +\infty$, not faster than an exponential $\exp(-2\lambda_j x)$, and the second term in (2.19) equals to

$$-\sum_{j=1}^n (s_j' - s_j)\exp(-\lambda_j x)\int_x^\infty \exp(-\lambda_j t)a(x,t)dt = -\sum_{j=1}^n (s_j' - s_j)\exp(-\lambda_j x)b_j(x), \qquad (2.20)$$

where
$$b_j(x) := \int_x^\infty \exp(-\lambda_j t) a(x,t) dt. \qquad (2.21)$$

For $x > a$, equation (2.17) becomes
$$a(x,y) = -\int_x^\infty M(y+t) a(x,t) dt - M(x+y), \quad y \geq x > a, \qquad (2.22)$$

because of (2.18). Equation (2.22) has a degenerate kernel and can be solved analytically. Let us write (2.22) as
$$a(x,y) = -\sum_{j=1}^n \gamma_j \exp(-\lambda_j y) b_j(x) - \sum_{j=1}^n \gamma_j \exp(-\lambda_j y) \exp(-\lambda_j x), \quad y \geq x > a, \qquad (2.23)$$

where
$$\gamma_j := s'_j - s_j. \qquad (2.24)$$

Then
$$b_j(x) = -\sum_{j=1}^n \gamma_j a_{ij}(x) b_j(x) - \sum_{j=1}^n \gamma_j a_{ij}(x) \exp(-\lambda_j x), \qquad (2.25)$$

where
$$a_{ij}(x) := \int_x^\infty \exp[-(\lambda_j + \lambda_i) y] dy = \frac{\exp[-(\lambda_j + \lambda_i) x]}{\lambda_j + \lambda_i}. \qquad (2.26)$$

System (2.25) with matrix (2.26) for large x has diagonally dominant matrix
$$\delta_{ij} + \gamma_j a_{ij}(x). \qquad (2.27)$$

Therefore system (2.25) is uniquely solvable by iterations and asymptotically one has
$$b_i(x) \sim \sum_{j=1}^n \gamma_j a_{ij}(x) \exp(-\lambda_j x), \quad x \to +\infty \qquad (2.28)$$

From (2.28) and (2.23) it follows that
$$a(x,x) \sim -\sum_{j=1}^n \gamma_j \exp(-\lambda_j x) \sum_{m=1}^n \gamma_m a_{jm}(x) \exp(-\lambda_m x) - \sum_{j=1}^n \lambda_j \exp(-2\lambda_j x)$$
$$= -\sum_{j,m=1}^n \frac{\gamma_j \gamma_m}{\lambda_j + \lambda_m} \exp[-2(\lambda_m + \lambda_j) x] - \sum_{j=1}^n \gamma_j \exp(-2\lambda_j x). \qquad (2.29)$$

Thus, if $q(x)$ satisfies (1.0), that is, $-2\frac{dA(x,x)}{dx} = 0$ for $x > a$, and $0 < \lambda_1 < \lambda_2 < \ldots \lambda_n$, then
$$q_1(x) := -2\frac{dB(x,x)}{dx} = -2\frac{da(x,x)}{dx}$$
$$= 2\frac{d}{dx}\left\{\frac{\gamma_1^2}{2\lambda_1} \exp(-4\lambda_1 x) + \gamma_1 \exp(-2\lambda_1 x)\right\}[1+o(1)], \quad x \to +\infty \qquad (2.30)$$

It follows from (2.30) that $q_1(x)$ decays not faster than $O(\exp(-2\lambda_1 x))$ as $x \to +\infty$ for any choice of s'_j such that $s'_1 - s_1 \neq 0$ and if $s'_j = s_j$, $1 \leq j \leq i < n$, then $q_1(x)$ decays not faster than $O(\exp(-2\lambda_{j+1} x))$ as $x \to +\infty$.

Step 3 is completed.

If $q(x)$ satisfies (1.0′) the argument is similar. For such potentials one uses the estimates (see [7, pp. 178, 209]):

$$|A(x,y)| \leq c\sigma\left(\frac{x+y}{2}\right), \quad \sigma(x) := \int_x^\infty |q(s)|ds, \quad (2.31)$$

$$\left|\frac{\partial A(x_1,x_2)}{\partial x_j} + \frac{1}{4}q\left(\frac{x_1+x_2}{2}\right)\right| < c\sigma(x_1)\sigma(\frac{x_1+x_2}{2}), \quad j = 1,2, \quad (2.32)$$

$$|F(2x)| \leq c\sigma(x). \quad (2.33)$$

These estimates show that $F(x)$, $A(x,x)$ and $A(x,y)$, for $y \geq x$, are majorized by $c\sigma(x)$, so that if $q(x)$ satisfies (1.0′) all these functions also satisfy estimates (2.31)–(2.33) with $\sigma(x) \leq c\exp(-|x|^\gamma)$ with $\gamma > 1$, where $c = \text{const} > 0$ denotes different constants.

Integral equation (2.17) yields under the assumption (1.0′) the following equation

$$a(x,y) + \int_x^\infty M(y+t)a(x,t)dt = -M(x+y) + O(\exp(-|x|^\gamma)) \quad (2.34)$$

This equation has a degenerate kernel and can be solved as before. Its solution has the property that $-2\frac{da(x,x)}{dx}$ decays as $x \to +\infty$, not faster than an exponential unless $s'_j = s_j$ for all j, $1 \leq j \leq n$.

Theorem 1.1 is proved. □

3. ALGORITHM FOR RECOVERY OF THE SPHERICALLY SYMMETRIC POTENTIALS FROM THE PHASE-SHIFT OF THE S-WAVE

In this section we assume that $q(x)$ is compactly supported and without loss of generality, that $a = 1$. The basic idea is a variant of the one in [10, p. 296] (which is originally proposed in [22]): one uses the scattering data to get Cauchy data for the transformation kernel $K(x,y)$, and these Cauchy data are used for a derivation of a nonlinear Volterra equation for $q(x)$.

To start with, define $K(x,y)$ as in [7] and [10, p. 251]:

$$\varphi(x,k) = \varphi_0 + \int_0^x K(x,y)\varphi_0 dy, \quad \varphi_0 := \varphi_0(x,k) := \frac{\sin kx}{k}. \quad (3.1)$$

The kernel $K(x,y)$ is the unique solution to the Goursat problem [7], [10, p. 251]):

$$K_{xx} - q(x)K = K_{yy}, \quad -x \leq y \leq x; \quad K(x,y) = -K(x,-y), \quad (3.2)$$

$$K(x,x) = \frac{1}{2}\int_0^x q(s)ds, \quad (3.3)$$

$$K(x,0) = 0, \tag{3.4}$$

so

$$q(x) = 2\frac{dK(x,x)}{dx}. \tag{3.5}$$

If the Cauchy data

$$\{K(1,y), K_x(1,y)\}_{-1\le y\le 1} \tag{3.6}$$

are known, then one gets the following equation for $q(x)$, which is derived in section 4 (see equation (4.6) in section 4):

$$q = T(q), \tag{3.7}$$

where

$$T(q) = -2\int_x^1 K(y, 2x-y)q(y)dy + f(x) \tag{3.8}$$

and

$$f(x) := 2[K_y(1, 2x-1) + K_x(1, 2x-1)]. \tag{3.9}$$

The kernel $K(x,y)$ depends on $q(x)$, so (3.7) is a nonlinear Volterra equation. This equation was studied in [10, pp. 299–302]. Let us check that the function $f(x)$ in (3.8), defined by formula (3.9), is uniquely determined by the phase shift $\delta(k)$ if $q(x)$ vanishes for $x > 1$.

Theorem 3.1. *The functions $K(1,y)$ and $K_x(1,y)$ for $-1 \le y \le 1$ are uniquely determined by $\delta(k)$ if $q(x) = 0$ for $x > 1$.*

Proof of Theorem 3.1: In the course of the proof it will be shown how to calculate $K(1,y)$ and $K_x(1,y)$ from the data $\delta(k)$. Let us recall that if $q(x) = 0$ for $x > 1$, then $\delta(k)$ determines uniquely the Jost function $f(k)$ as the unique solution to the Riemann problem (2.1), namely the solution which satisfies the condition $f(\infty) = 1$ in $\overline{\mathbb{C}}_+$, and has exactly n simple purely imaginary zeros in $\mathbb{C}_+$.

Let us assume that Step 1 in section 2 is completed, that is, $f(k)$ is found from $\delta(k)$. Since the function (3.1) has the asymptotics

$$\varphi = \frac{|f(k)|}{k}\sin[kx + \delta(k)] + o(1) \text{ as } x \to +\infty, \tag{3.10}$$

one concludes that for $x > 1$, that is, in the region where $q(x) = 0$, (3.10) implies

$$\varphi(x) = \frac{|f(k)|}{k}\sin[kx + \delta(k)], \quad x > 1. \tag{3.11}$$

In particular, (3.1) and (3.11) imply

$$\int_0^1 K(1,y)\sin ky\, dy = -\sin k + |f(k)|\sin[k + \delta(k)], \tag{3.12}$$

and

$$\int_0^1 K_x(1,y) \sin ky \, dy = -K(1,1) \sin k - k \cos k + k|f(k)| \cos[k + \delta(k)]. \quad (3.13)$$

Since $\delta(k)$ and $|f(k)|$ are known for all $k > 0$, one can take $k = n\pi$, and get

$$\int_0^1 A(y) \sin(n\pi y) dy = A_n, \quad A(y) := K(1,y), \quad (3.14)$$

$$\int_0^1 B(y) \sin(n\pi y) dy = B_n, \quad B(y) := K_x(1,y), \quad (3.15)$$

$$A_n := |f(n\pi)| \sin[n\pi + \delta(n\pi)] = |f(n\pi)|(-1)^n \sin \delta(n\pi), \quad (3.16)$$

$$B_n := -n\pi(-1)^n + n\pi|f(n\pi)| \cos[n\pi + \delta(n\pi)] = n\pi(-1)^n[-1 + |f(n\pi)| \cos \delta(n\pi)]. \quad (3.17)$$

Note that $|A_n| = O(\frac{1}{n})$ as $n \to \infty$ and, if q is piecewise C^1, then $\{B_n\} = O(\frac{1}{n})$ also. If $q \in L^2(0,1)$ then the sequence B_n belongs to ℓ^2.

Equations (3.14) and (3.15) determine uniquely both functions $A(y)$ and $B(y)$ since the system of functions $\{\sin(n\pi y)\}_{n=1,2,\ldots}$ is a complete orthonormal system in $L^2[0,1]$. In fact, $A(y)$ and $B(y)$ can be written as the Fourier series:

$$A(y) = 2 \sum_{n=1}^{\infty} A_n \sin(n\pi y), \quad B(y) = 2 \sum_{n=1}^{\infty} B_n \sin(n\pi y). \quad (3.18)$$

Theorem 3.1 is proved. □

Let us now describe an iterative process for finding the potential $q(x)$ from the data $K(1,y)$ and $K_x(1,y)$. In section 4 a version of this iterative process is studied in detail. If $A(y) = K(1,y)$ and $B(y) = K_x(1,y)$ are found, then $f(x)$ can be calculated by formula (3.9) and one can try to solve equation (3.8) by an iterative process considered in [RuSa1,RuSa2]. Namely, take

$$q_0(x) = f(x), \quad q_{n+1}(x) := 2 \frac{dK_n(x,x,q_n)}{dx}, \quad (3.19)$$

assume that $q_n(x)$ is known and find $K_n(x,y,q_n)$ as the unique solution to the problem (3.20)–(3.21):

$$\frac{\partial^2 K_n(x,y;q_n)}{\partial x^2} - q_n(x) K_n(x,y;q_n) = \frac{\partial^2 K_n(x,y;q_n)}{\partial y^2}, \quad -1 \leq -x \leq y \leq x \leq 1 \quad (3.20)$$

$$K_n(1,y) = A(y), \quad \frac{\partial K_n(1,y)}{\partial x} = B(y). \quad (3.21)$$

Problem (3.20)–(3.21) is a Cauchy problem for the wave equation. Its solution yields $q_{n+1}(x)$ by the second formula (3.19). Then problem (3.20)–(3.21) is solved with $q_{n+1}(x)$ in place of $q_n(x)$ and $K_{n+1}(x,y;q_{n+1})$ (in place of $K_n(x,y;q_n)$) is found.

In [22] a proof of convergence of this iterative process is outlined.
Let us discuss a possible method for finding λ_j from $\delta(k)$.
Note that

$$\exp[2i\delta(k)] = \frac{f(-k)}{f(k)}, \tag{3.22}$$

and

$$f(-k) = \overline{f(\overline{k})}, \quad k \in \mathbb{C}, \tag{3.23}$$

where the bar stands for complex conjugate. Formula (3.23) is an immediate consequence of the reality condition

$$f(-k) = \overline{f(k)}, \quad -\infty < k < \infty, \tag{3.24}$$

and of the fact that $f(k)$ is an entire function of k.

Indeed, if $f(k) = \sum_{j=0}^{\infty} f_j k^j$, f_j are constants, then (3.24) implies

$$f_j(-1)^j = \overline{f_j}, \quad \forall j, \tag{3.25}$$

that is:

$$f_{2m} = \overline{f_{2m}} := b_{2m}, \quad f_{2m+1} = i f_{2m+1}, \tag{3.26}$$

where f_m are real numbers, so

$$f(k) = \sum_{m=0}^{\infty} b_{2m} k^{2m} + i \sum_{m=0}^{\infty} b_{2m+1} k^{2m+1} \tag{3.27}$$

Formula (3.23) follows from (3.27).

Consider the function

$$F_S(x) = \frac{1}{2\pi} \int_{-\infty}^{\infty} \left[1 - \frac{f(-k)}{f(k)} \right] \exp(ikx) dk. \tag{3.28}$$

Since $q(x) = 0$ for $x > a$, it follows (see [10, p. 253]) that $A(0,y) = 0$ for $y > 2a$.
In what follows we assume without loss of generality that $a = 1$ and $q(x) = 0$ for $x > 1$.
Then it follows that

$$f(k) = 1 + \int_0^2 A(0,y) \exp(iky) dy. \tag{3.29}$$

Therefore,

$$f(k) - f(-k) = \int_0^2 A(0,y) [\exp(iky) - \exp(-iky)] dy. \tag{3.30}$$

Let us assume that $x > 2$ and $\lambda_j \neq 1$, $\forall j$. Then, using the residue theorem at the last step, one gets:

$$F_S(x) = \frac{1}{2\pi} \int_{-\infty}^{\infty} \frac{dk}{f(k)} \int_0^2 A(0,y) \left(\exp[ik(x+y)] - \exp[ik(x-y)] \right) dy$$

$$= \int_0^2 dy A(0,y) \left(-\frac{d^2}{dx^2} + 1 \right) \frac{1}{2\pi} \int_{-\infty}^{\infty} \frac{dk}{f(k)(1+k^2)} \left[e^{ik(x+y)} - e^{ik(x-y)} \right]$$

$$= \left(-\frac{d^2}{dx^2} + 1 \right) \int_0^2 dy A(0,y) i \left(\sum_{j=1}^J \frac{e^{-\lambda_j(x+y)} - e^{-\lambda_j(x-y)}}{\dot{f}(i\lambda_j)(1-\lambda_j^2)} + \frac{e^{-(x+y)} - e^{-(x-y)}}{f(i)2i} \right)$$

$$= i \left(-\frac{d^2}{dx^2} + 1 \right) \sum_{j=1}^J \frac{e^{-\lambda_j x}}{\dot{f}(i\lambda_j)(1-\lambda_j^2)} \int_0^2 dy A(0,y) \left[e^{-\lambda_j y} - e^{\lambda_j y} \right], \quad x > 2.$$

Therefore

$$F_S(x) = -i \sum_{j=1}^J \frac{e^{-\lambda_j x}}{\dot{f}(i\lambda_j)} \int_0^2 A(0,y) \left(e^{\lambda_j y} - e^{-\lambda_j y} \right) dy, \quad x > 2. \tag{3.31}$$

It follows now that λ_j can be uniquely determined from the asymptotics of $F_S(x)$ as $x \to +\infty$.

Numerically the problem consists in finding exponents $\lambda_j > 0$, $0 < \lambda_1 < \lambda_2 < \ldots \lambda_J$, from the knowledge of the function

$$F_S(x) = \sum_{j=1}^J \gamma_j e^{-\lambda_j x}, \quad \gamma_j := \frac{-i \int_0^2 A(0,y) \left[e^{\lambda_j y} - e^{-\lambda_j y} \right] dy}{\dot{f}(i\lambda_j)} \tag{3.32}$$

for all $x > 0$. The constants γ_j and λ_j are not known a priori, and the number J of bound states is known. Indeed, since $f(k) := |f(k)| e^{-i\delta(k)}$, since $\delta(\infty) = 0$ by definition, and $\delta(-k) = -\delta(k)$ for real k, one has

$$J = \mathrm{ind} f(k) := \frac{1}{2\pi} \Delta_{(-\infty,\infty)} \arg f(k) = -\frac{1}{2\pi} \Delta_{(-\infty,\infty)} \delta(k)$$

$$= -\frac{1}{2\pi} [\delta(+\infty) - \delta(+0) + \delta(-0) - \delta(-\infty)] = \frac{1}{\pi} \delta(+0), \text{ if } f(0) \neq 0, \tag{3.33}$$

and

$$\mathrm{ind} f = J + \frac{1}{2}, \quad \text{if } f(0) = 0, \text{ since } \dot{f}(0) \neq 0. \tag{3.34}$$

Note that $\mathrm{ind} S(k) = \mathrm{ind} \frac{f(-k)}{f(k)} = \mathrm{ind} \overline{f(k)} - \mathrm{ind} f(k) = -2 \mathrm{ind} f(k)$. Thus

$$\mathrm{ind} S(k) = \begin{cases} -2J & \text{if } f(0) \neq 0, \\ -2J - 1 & \text{if } f(0) = 0. \end{cases} \tag{3.35}$$

A method for finding λ_j is known, it is Prony's method described, for example, in [14] and used in [15–19].

The problem of finding λ_j from $F_S(x)$ is an ill-posed one: since the function $e^{-\lambda_j x}$ decay rapidly as $x \to +\infty$, a small noise in the data $F_S(x)$ complicates extraction of λ_j from $F_S(x)$ [16–18].

4. AN ITERATIVE SOLUTION OF THE INVERSE PROBLEM

In this section we derive equation (3.7) and study its iterative solution.

To derive (3.7), following the Riemann method, let us integrate (3.2) over the triangle D bounded by the straight lines $s = 1, t - x = s - x, t - x = x - s$. This yields

$$\int_D q(s) K(s,t) ds\, dt = \int_D (K_{ss} - K_{tt}) ds\, dt = \int_{\partial D} K_s dt + K_t ds := I, \quad (4.1)$$

where Green's formula was used at the last step, ∂D is the boundary of D which is passed counterclockwise. Denote the vertices of the triangle by $A = (1,1), B = (1, 2x-1), C = (x,x)$. The lines CA has the equation $t = s$, the line CB has the equation $t - x = x - s$, and the line BA has the equation $s = 1$. The integral I in (4.1) can be written as

$$I = \int_{2x-1}^1 K_s(1,t) dt - \int_x^1 d_s K(s, 2x-s) - \int_x^1 d_s K(s,s)$$

$$= \int_{2x-1}^1 K_s(1,t) dt - K(1, 2x-1) + K(x,x) - K(1,1) + K(x,x). \quad (4.2)$$

From (4.1) and (4.2) one gets:

$$K(x,x) = \frac{1}{2} \int_x^1 ds\, q(s) \int_{2x-s}^s K(s,t) dt + \frac{K(1, 2x-1) + K(1,1)}{2} - \frac{1}{2} \int_{2x-1}^1 K_s(1,t) dt. \quad (4.3)$$

Similarly, one derives

$$K(x,y) = \frac{1}{2} \int_x^1 dsq(s) \int_{y+x-s}^{y-x+s} K(s,t) dt + \frac{K(1, y+x-1) + K(1, y-x+1)}{2}$$

$$- \frac{1}{2} \int_{y+x-1}^{y-x+1} K_s(1,t) dt, \quad 0 \le |y| \le x \le 1.. \quad (4.4)$$

From (4.3) and (3.5) it follows that

$$q(x) = -2 \int_x^1 q(s) K(s, 2x-s) ds + 2 K_y(1, 2x-1) + 2 K_x(1, 2x-1) := T(q), \quad (4.5)$$

which is equation (3.7).

Let us now assume that $f(x)$, defined by (3.9), is known, and consider equation (3.8):

$$q(x) = A(q) + f, \quad A(q) := -2 \int_x^1 q(s) K(s, 2x-s) ds. \quad (4.6)$$

Let us consider (4.6) as an equation with the nonlinear operator

$$T(q) := f - 2 \int_x^1 q(s) K(s, 2x-s; q) ds, \quad (4.7)$$

where $K(x,y;q)$ is the unique solution to the equation (4.4) which we write as

$$K(x,y) = g(x,y) + \frac{1}{2} \int_{D_{xy}} q(s) K(s,t) ds\, dt := g + B(q) K. \quad (4.8)$$

Here D_{xy} is the triangle on the (s,t) plane bounded by the straight lines:

$$s = 1, \quad t = y+s-x, \quad t = y-(s-x), \tag{4.9}$$

$B(q)K$ is the integral operator in (4.8), and

$$g(x,y) := \frac{K(1,y+x-1)+K(1,y-x+1)}{2} - \frac{1}{2}\int_{y+x-1}^{y-x+1} K_s(1,t)dt, \quad 0 \leq |y| \leq x \leq 1. \tag{4.10}$$

The function $g(x,y)$ is uniquely and numerically efficiently determined by the phase shift $\delta(k)$, as was proved in Theorem 3.1.

Let us assume that

$$q(x) \in L^\infty(0,1) \text{ and } \sup_{0 \leq x \leq 1} |q(x)| := \mathbf{q}$$

We prove that equation (4.8) is uniquely solvable by iterations and its solution $K(x,y)$ is a differentiable function provided that $q(x) \in L^\infty(0,1)$. Let us define the Banach space X_m of $L^\infty(0,1)$ functions with the norm

$$\|q\|_m := \sup_{0 \leq x \leq 1} \left\{ e^{-m(1-x)}|q(x)| \right\}, \tag{4.11}$$

where $m > 0$ is a sufficiently large number which we fix later. For any fixed $m > 0$ the norm (4.11) is equivalent to $L^\infty(0,1)$ norm. The reason for choosing norm (4.11) is simple: for $m > 0$ sufficiently large ($m \gg 1$) we prove that T considered on a convex closed set

$$M_R := \left\{ q : \sup_{0 \leq x \leq 1} |q(x)| \leq R \right\}$$

is a contraction map:

$$\|T(q) - T(p)\|_m < \gamma_m \|q-p\|_m, \quad 0 < \gamma_m < 1, \quad m \gg 1, \quad p,q \in M_R. \tag{4.12}$$

Estimate (4.12) alone is not sufficient for the contraction mapping principle to be applicable to equation (4.6) because we did not prove that the map T sends the set M_R into itself. However, estimate (4.12) implies that equation (4.6) has a unique solution in X_m. Indeed, $R > 0$ can be chosen arbitrary large, and if equation (4.6) has two solutions, then taking R such that both solutions belong to M_R, one can use (4.12) to conclude that their difference is zero.

Since the solution of (4.6) is unique, and estimate (4.12) holds, one may try to compute this solution by iterations:

$$q_{n+1}(x) = T(q_n), \quad q_0 = 0, \tag{4.13}$$

where the function $K = K(s,t,q_n)$ in equation (4.6) is the unique solution to (4.8) with $q = q_n(s)$, and the functions $f(x)$ in (4.6) and $g(x,y)$ in (4.8) do not depend on the index n. In theorem 4.2 an iterative process (4.43) is suggested for a simultaneous computation of $q(x)$ and $K(x,y)$ and convergence of this process is proved. A similar proof can be given for convergence of the process (4.13), but we do not discuss it because one has to calculate the kernel $K(x,y)$ in order to use process (4.13), and process (4.43) calculates both functions, $q(x)$ and $K(x,y)$, simultaneously.

To realize this plan, we first prove the unique solvability of equation (4.8).

Theorem 4.1. *Equation (4.8) is uniquely solvable in $L^\infty(D)$, where $D := \{x,y : 0 \leq |y| \leq x \leq 1\}$. The iterative process*

$$K_{n+1} = B(q)K_n + g, \quad K_0 = g, \qquad (4.14)$$

converges in X_m to the unique solution of (4.8).

If $q(x) \in L^\infty(0,1)$, then the kernel $K(x,y) = K(x,y;q)$ of the transformation operator is a differentiable function which is the unique solution to the equation (31) in [10, p. 252]:

$$K(x,y) = \frac{1}{2}\int_{\frac{x-y}{2}}^{\frac{x+y}{2}} q(s)ds + \int_{\frac{x-y}{2}}^{\frac{x+y}{2}} ds \int_0^{\frac{x-y}{2}} q(s+t)K(s+t, s-t)dt. \qquad (4.14')$$

Equation (4.14′) is uniquely solvable by iterations. This is proved, for example, in [7, p. 14], where the following estimate is obtained [7, p. 14]:

$$|K(x,y)| \leq \frac{1}{2}w\left(\frac{x+y}{2}\right)\exp\left[Q(x) - Q\left(\frac{x+y}{2}\right) - Q\left(\frac{x-y}{2}\right)\right], \quad 0 \leq |y| \leq x, \quad (4.15)$$

where

$$w(x) := \max_{0 \leq s \leq x}\left|\int_0^s q(t)dt\right|, \quad Q(x) := \int_0^x ds \int_0^s |q(t)|dt. \qquad (4.16)$$

Let $0 \leq |y| \leq x \leq 1$, and $\int_0^1 |q(s)|ds := c_q \leq \mathbf{q}$, then (4.15) implies

$$|K(x,y)| \leq \frac{1}{2}c_1 \exp(c_q) := C_q, \qquad (4.17)$$

where $c_1 > 0$ is a constant.

Estimate (4.17) and equation (4.14′) imply:

$$\int_{-1}^1 |K_x(1,t)|dt \leq c_q(1+C_q). \qquad (4.18)$$

Therefore, (4.17) and (4.18), (3.9) and (4.10) yield

$$|g(x,y)| \leq C_q + \frac{c_q(1+C_q)}{2}; \quad |f(x,y)| \leq \mathbf{q} + 6\mathbf{q}C_q. \qquad (4.19)$$

Let us summarize the results:

Lemma 4.2. *If $\max_{0 \leq x \leq 1} |q(x)| \leq \mathbf{q}$, then*

$$\max |g(x,y)| \leq C_q + \frac{c_q(1+C_q)}{2}, \quad \max |f(x,y)| \leq \mathbf{q} + 6\mathbf{q}C_q. \qquad (4.20)$$

Now let us prove

Lemma 4.3. *If $\mathbf{q} < \infty$, then equation (4.8) is uniquely solvable in $L^\infty(D)$ by iterations and its solution is*

$$K(x,y) = \sum_{j=0}^\infty B^j(q)g. \qquad (4.21)$$

The series (4.21) converges in $C[0,1]$, and

$$|B^j(q)g| \leq g_0 \frac{\mathbf{q}^j(1-x)^{2j}}{(2j)!}, \quad g_0 := \max_{0 \leq |y| \leq x \leq 1} |g(x,y)|. \qquad (4.22)$$

Proof of Lemma 4.3: The proof of (4.22) is by induction. Equation (4.21) is an immediate consequence of (4.22). Equations (4.21) and (4.22) imply the unique solvability of (4.8) by iterations. Estimate (4.22) shows that the spectral radius of the linear operator $B(q)$ in $L^\infty(D)$ equals zero, so $B(q)$ is a Volterra operator.

So the proof of Lemma 4.3 is complete as soon as (4.22) is proved.

If $j = 0$ then (4.22) is obvious. Suppose (4.22) is already proved for $0 \leq i \leq j$, and let us prove it for $i = j+1$. One has, with $B := B(q)$,

$$|B^{j+1}g| = \left|\frac{1}{2}\int_{D_{xy}} q(s)(B^j g)(s,t)dsdt\right| \leq \frac{1}{2}\int_x^1 ds|q(s)|\int_{x+y-s}^{x-y+s} \frac{g_0 q^j}{(2j)!}(1-s)^{2j} dt$$

$$\leq \frac{g_0 q^{j+1}}{(2j)!}\frac{1}{2}\int_x^1 ds(1-s)^{2j}(2s-2x) = \frac{g_0 q^{j+1}}{(2j)!}(1-x)^{2j+2}\left(\frac{1}{2j+1} - \frac{1}{2j+2}\right)$$

$$= \frac{g_0 q^{j+1}(1-x)^{2j+2}}{(2j+2)!}. \tag{4.23}$$

Thus (4.22) is proved. Lemma 4.3 is proved. □

Let us now prove the following lemma:

Lemma 4.4. *If $p, q \in M_R$, then*

$$\|T(p) - T(q)\|_m \leq \gamma_m \|p-q\|_m, \quad 0 < \gamma_m < 1, \tag{4.24}$$

where $\|\cdot\|_m$ is defined in (4.11) and $m = m(R)$ is chosen sufficiently large.

Proof of Lemma 4.4: Equations (4.7) and (4.8) form a system of equations for $q(x)$ and $K(x,y)$. Equation (4.8) is a linear with respect to $K(x,y)$ Volterra-type equation, and if $K := K(x,y;q)$ is found from (4.8) and is substituted into (4.6), then (4.6) becomes a single nonlinear Volterra-type equation for $q(x)$.

In Lemma 4.3 the solution to (4.8) was investigated.

Let us now prove (4.24). If (4.24) is proved, then it follows that equation (4.6) has at most one solution in M_R, but since $R > 0$ can be taken arbitrary large, this implies that equation (4.6) has at most one solution in $L^\infty(0,1)$.

However, since the index m for which $0 \leq \gamma_m < 1$, depends on R, and we do not prove that the map T maps a ball $\{q : \sup_{0 \leq x \leq 1} |q(x)| \leq R\}$ into itself for a certain R, we cannot claim that the contraction mapping principle is applicable and that the iterative process (3.19) - (3.21) converges.

To prove (4.24), consider

$$T(p) - T(q) = \sum_{j=1}^\infty \int_x^1 \left[(B^j(p)g)(s,2x-s)p(s) - (B^j(q)g)(s,2x-s)q(s)\right] ds. \tag{4.25}$$

The idea of the proof is to use the Volterra properties of the operators B and T. One has

$$p(s)\left(B^j(p)g\right) - q(s)\left(B^j(q)g\right)(s,2x-s) = (p-q)\left(B^j(p)g\right) + q(s)\left[B^j(p) - B^j(q)\right]g. \tag{4.26}$$

From (4.22) it follows that

$$\sup_{0\leq x\leq 1}\left\{e^{-m(1-x)}\int_x^1 |p-q|\,|(B^j(p)g)(s,2x-s)|\,ds\right\}$$

$$\leq \frac{g_0\mathbf{p}^j}{(2j)!}\sup_{0\leq x\leq 1}\left\{e^{-m(1-x)}(1-x)^{2j}\int_x^1 e^{m(1-s)}|p-q|e^{-m(1-s)}\,ds\right\}$$

$$\leq \frac{g_0\mathbf{p}^j}{(2j)!}\sup_{0\leq x\leq 1}\left\{e^{-m(1-x)}(1-x)^{2j}\frac{e^{m(1-x)}-1}{m}\right\}\|p-q\|_m$$

$$\leq \frac{g_0\mathbf{p}^j}{(2j)!\,m}\|p-q\|_m. \tag{4.27}$$

Let us use the following representation:

$$B^j(p)-B^j(q) = [B(p)-B(q)]B^{j-1}(p)+B(q)[B(p)-B(q)]B^{j-2}(p)$$
$$+B^2(q)[B(p)-B(q)]B^{j-3}(p)+\cdots+B^{j-1}(q)[B(p)-B(q)]$$
$$= \sum_{i=0}^{j-1} B^i(q)[B(p)-B(q)]B^{j-i-1}(p). \tag{4.28}$$

Using (4.22) one gets

$$\left|(B^i(q)[B(p)-B(q)]B^{j-i-1}(p)g)(s,2x-s)\right|$$
$$\leq g_0\mathbf{p}^{j-i-1}\mathbf{q}^i\int_s^1 |p(\sigma)-q(\sigma)|\,ds\,\frac{(1-s)^{2j-2}2i(2i-1)}{(2j-2)!}. \tag{4.29}$$

Thus

$$\left\|\int_x^1 ds\,q(s)\left(B^i(q)[B(p)-B(q)]B^{j-i-1}(p)g\right)(s,2x-s)\right\|_m$$

$$\leq \frac{g_0\mathbf{p}^{j-i-1}\mathbf{q}^i 2i(2i-1)}{(2j)!}\sup_{0\leq x\leq 1}\left\{e^{-m(1-x)}\int_x^1 ds\int_s^1 |p(\sigma)-q(\sigma)|d\sigma(1-s)^{2j-2}\right\}$$

$$\leq \frac{g_0\mathbf{p}^{j-1}\mathbf{q}^i 2i(2i-1)}{(2j-2)!}\sup_{0\leq x\leq 1}\left\{e^{-m(1-x)}\int_x^1 d\sigma|p(\sigma)-q(\sigma)|\int_x^\sigma ds(1-s)^{2j-2}\right\}$$

$$\leq \frac{g_0\mathbf{p}^{j-1}\mathbf{q}^i 2i(2i-1)}{(2j-1)!}\sup_{0\leq x\leq 1}\left\{e^{-m(1-x)}\int_x^1 d\sigma|p(\sigma)-q(\sigma)|(1-x)^{2j-1}\right\}$$

$$\leq \frac{g_0\mathbf{p}^{j-1}\mathbf{q}^i 2i(2i-1)}{(2j-1)!\,m}\|p-q\|_m. \tag{4.30}$$

From (4.25)–(4.30) one gets

$$\|T(p)-T(q)\|_m \leq \frac{1}{m}\sum_{j=1}^\infty\left(\frac{g_0\mathbf{p}^j}{(2j)!}+\sum_{i=0}^{j-1}\frac{g_0\mathbf{p}^{j-1}\mathbf{q}^i 2i(2i-1)}{(2j-1)!}\right)\|p-q\|_m$$

$$\leq \frac{g_0\|p-q\|_m}{m}\left(\sum_{j=1}^\infty\frac{\mathbf{p}^j}{(2j)!}+4\sum_{j=1}^\infty\frac{\mathbf{p}^{j-1}}{(2j-1)!}\sum_{i=0}^{j-1}\mathbf{q}^i i^2\right)$$

$$\leq \gamma_m\|p-q\|_m, \tag{4.31}$$

where

$$\gamma_m := \frac{1}{m}\left[\sum_{j=1}^{\infty}\frac{\mathbf{p}^j}{(2j)!} + 4\sum_{j=1}^{\infty}\frac{\mathbf{p}^{j-1}}{(2j-1)!}\sum_{i=0}^{j-1}\mathbf{q}^i i^2\right]$$
$$\leq \frac{1}{m}\left[e^{\sqrt{\mathbf{p}}} + 4\sum_{j=1}^{\infty}\frac{\mathbf{p}^{j-1}(1+\mathbf{q})^{j-1}j^3}{(2j-1)!}\right] \leq \frac{c(\mathbf{p},\mathbf{q})}{m}. \tag{4.32}$$

Here

$$c(\mathbf{p},\mathbf{q}) := e^{\sqrt{\mathbf{p}}} + 4\sum_{j=1}^{\infty}\frac{\mathbf{p}^{j-1}(1+\mathbf{q})^{j-1}j^3}{(2j-1)!}, \quad \mathbf{p} := \sup_{0\leq x\leq 1}|p(x)|, \quad \mathbf{q} := \sup_{0\leq x\leq 1}|q(x)|. \tag{4.33}$$

Assume that

$$\mathbf{p} \leq R, \quad \mathbf{q} \leq R, \tag{4.34}$$

and choose m such that

$$\frac{c(\mathbf{p},\mathbf{q})}{m} < 1. \tag{4.35}$$

Then (4.12) holds for any $p(x)$ and $q(x)$ for which (4.34) holds. Lemma 4.4 is proved. $\square$

For exact data f and g in (4.7) and (4.8), one knows that the system of equations (4.7), (4.8) has a solution $q(x)$, so there exists a number R such that $\mathbf{q} < R$. From Lemma 4.4 it follows that this solution is unique in the whole space $L^{\infty}(0,1)$.

Define

$$u := \begin{pmatrix} q \\ K \end{pmatrix}, \quad h := \begin{pmatrix} f \\ g \end{pmatrix}, \quad W(u) := \begin{pmatrix} -2\int_x^1 q(s)K(s,2x-s)ds \\ \frac{1}{2}\int_{D_{xy}} q(s)K(s,t)ds\,dt \end{pmatrix} \tag{4.36}$$

and consider the system of equations (4.6)–(4.8) as an operator equation

$$u = V(u) := W(u) + h \tag{4.37}$$

in the space $L(x_0)$ of $L^{\infty}(x_0,1) \times L^{\infty}(x,y: 0 < |y| \leq x \leq 1, x_0 \leq x \leq 1)$ functions with the norm $||u|| = \sup_{x_0 \leq x \leq 1}|q(x)| + \sup_{\substack{x_0 \leq x \leq 1 \\ |y|\leq x}}|K(x,y)|$.

Lemma 4.5. *Let $||h|| \leq R$. Then the map V maps the ball $B(2R) := \{U: U \in L_{x_0}, ||U|| \leq 2R\}$ into itself and is a contraction map on $B(2R)$ provided that*

$$1 - x_0 < \frac{1}{12R}. \tag{4.38}$$

Proof: Note that if a and b are positive numbers and $a + b \leq 2R$, then $ab \leq R^2$. Therefore, using (4.36) and (4.38), one gets:

$$||V(u)|| \leq ||h|| + 2(1-x_0)R^2 + (1-x_0)R^2 \leq R + 3R^2(1-x_0) < 2R, \tag{4.39}$$

so V maps $B(2R)$ into itself if (4.38) holds. Moreover,

$$\begin{aligned}
||V(u) - V(\tilde{u})|| &= ||W(u) - W(\tilde{u})|| \\
&\leq \sup_{x_0 \leq x \leq 1} \Big\{ 2 \int_x^1 |q(s) - \tilde{q}(s)| |K(s, 2x-s)| ds + 2 \int_x^1 |\tilde{q}(s)| |K - \tilde{K}| ds \\
&\quad + \int_x^1 ds |q - \tilde{q}| \sup_{|t| \leq s} |K(s,t)| (s-x) \\
&\quad + \int_x^1 ds |\tilde{q}(s)| \sup_{|t| \leq s} |K(s,t) - \tilde{K}(s,t)| (s-x) \Big\} \\
&\leq ||u - \tilde{u}|| (1 - x_0) [8R + 4R] = 2R(1 - x_0) ||u - \tilde{u}||,
\end{aligned} \qquad (4.40)$$

V is a contraction on $B(2R)$ if (4.38) holds.

From Lemma 4.5 it follows that the system (4.6)–(4.8) is uniquely solvable by iterations for any $h \in B(R)$ and its solution belongs to $B(2R)$ if condition (4.38) holds. This means that the iterative process defines the unique solution $\{\tilde{q}(x), \tilde{K}(x,y)\}$ for $x_0 \leq x \leq 1$, $0 < x_0 < \frac{1}{12R}$, $|y| \leq x$.

Since the pair $\{q(x), K(x,y)\}$ solves the system (4.6)–(4.8), we conclude that $\tilde{q} = q(x)$, $\tilde{K} = K(x,y)$, where $q(x)$ and $K(x,y)$ is the pair underlying the exact data $\{f, g\}$, that is, $q(x)$ is the original potential which generated the data $\delta(K)$ and, therefore, the data $\{f, g\}$, and the transformation kernel $K(x,y) = K(x, y; q)$.

Define now $\{f_0, g_0\}$, where (cf (3.9) and (4.10))

$$f_0 := 2[K_y(x_0, 2x - x_0) + K_x(x_0, 2x - x_0)], \qquad (4.41)$$

$$g_0 := \frac{K(x_0, y + x - x_0) + K(x_0, y - x + x_0)}{2} - \frac{1}{2} \int_{y+x-x_0}^{y-x+x_0} K_s(x_0, t) dt. \qquad (4.42)$$

The crucial point now is this: since $\tilde{K}(x,y)$ and $\tilde{q}(x)$ which are obtained locally, in the interval $x_0 \leq x < 1$, are the unique $q(x)$ and $K(x,y)$ the data $\{f_0, g_0\}$ lies in the same ball $B(2R)$ as the original data $\{f, g\}$, since the data $\{f_0, g_0\}$ depend only on the unique underlying $q(x)$. Therefore, one can repeat the argument, construct the solution $\{\tilde{q}, \tilde{K}\}$, $\tilde{q} = q(x)$, $\tilde{K} = K(x,y)$, on the interval (x_1, x_0) by iterations, $x_0 - x_1 < \frac{1}{12R}$, with the same R. Thus, in a finite number of steps ($\leq 12R$) one can construct the solution $q(x)$ (and the corresponding $K(x,y)$) for any $0 \leq x \leq 1$.

We have proved

Theorem 4.6. *If the original data $\delta(k)$ corresponds to a $q(r) \in L^\infty(0,1)$, $q(r) = 0$ for $r > 1$, then one can uniquely reconstruct $q(r)$ from $\delta(k)$, known for $0 < k < \infty$, by*

1. *finding the data (3.6) according to Theorem 3.1;*

2. *solving the system (4.6)–(4.8), or, equivalently, equation (4.37), by iterations:*

$$u_{n+1} = W(u_n) + h, \quad u_0 = h. \qquad (4.43)$$

Iterative process (4.43) converges in L_{x_0} at least for $x_0 \leq x \leq 1$, $1 - \frac{1}{12R} \leq x_0 \leq 1$, where $R = ||h||$. This iterative process yields $q(x)$ and $K(x,y)$ for $x_0 \leq x \leq 1$, $|y| \leq x$, and if one uses the data $\{K(x_0,y), K_x(x_0,y)\}$ then one can repeat the iterative process on the interval of the same length $\left(< \frac{1}{12R}\right)$ starting from the point $x_0 = 1 - \frac{1}{12R}$ with the data $h_0 := \begin{pmatrix} f_0 \\ g_0 \end{pmatrix}$, where $\{f_0, g_0\}$ are given by (4.41) and (4.42), and obtain $q(x)$ and $K(x,y)$ on the interval $\left(x_0 - \frac{1}{12R}, x_0\right)$. In this way one constructs the solution $q(r)$ for all $0 < r < 1$ in a finite number of steps.

REFERENCES

1. T. Aktosun, "Bound states and inverse scattering for the Schrödinger equation in one dimension," *J. Math. Phys*, **35**, No. 12, 6231–6236 (1994).
2. F. Gakhov, *Boundary Value Problems*, Pergamon, New York (1966).
3. F. Gesztesy and B. Simon, "Inverse spectral analysis with partial information on the potential I.," *Helv. Phys. Acta*, **70**, 66–71 (1997).
4. B. Grebert and R. Weder, "Reconstructon of a potential on the line which is a priori known on the half-line," *SIAM J. Appl. Math*, **55**, No. N1, 242–254 (1995).
5. M. Klibanov and P. Sacks, "Use of partial knowledge of the potential in the phase problem of inverse scattering," *J. Comput. Phys*, **112**, 273–281 (1994).
6. _____, "Phaseless inverse scattering and the phase problem in optics," *J. Math. Phys.*, **33**, 3813–3821 (1992).
7. V. Marchenko, *Sturm–Liouville Operators and Applications*, Birkhäuser, Basel (1986).
8. S. Mikhlin, ed., *Linear Equations of Mathematical Physics (in Russian)*, Nauka, Moscow (1964).
9. R. Newton, "Remarks on scattering theory," *Phys. Rev.*, **101**, 1588–1596 (1956).
10. A. G. Ramm, *Multidimensional Inverse Scattering Problems*, Longman, New York (1992) [Expanded Russian edition MIR, Moscow (1994)].
11. _____, "Recovery of the potential from fixed energy scattering data," *Inverse Problems*, **4**, 877–886 (1988); **5**, 255 (1989).
12. _____, "Conditions under which the scattering matrix is analytic," *Sov. Physics Doklady*, **157**, 1073–1075 (1964).
13. _____, "Inverse scattering on half-line," *J. Math. Anal. App.*, **133**, No. 2, 543–572 (1988).
14. _____, *Scattering by Obstacles*, D. Reidel, Dordrecht, 1–442 (1986).
15. _____, "Theoretical and practical aspects of singularity and eigenmode expansion methods," *IEEE A-P*, **28**, No. N6, 897–901 (1980).
16. _____, "Extraction of resonances from transient fields," *IEEE A-P Trans.*, **33**, 223–226 (1985).
17. _____, "Calculation of resonances and their extraction from transient fields," *J. Math. Phys.*, **26**, No. 5, 1012–1020 (1985).
18. _____, "On the singularity and eigenmode expansion methods," *Electromagnetics*, **1**, No. N4, 385–394 (1981).
19. _____, "Mathematical foundations of the singularity and eigenmode expansion methods," *J. Math. Anal. Appl.*, **86**, 562–591 (1982).
20. A. G. Ramm and P. Stefanov, "Fixed-energy inverse scattering for non-compactly supported potentials," *Math. Comp. Modelling*, **18**, No. N1, 57–64 (1993).
21. W. Rundell and P. Sacks, "On the determination of potentials without bound state data," *Jour. of Comp. and Appl. Math.*, **55**, 325–347 (1994).
22. _____, "Reconstruction techniques for classical Sturm–Liouville problems," *Math. Comp.*, **58**, 161–183 (1992).

A TURNING POINT PROBLEM ARISING IN CONNECTION WITH A LIMITING ABSORPTION PRINCIPLE FOR SCHRÖDINGER OPERATORS WITH GENERALIZED VON NEUMANN–WIGNER POTENTIALS

Peter Rejto and Mario Taboada

School of Mathematics
University of Minnesota
Minneapolis, MN
Department of Mathematics
Old Dominion University
Norfolk, VA

1. INTRODUCTION

In connection with our study of "A Limiting Absorption Principle for Schrödinger Operators with Generalized Von Neumann–Wigner Potentials" [7, 8] we needed to solve a turning point problem for a family of ordinary differential equations. This family arose by separation of variables from the partial differential equation $(-\Delta + p_0 - \lambda)f = 0$, where Δ is the Laplacian, p_0 is a spherically symmetric potential function given by definition (2.1) and $\lambda > 0$ is the spectral parameter. Following the notation of [7, 8] we parametrize this family by $j = 0, 1, 2, \ldots$. It is a key feature of this family of differential equations that for each j the corresponding equation has at least one turning point and that as j tends to infinity so do the turning points. Hence our family of operators is similar to the one of Langer [6, 9]. It is different from the one of Langer inasmuch as we do not need a uniform expansion for the solution which is normalized near zero. All that we need is a lower estimate for the absolute values of the solutions at a family of points $\sigma^{(j)}$ which satisfy the inequality of conclusion (2.15) of Theorem Theorem 2.1. Hence these points are to the left of the turning points. On account of these weaker conclusions, we can allow to add an oscillatory potential to the one of Langer.

In Section 2, in definition (2.3) we describe our family of ordinary differential equations in more specific terms. Then in Theorem 2.1, which is our main theorem, we formulate our results. In this theorem we treat a family of solutions of the basic equations (2.3), f_m. Specifically we assume that at the points of the sequence σ_j of definition (2.8) the Cauchy data of the solutions f_m satisfy the asymptotic formulae (2.9) and (2.10). Then we conclude that there is a sequence $\sigma^{(j)}$ such that: they satisfy the inequality of conclusion (2.15) and at the points $\sigma^{(j)}$ the absolute values of the solutions satisfy the lower estimate of conclusion (2.16).

In Section 3 we start the proof of Theorem 2.1. For this purpose we define a family of intervals, $\mathcal{J}_{m,j}$, which contain the turning points. More specifically, they satisfy assumption (3.1). That is to say, the left endpoints are such that $\inf \mathcal{J}_{m,j} = \sigma_j$, where the sequence σ_j is given by definition (2.8) and the right endpoints, $\sup \mathcal{J}_{m,j}$, are such that conclusion (2.15) of Theorem 2.1 holds for them in place of $\sigma^{(j)}$. Then, we construct a family of approximate solutions $k_{m,j}$ over this family of intervals. First we follow an adaptation of the method of Langer [6, 9], which was done in [4],[Section 5] and in formulae (3.5) and (3.10) we define the adjusted Airy functions y_m and z_m. Note that such an adjustment is needed since the turning points of our family of differential equations tend to infinity while the Airy equation has a single turning point at zero. Then we define our approximations $k_{m,j}$ in formula (3.11) as linear combinations of the functions y_m and z_m. The coefficients in this linear combinations are determined from the requirement that at the left endpoints, $\inf \mathcal{J}_{m,j} = \sigma_j$, the approximate solutions $k_{m,j}$ and the solutions f_m of Theorem 2.1 should have the same Cauchy data. Having defined these approximate solutions, in Theorem 3.1 we show that conclusion (2.16) of Theorem 2.1 holds with the approximate solutions $k_{m,j}$ in place of the solutions f_m and with the sequence $\sup \mathcal{J}_{m,j}$ in place of the sequence of conclusion (2.15). We start the proof of Theorem 3.1 by showing that conclusion (3.13) holds for the first term of definition (3.11) in place of the approximate solutions of definition (3.11). This is the statement of Proposition 3.2. To prove Proposition 3.2 we asymptotically evaluate two of the factors of the conclusion. Specifically, in Lemma 3.3 we asymptotically evaluate the third factor of the conclusion of Proposition 3.2. The proof of Lemma 3.3 is modeled on the one of [4],[Proposition 5.3] and it makes essential use of [4],[Lemma 5.4]. In Lemma 3.4 we asymptotically evaluate the absolute value of the second factor of the conclusion of Proposition 3.2. There is no need to estimate the first factor of the conclusion of Proposition 3.2, since conclusion (2.16) of the main Theorem 2.1 is stated in terms of this factor. We complete the proof of Theorem 3.1 by showing that the second term of definition (3.11) is asymptotically zero compared to the first term. This is the statement of Proposition 3.5. We prove Proposition 3.5 with the help of Lemmas 3.6 and 3.7.

In Section 4 we complete the proof of Theorem 2.1. We see from Theorem 3.1 and from assumption (3.1) that it suffices to show that at the right endpoints, $\sup \mathcal{J}_{m,j}$, the approximate solutions $k_{m,j}$ and the solution f_m are asymptotically equal. This is the statement of Theorem 4.1. As a first step of the proof of Theorem 4.1 we introduce a weighted norm. As a second step of the proof of Theorem 4.1 we show that the weighted norm of the difference $f_m - k_{m,j}$ tends to zero as j tends to ∞. This is the statement of Proposition 4.2 We start the proof of Proposition 4.2 by formulating an upper bound for the conclusion of Proposition 4.2. This is done in Lemma 4.3. To prove Lemma 4.3 we formulate an integral equation for f_m in terms of the approximate solutions $k_{m,j}$. Then we show that the general Corollary A.2 of the Appendix applies to the integral equation of Lemma 4.3. To verify the assumptions of Corollary A.2 we make essential use of the estimates of [4, Section 6]. We continue the proof of Proposition 4.2 by showing that the second factor of the conclusion of Lemma 4.3 tends to zero as j tends to ∞. This is the statement of Lemma 4.4. We complete the proof of Proposition 4.2 by showing that the fourth factor of the conclusion of Lemma 4.3 remains bounded as j tends to ∞. This is the statement of Lemma 4.5.

In the Appendix first, we formulate the Love–Erdelyi–Olver bound for the norm of a

Volterra operator acting on a space of continuous functions with a weighted norm. Second, in Lemma A.1 we state a resolvent estimate. Third, we state Corollary A.2. This corollary is an adaptation of Lemma A.1 to an integral equation which arises from two approximate solutions and from an error potential of a second order ordinary differential equation. In other words, each of the two approximate solutions satisfies the same second order differential equations and the potential of this differential equation is the sum of an original potential and of an error potential.

For other classes of differential equations with a turning point we refer to the books of Wasov [12], and Olver [6] and Sibuya [11].

It is a pleasure to thank professors Devinatz and Sibuya for valuable conversations. At the same time it is a pleasure to thank the organizers of the special Waves and Scattering year at the IMA for making this work possible.

2. FORMULATION OF THE RESULT

To formulate Theorem 2.1, which is our main theorem, we need some notations. Let the oscillating potential p_0 be given by

$$p_0(\rho) = c\frac{\sin b\rho}{\rho^\beta}, \quad \rho \in \mathcal{R}^+ \tag{2.1}$$

and define the corresponding separated potentials by using definition (2.1) define

$$p_{0,j}(\rho) = j(j+1)\rho^{-2} + p_0(\rho), \quad \rho \in \mathcal{R}^+ = (0,\infty), j = 0,1,2,... \tag{2.2}$$

Then for a given parameter $\lambda \in \mathcal{R}^+$ our family of basic equations is given by

$$f''(\rho) + (\lambda - p_{0,j}(\rho))f(\rho) = 0, \tag{2.3}$$

Note that for $j \in \mathbb{Z}^+, j \neq 0$ and $\lambda \in \mathcal{R}^+$,

$$\lim_{\rho \to 0}(\lambda - p_{0,j}(\rho)) = -\infty \text{ and } \lim_{\rho \to \infty}(\lambda - p_{0,j}(\rho)) = \lambda.$$

Hence, there is at least one point where the coefficient of the basic equation (2.3) changes sign, and so does the qualitative behavior of the solution. Such points are called transition points, [9], [1] or turning points, [12]. Actually, we need only approximate turning points. To describe them, define

$$\nu = \nu_j = (j(j+1) + 1/4)^{1/2} = j + 1/2, j = 0,1,2,... \tag{2.4}$$

Then we see from definitions (2.4), (2.2) and (2.1) that such approximate turning points are given by

$$\tau = \lambda^{-1/2}\nu. \tag{2.5}$$

Note that for each ν such a point is a zero of the function $\lambda - \psi_\nu$, where

$$\psi_\nu(\rho) = \nu^2 \rho^{-2}. \tag{2.6}$$

We shall also need that p_0 and its derivative can be estimated in in terms of this function:

$$|p_0| = O(\nu^{-\beta})\psi_\nu^{\frac{\beta}{2}} \text{ and } |p_0'| = O(\nu^{-\beta})\psi_\nu^{\frac{\beta}{2}}, \text{ on } \mathcal{R}^+, \text{ uniformly in } \nu > 1. \quad (2.7)$$

In the following theorem, which is our main theorem, we study the solutions of the basic equation near these approximate turning points. More specifically, for each κ and $\lambda > 0$ define the sequence

$$\sigma_j = \lambda^{-1/2}\nu_j \cdot (1 - \nu_j^{\kappa-1}), j = 0,1,2,\ldots \quad (2.8)$$

We shall assume that the given solution of the basic equation (2.3) is such that

$$f_m(\sigma_j) \sim (4\pi)^{-1/2} \exp\left(\frac{i\pi}{4}\right)(\lambda - \psi_\nu(\sigma_j))^{-1/4} \cdot \exp\left[-i\int_{\lambda^{-1/2}\nu}^{\sigma_j}(\lambda - \psi_\nu(\sigma))^{1/2}d\sigma\right]. \quad (2.9)$$

and

$$f_m'(\sigma_j)f_m(\sigma_j)^{-1} \sim -i(\lambda - \psi_\nu(\sigma_j))^{1/2}, \text{ for } j \to \infty. \quad (2.10)$$

More specifically, in these asymptotic formulae, we take $0 \leq \kappa \leq 1$ and take the branch of the square root function to be the principal one:

$$\text{Re}(z^{1/2}) > 0 \text{ for } z \in \mathcal{C}\setminus(-\infty,0], \text{ so that } \text{Re}(-iz^{1/2}) > 0, \text{ for } \text{Im} z > 0, \text{Re} z < 0. \quad (2.11)$$

Then we define,

$$-i\int_{\lambda^{-1/2}\nu}^{\sigma_j}(\lambda - \psi_\nu(\sigma))^{1/2}d\sigma = -i\lim_{\epsilon \to +0}\int_{(\lambda+i\epsilon)^{-1/2}\nu}^{\sigma_j}((\lambda + i\epsilon - \psi_\nu(\sigma))^{1/2})d\sigma. \quad (2.12)$$

In the theorem that follows we formulate a lower estimate for the absolute value of f_m to the right of the approximate turning points.

Theorem 2.1. *Let the constant β in definition (2.1) satisfy*

$$\frac{2}{3} < \beta \leq 1, \quad (2.13)$$

let the given constant κ satisfy

$$\frac{1}{3} < \kappa < 2\beta - 1 \quad (2.14)$$

and let the corresponding sequence σ_j be given by definition (2.8). Next let f_m be a given solution of the basic equation (2.3) for which the asymptotic formulae (2.9) and (2.10) hold, where the function ψ_ν is given by definition (2.6). Then, there is a sequence $\sigma^{(j)}$ with the following property:

$$\lambda^{-1/2}\nu_j \cdot (1 + \nu_j^{\kappa-1}) \leq \sigma^{(j)} \leq \lambda^{-1/2}\nu_j \cdot (1 + 2\nu_j^{\kappa-1}), \text{ for } j = 0,1,2,\ldots, \lambda > 0, \quad (2.15)$$

and

$$\liminf_{j\to\infty}|\lambda - \psi_\nu(\sigma^{(j)})|^{1/4} \cdot |f_m(\sigma^{(j)})| > 0. \quad (2.16)$$

Before proving Theorem 2.1 we make some observations in addition to the ones of the Introduction.

First, we note that assumption (2.13) assures that the set of constants κ satisfying assumption (2.14) is not empty.

Second, we note that assumption (2.14) implies assumption (3.1) of [7] and conclusion (2.14) implies assumption (3.2) of [7]. It is an elementary fact that the solution f_ℓ of [4] satisfies assumptions (2.9) and (2.10). Hence conclusion (2.16) also holds for f_ℓ in place of f_m. That is to say, assumption (3.3) of [7] holds. Thus *the main Theorem 2.1 implies the* ADDITIONAL HYPOTHESIS of [7].

Third, we observe that although the main Theorem 2.1 covers the special case of $\beta = 1$, it can be replaced by the results of [4]. More specifically, we claim that applying [4],[Theorems 5.1 and 7.1] with $\delta = 2$ and $\delta = 3$ allows us to conclude that the ADDITIONAL HYPOTHESIS holds with $\beta = 1$ and $\kappa = 1$. Indeed, combining such an application with the Sturm Oscillation Theorem [13] we find a sequence $\sigma^{(j)}$ for which conclusion (2.15) holds with $\kappa = 1$ and is such that

$$\liminf_{j \to \infty} |f_\ell(\sigma^{(j)})| > 0. \tag{2.17}$$

We see from definition (2.6) that the function $\lambda - \psi_\nu$ is increasing and so, conclusion (2.15) yields

$$\lambda - \psi_\nu(\sigma^{(j)}) \geq \lambda - \psi_\ell(2\lambda^{-1/2}\nu_j)$$

Another application of definition (2.6) yields

$$\lambda - \psi_\ell(2\lambda^{-1/2}\nu_j) = \frac{1}{2}\lambda \tag{2.18}$$

and so,

$$\lambda - \psi_\nu(\sigma^{(j)}) \geq \frac{1}{2}\lambda. \tag{2.19}$$

Combining the lower estimates (2.19), (2.17) and using that by assumption $\lambda > 0$ we arrive at assumption (3.3) of [7]. Thus the solution f_ℓ of [4] satisfies the ADDITIONAL HYPOTHESIS of [7].

Finally, we note that assumptions (2.13) and (2.14) imply that the sequence of definition (2.8) is to the left of the turning points. This implication hinges on the inequality,

$$\sup_{0 < \rho < \sigma_j} (\lambda - p_{0,j})(\rho)) < 0, \tag{2.20}$$

which holds for large enough j.

3. CONSTRUCTION OF APPROXIMATE SOLUTIONS

In this section we start the proof of Theorem 2.1. For this purpose we construct approximate solutions for the solutions f_m of Theorem 2.1 over a family of intervals which contain the turning points and whose endpoints are sufficiently far from the turning points.

More specifically, let the sequence σ_j be given by definition (2.8) and we assume that the intervals $\mathcal{J}_{m,j}$ are such that

$$\inf \mathcal{J}_{m,j} = \sigma_j \text{ and } \lambda^{-1/2} \nu_j \cdot (1 + \nu_j^{\kappa-1}) \leq \sup \mathcal{J}_{m,j} \leq \lambda^{-1/2} \nu \cdot (1 + 2\nu_j^{\kappa-1}), j = 0, 1, 2, \ldots, \lambda \neq 0. \tag{3.1}$$

Note that the second inequality of assumption (3.1) with $\sigma^{(j)}$ in place of $\sup \mathcal{J}_{m,j}$ yields conclusion (2.15) of Theorem 2.1. Since the simplest differential equation which has a turning point is the Airy equation, we do this construction with the help of the Airy functions [E],[O],[W]. In fact, in [4, Section 5] we adapted this turning point analysis to our basic equation (2.3).

We start this construction by describing the approximations of [4]: First we define

$$g(z) = \int_1^z (1 - y^{-2})^{1/2} dy, \text{ for } z \in \mathcal{C}\setminus(-\infty, 1], \tag{3.2}$$

where the path of integration is taken along the line segment from 1 to z and as in (2.9), the square root in the integrand is the principal branch. Second, we define, at least formally, the multi-valued functions,

$$\zeta(z) = \left(\frac{3}{2} g(z)\right)^{2/3}, \text{ for } z \in \mathcal{C}\setminus(-\infty, 1]. \tag{3.3}$$

Third, with the help of definitions (3.3), (2.5) and (2.4) we define

$$\varphi(z) = \nu^{2/3} \zeta(\tau^{-1} z), \text{ for } z \notin \tau(-\infty, 1]. \tag{3.4}$$

We continue this construction with additional definitions. First we define the approximate solution,

$$y_m(\rho) = \varphi'(\rho)^{-1/2} \text{Ai}(-\varphi(\rho)), \tag{3.5}$$

where Ai is the Airy function in the usual notation [1], [6], [12]. In other words, it satisfies the differential equation,

$$\text{Ai}''(z) - z \text{Ai}(z) = 0 \text{ for } z \in \mathcal{C} \tag{3.6}$$

and the asymptotic formula,

$$\text{Ai}(z) \sim (4\pi)^{-1/2} z^{-1/4} \exp\left(-\frac{2}{3} z^{3/2}\right), \tag{3.7}$$

$$\text{for } z \in \mathcal{S}(-\pi, \pi), \ |z| \to \infty. \tag{3.8}$$

Here, as usual, for any given ϑ and ϑ' in $\mathcal{R}$ we define the open sector

$$\mathcal{S}(\vartheta, \vartheta') = \{z : \vartheta < \arg z < \vartheta', z \neq 0\}. \tag{3.9}$$

and the fractional powers are analytic continuations of the principal ones in the simply connected sector $\mathcal{S}(-\pi, \pi)$. Second, we define another approximate solution,

$$z_m(\rho) = \varphi'(\rho)^{-1/2} \text{Ai}\left(-\exp\left(\frac{2\pi i}{3}\right) \varphi(\rho)\right). \tag{3.10}$$

Third, we define

$$k_{m,j}(\rho) = \alpha_{m,j} y_m(\rho) + \beta_{m,j} z_m(\rho), \qquad \rho \in \mathcal{J}_{m,j}. \tag{3.11}$$

where these connection coefficients are determined from the requirement that at the left endpoint $\inf \mathcal{J}_{m,j}$, the exact solution f_m and the approximation $k_{m,j}$ should have the same Cauchy data. This leads to the system of equations,

$$\begin{aligned}\alpha_{m,j} y_m(\inf \mathcal{J}_{m,j}) + \beta_{m,j} z_m(\inf \mathcal{J}_{m,j}) &= f_m(\inf \mathcal{J}_{m,j}) \\ \alpha_{m,j} y'_m(\inf \mathcal{J}_{m,j}) + \beta_{m,j} z'_m(\inf \mathcal{J}_{m,j}) &= f'_m(\inf \mathcal{J}_{m,j}).\end{aligned} \tag{3.12}$$

In the following theorem we show that conclusion (2.16) of Theorem 2.1 holds for the approximate solutions $k_{m,j}$ in place of f_m.

Theorem 3.1. *Let the assumptions and notations of Theorem 2.1 hold and let the approximate solutions $k_{m,j}$ be given by definitions (3.11) and (3.12). Then, there is a family of intervals satisfying assumption (3.1) such that*

$$\liminf_{j \to \infty} |\lambda - \psi_\nu(\sup \mathcal{J}_{m,j})|^{1/4} \cdot |k_{m,j}(\sup \mathcal{J}_{m,j})| > 0. \tag{3.13}$$

We start the proof of Theorem 3.1 by showing that conclusion (3.13) holds for the first terms of definition (3.11) in place of the approximate solutions of definition (3.11). This is the statement of the proposition that follows.

Proposition 3.2. *Let the assumptions of Theorem 3.1 hold. Then, there is a family of intervals satisfying assumption (3.1) such that*

$$\liminf_{j \to \infty} |\lambda - \psi_\nu(\sup \mathcal{J}_{m,j})|^{1/4} \cdot |\alpha_{m,j}| \cdot |y_m(\sup \mathcal{J}_{m,j})| > 0. \tag{3.14}$$

We start the proof of Proposition 3.2 by asymptotically evaluating $y_m(\sup \mathcal{J}_{m,j})$ for any family of intervals satisfying assumption (3.1). This is done in the lemma that follows. In it, for brevity, we define

$$h_\tau^\pm(\rho) = \exp(\pm i \nu g(\rho \tau^{-1})). \tag{3.15}$$

Lemma 3.3. *Let the assumptions of Theorem 3.1 hold and let the functions $h_\tau^\pm(\rho)$ be given by definitions (3.15). Then, for each family of intervals satisfying assumption (3.1),*

$$y_m(\sup \mathcal{J}_{m,j}) \sim (4\pi)^{-1/2} \exp\left(\frac{i\pi}{4}\right) (\lambda - \psi_\nu(\sup \mathcal{J}_{m,j}))^{-1/4} \{h_\tau^- - i h_\tau^+\}(\sup \mathcal{J}_{m,j}), \text{ for } j \to \infty. \tag{3.16}$$

We start the proof of Lemma 3.3 by showing that

$$\lim_{j \to \infty} |\varphi(\sup \mathcal{J}_{m,j})| = \infty. \tag{3.17}$$

To prove relation (3.17) we note that according to definition (3.2) the function g is increasing on the interval $(1, \infty)$ and that definition (2.5) and assumption (3.1) yield the inequality

$$\tau^{-1} \sup \mathcal{J}_{m,j} \geq 1 + \nu_j^{\kappa-1}.$$

Hence

$$g(\sup \mathcal{I}_{m,j}\tau^{-1}) \geq \int_1^{1+\nu_j^{\kappa-1}} (1-y^{-2})^{1/2} dy.$$

It is an elementary fact that

$$\int_1^{1+\nu_j^{\kappa-1}} (1-y^{-2})^{1/2} dy > \frac{1}{3}\nu_j^{\frac{3}{2}(\kappa-1)}, \text{ for } \nu_j^{(\kappa-1)} \leq 1.$$

Combining the previous two inequalities with definitions (3.4), (3.3) and (3.2) we find

$$\varphi(\sup \mathcal{I}_{m,j}) \geq \left(\frac{1}{2}\right)^{2/3} \cdot \nu_j^{\kappa-\frac{1}{3}}, \text{ for } \nu_j^{(\kappa-1)} \leq 1.$$

Combining the previous inequality, in turn, with the first half of assumption (2.14) of Theorem 2.1 and with definition (2.4) we find relation (3.17).

We continue the proof of Lemma 3.3 by showing that

$$-\varphi(\sup \mathcal{I}_{m,j}) \in \overline{\mathcal{S}(-\pi, -\pi/2)}, \tag{3.18}$$

where $\overline{\mathcal{S}(-\pi, -\pi/2)}$ denotes the closure of the open sector of definition (3.9). Indeed, [4], [conclusion (5.41) of Lemma 5.4] implies

$$-\varphi((\tau, \infty)) \subset \overline{\mathcal{S}(-\pi, -\pi/2)}. \tag{3.19}$$

We see from assumption (3.1) and from definition (2.5) that the point $\sup \mathcal{I}_{m,j}$ is in the the interval (τ, ∞) and so, the previous relation yields relation (3.18).

We complete the proof of Lemma 3.3 by showing that relations (3.18) and (3.17) imply it. To see this note that these two relations together allow us to apply a version of the asymptotic formula (3.7) to this sequence. This version, [1],[Formula (12) of Sec. 10.4], says that

$$\text{Ai}(z) \sim (4\pi)^{-1/2} z^{-1/4} \left\{ \exp\left(-\frac{2}{3}z^{3/2}\right) - i\exp\left(\frac{2}{3}z^{3/2}\right) \right\}, \tag{3.20}$$

$$\text{for } z \in \mathcal{S}\left(-\frac{5\pi}{3}, -\frac{\pi}{3}\right), \quad |z| \to \infty. \tag{3.21}$$

Here, the fractional powers are defined with the help of $-\frac{5\pi}{3} < \arg z < -\frac{\pi}{3}$. Furthermore, the convergence is uniform in any closed subsector of the sector of assumption (3.21). Application of the asymptotic formula (3.20) to the sequence $z = -\varphi(\sup \mathcal{I}_{m,j})$ yields,

$$\text{Ai}(-\varphi(\sup \mathcal{I}_{m,j})) \sim (4\pi)^{-1/2} [-\varphi(\sup \mathcal{I}_{m,j})]^{-1/4} \cdot$$
$$\cdot \left\{ \exp\left(-\frac{2}{3}[-\varphi(\sup \mathcal{I}_{m,j})]^{3/2}\right) - i\exp\left(\frac{2}{3}[-\varphi(\sup \mathcal{I}_{m,j})]^{3/2}\right) \right\}, \text{ for } j \to \infty. \tag{3.22}$$

According to [3, conclusion (5.6) of Lemma 5.1],

$$[-\varphi(\rho)]^{3/2} = \frac{3}{2} i\nu g(\rho\tau^{-1}), \text{ for } \rho \in \mathcal{R}^+ \tag{3.23}$$

A Turning Point Problem

and so, we see from definitions (3.15) that

$$\exp\left(\pm\frac{2}{3}[-\varphi(\sup \mathfrak{I}_{m,j})]^{3/2}\right) = h_\tau^\pm(\sup \mathfrak{I}_{m,j}). \tag{3.24}$$

Inserting formulae (3.24) and the asymptotic formula (3.22) into the definition (3.5) we obtain

$$y_m(\sup \mathfrak{I}_{m,j}) \sim (4\pi)^{-1/2} \varphi'(\sup \mathfrak{I}_{m,j})^{-1/2} [-\varphi(\sup \mathfrak{I}_{m,j})]^{-1/4} \{h_\tau^- - ih_\tau^+\}(\sup \mathfrak{I}_{m,j}), \text{ for } j \to \infty. \tag{3.25}$$

According to [3],[Lemma 5.2],

$$[\varphi'(\rho)^2 \varphi(\rho)]^{1/4} = \exp\left(\frac{\pi i}{4}\right) [\varphi'(\rho)]^{1/2} [-\varphi(\rho)]^{1/4}, \quad \rho \in \mathcal{R}^+, \tag{3.26}$$

where the fractional powers are taken as the principal branches. It is an elementary consequence of definitions (3.4), (2.6) and (2.5) that

$$\varphi'(\rho)^2 \varphi(\rho) = \lambda - \psi_\nu(\rho). \tag{3.27}$$

Inserting formula (3.26) and the differential equation (3.27) into the asymptotic formula (3.25) and using definition (2.6) we arrive at conclusion (3.16). This completes the proof of Lemma 3.3.

We continue the proof of Proposition 3.2 by choosing a sequence $\sup \mathfrak{I}_{m,j}$ for which conclusion (2.16) of Theorem 2.1 holds and is such that

$$|\{h_\tau^- - ih_\tau^+\}(\sup \mathfrak{I}_{m,j})| = 2. \tag{3.28}$$

Since,

$$\left|\exp\left(-i\frac{\pi}{4}\right) - i\exp\left(i\frac{\pi}{4}\right)\right| = 2,$$

we see from definition (3.15) that it suffices to show that there is a j_0 such that to each $j > j_0$ there is a point ρ_j with the following property:

$$\lambda^{-1/2} \nu_j \cdot (1 + \nu_j^{\kappa-1}) \leq \rho_j \leq \lambda^{-1/2} \nu \cdot (1 + 2\nu_j^{\kappa-1}), \tag{3.29}$$

and

$$\nu_j g(\rho_j \tau^{-1}) \equiv \frac{\pi}{4}, \mod 2\pi. \tag{3.30}$$

Indeed, definitions (3.2) and (2.5) yield,

$$\nu_j g(\lambda^{-1/2} \nu_j \cdot (1 + 2\nu_j^{\kappa-1})\tau^{-1}) - \nu_j g(\lambda^{-1/2} \nu_j \cdot (1 + \nu_j^{\kappa-1})\tau^{-1}) = \nu_j \int_{1+\nu_j^{\kappa-1}}^{1+2\nu_j^{\kappa-1}} (1 - y^{-2})^{1/2} dy. \tag{3.31}$$

Now it is an elementary fact that

$$\int_{1+\nu_j^{\kappa-1}}^{1+2\nu_j^{\kappa-1}} (1 - y^{-2})^{1/2} dy > \frac{1}{3} \int_{1+\nu_j^{\kappa-1}}^{1+2\nu_j^{\kappa-1}} (1 - y)^{1/2} dy,$$

and that the integral on the right of the previous inequality is of the order of $v_j^{\frac{3}{2}(\kappa-1)}$. We see from the first half of assumption (2.14) of Theorem 2.1 that

$$\lim_{j\to\infty} v_j \cdot v_j^{\frac{3}{2}(\kappa-1)} = \infty.$$

Inserting these two elementary facts into formula (3.31) we find

$$\lim_{j\to\infty} \left[v_j g(\lambda^{-1/2} v_j \cdot (1+2v_j^{\kappa-1})\tau^{-1}) - v_j g(\lambda^{-1/2} v_j \cdot (1+v_j^{\kappa-1})\tau^{-1}) \right] = \infty.$$

Hence, for large enough j the increment of the function $v_j g(\rho_j \tau^{-1})$ of the variable ρ over the interval of relation (3.29) is at least 2π. Then relations (3.30) and (3.29) follow by continuity. Finally, inequality (3.29) allows us to define

$$\sup \mathcal{J}_{m,j} = \rho_j, \tag{3.32}$$

so that the second half of assumption (3.1) should hold. Thus formula (3.28) follows.

We complete the proof of Proposition 3.2 by asymptotically evaluating the absolute values of the constants $\alpha_{m,j}$. This is done in the lemma that follows.

Lemma 3.4. *Let the assumptions of Theorem 3.1 hold and let the constants $\alpha_{m,j}$ be given by the system of linear equations (3.12). Then*

$$\lim_{j\to\infty} |\alpha_{m,j}| = (4\pi)^{1/2}. \tag{3.33}$$

As a first step of the proof of Lemma 3.4 we show the asymptotic formula:

$$\alpha_{m,j} \sim \pi^{1/2} \exp\left(\frac{5\pi i}{12}\right) \cdot$$

$$\cdot \exp\left[-i \int_{\lambda^{-1/2}v}^{\sigma_j} (\lambda - \psi_\nu(\sigma))^{1/2} d\sigma\right] \left[(\lambda - \psi_\nu)^{-1/4} z'_m + i(\lambda - \psi_\nu)^{1/4} z_m \right] (\sigma_j),$$

$$\text{for } j \to \infty. \tag{3.34}$$

Indeed, the determinant of the system (3.12) is given by the Wronskian $W(y_m, z_m)$. It is an elementary consequence of definitions (3.10) and (3.5) that

$$W(y_m, z_m) = -W\left(\text{Ai}(z), \text{Ai}\left(\exp\left(\frac{2\pi i}{3}\right)z\right)\right).$$

Using the well known value of Wronskian on the right [6], [p.48; Ex.8.1] we find

$$W(y_m, z_m) = -\frac{1}{2\pi} \exp\left(\frac{-\pi i}{6}\right). \tag{3.35}$$

Hence application of Cramer's rule to the system (3.12) yields

$$\alpha_{m,j} = 2\pi \exp\left(\frac{\pi i}{6}\right) [f_m z'_m - f'_m z_m](\sigma_j). \tag{3.36}$$

A Turning Point Problem

Now inserting assumptions (2.10) and (2.9) of Theorem 2.1 into formula (3.36) we obtain the asymptotic formula (3.34).

As a second and main step of the proof of Lemma 3.4 we evaluate the second factor of the asymptotic formula (3.34). More specifically, we claim that for the sequence of definition (2.8),

$$\exp\left(-i\int_{\lambda^{-1/2}\nu}^{\sigma_j}(\lambda-\psi_\nu(\sigma))^{1/2}d\sigma\right)\left[i(\lambda-\psi_\nu)^{1/4}z_m+(\lambda-\psi_\nu)^{-1/4}z'_m\right](\sigma_j) \sim 2i, \text{ for } j \to \infty. \quad (3.37)$$

To prove the asymptotic formula (3.37) we asymptotically evaluate $z_m(\sigma_j)$ and $z'_m(\sigma_j)$:

$$z_m(\sigma_j) \sim (4\pi)^{-1/2}\exp\left(\frac{\pi i}{12}\right)\cdot(\lambda-\psi_\nu(\sigma_j))^{-1/4}\cdot\exp\left[i\int_{\lambda^{-1/2}\nu}^{\sigma_j}(\lambda-\psi_\nu(\sigma))^{1/2}d\sigma\right], \quad (3.38)$$

and

$$z'_m(\sigma_j) \sim i(\lambda-\psi_\nu(\sigma_j))^{1/2}\cdot z_m(\sigma_j), \text{ for } j \to \infty. \quad (3.39)$$

Similarly to the proof of Lemma 3.3, we start the proof of the asymptotic formula (3.38) by showing that relation (3.17) holds for the sequence of definition (2.8) in place of that sequence:

$$\lim_{j\to\infty}|\varphi(\sigma_j)| = \infty. \quad (3.40)$$

Indeed, definitions (2.8) and (2.5) yield

$$\sigma_j\tau^{-1} = (1-\nu_j^{\kappa-1})$$

Similarly to the proof of relation (3.17), we see that the previous relation implies relation (3.40).

Similarly to the proof of Lemma 3.3, we continue the proof of the asymptotic formula (3.38) by showing that

$$-\exp\left(\frac{2\pi i}{3}\right)\varphi(\sigma_j) \in \overline{S}\left(-\frac{\pi}{3},\frac{2\pi}{3}\right), \quad \text{for } j \in \mathbb{Z}^+. \quad (3.41)$$

Indeed, [4], [relation (5.48)] implies

$$\varphi((0,\infty)) \subset \overline{S}(0,\pi).$$

and so, relation (3.41) follows. Since the closed sector of relation (3.41) is contained in the open sector of relation (3.8), relation (3.40) allows us to apply the asymptotic formula (3.7) to the sequence $z = -\exp\left(\frac{2\pi i}{3}\right)\varphi(\sigma_j)$. This yields,

$$\text{Ai}\left(-\exp\left(\frac{2\pi i}{3}\right)\varphi(\sigma_j)\right) \sim (4\pi)^{-1/2}\left[-\exp\left(\frac{2\pi i}{3}\right)\varphi(\sigma_j)\right]^{-1/4}$$

$$\times \exp\left(-\frac{2}{3}\left[-\exp\left(\frac{2\pi i}{3}\right)\varphi(\sigma_j)\right]^{3/2}\right). \quad (3.42)$$

Similarly to formula (3.23) we see that

$$\left[-\exp\left(\frac{2\pi i}{3}\right)\varphi(\sigma_j)\right]^{3/2} = -i\frac{3}{2}vg(\sigma_j\tau^{-1}), \text{ for } \rho \in \mathcal{R}^+.$$

Setting $z = \sigma_j\tau^{-1}$ in definition (3.2), making the change of variable $\sigma(y) = y\tau$ in the resulting integral and using definitions (2.6) and (2.5) we see that

$$vg(\sigma_j\tau^{-1}) = \int_{\lambda^{-1/2}\nu}^{\sigma_j}(\lambda - \psi_\nu(\sigma))^{1/2}d\sigma. \tag{3.43}$$

Inserting the previous two formulae into the asymptotic formula (3.42) we obtain,

$$\text{Ai}\left(-\exp\left(\frac{2\pi i}{3}\right)\varphi(\sigma_j)\right) \sim (4\pi)^{-1/2}\left[-\exp\left(\frac{2\pi i}{3}\right)\varphi(\sigma_j)\right]^{-1/4}$$
$$\times \exp\left[i\int_{\lambda^{-1/2}\nu}^{\sigma_j}(\lambda - \psi_\nu(\sigma))^{1/2}d\sigma\right]. \tag{3.44}$$

Multiplying the asymptotic formula (3.44) by $\varphi'(\sigma_j)^{-1/2}$, inserting formula (3.26) and the differential equation (3.27) into the result and using definition (3.10), we arrive at the asymptotic formula (3.38). Since the asymptotic formula (3.7) holds uniformly in z in any open sector whose closure is contained in the open sector $\mathcal{S}(-\pi, \pi)$, the asymptotic formula (3.38) holds uniformly in λ in any compact subset of the open positive axis.

We start the proof of the asymptotic formula (3.39) by differentiating definition (3.10), which yields,

$$z'_m(\rho) = -\exp\left(\frac{2\pi i}{3}\right)\varphi'(\rho)^{1/2}\text{Ai}'\left(-\exp\left(\frac{2\pi i}{3}\right)\varphi(\rho)\right)$$
$$-\frac{1}{2}\varphi''(\rho)\varphi'(\rho)^{-3/2}\text{Ai}\left(-\exp\left(\frac{2\pi i}{3}\right)\varphi(\rho)\right). \tag{3.45}$$

We continue the proof of the asymptotic formula (3.39) by showing that it holds for the first term of formula (3.45) in place of $z'_m(\rho)$,

$$-\exp\left(\frac{2\pi i}{3}\right)\varphi'(\sigma_j)^{1/2}\text{Ai}'\left(-\exp\left(\frac{2\pi i}{3}\right)\varphi(\sigma_j)\right) \sim i(\lambda - \psi_\nu(\sigma_j))^{1/2} \cdot z_m(\sigma_j), \text{ for } j \to \infty. \tag{3.46}$$

To prove the asymptotic formula (3.46) we need the well known fact that the asymptotic formula (3.7) can be differentiated. More specifically, we need that assumption (3.8) implies,

$$\text{Ai}'(z) \sim -(4\pi)^{-1/2}z^{1/4}\exp\left(-\frac{2}{3}z^{3/2}\right). \tag{3.47}$$

Now replacing the asymptotic formula (3.7) by the asymptotic formula (3.47) in the proof of the asymptotic formula (3.44) we obtain

$$\text{Ai}'\left(-\exp\left(\frac{2\pi i}{3}\right)\varphi(\sigma_j)\right) \sim -(4\pi)^{-1/2}\left[-\exp\left(\frac{2\pi i}{3}\right)\varphi(\sigma_j)\right]^{1/4}$$
$$\times \exp\left[i\int_{\lambda^{-1/2}\nu}^{\sigma_j}(\lambda - \psi_\nu(\sigma))^{1/2}d\sigma\right]. \tag{3.48}$$

A Turning Point Problem

Multiplying the asymptotic formula (3.48) by $\varphi'(\sigma_j)^{1/2}$, inserting the differential equation (3.27) and formula (3.26) into the result and using the asymptotic formula (3.38), we arrive at the asymptotic formula (3.46).

We complete the proof of the asymptotic formula (3.39) by showing that the second term of formula (3.45) is asymptotically zero compared to the first one. We see from the asymptotic formulae (3.44) and (3.48) that this is implied by

$$\lim_{j\to\infty} |\varphi''(\varphi')^{-2}\varphi^{-1/2}|(\sigma_j) = 0. \tag{3.49}$$

We see from the differential equation (3.27) that

$$\varphi''(\varphi')^{-2}\varphi^{-1/2} = \varphi''\varphi^{1/2}(\lambda - \psi_\nu)^{-1}$$

and that

$$2\varphi''\varphi^{1/2}(\lambda - \psi_\nu)^{1/2} + (\varphi')^3 = -\psi_\nu'$$

The previous two formulae together show that

$$\varphi''(\varphi')^{-2}\varphi^{-1/2} = -\frac{1}{2}(\lambda - \psi_\nu)^{-3/2}\psi_\nu' \frac{1}{2}(\lambda - \psi_\nu)^{-3/2}(\varphi')^3. \tag{3.50}$$

Combining formula (3.50) with definition (2.8) and with the first half of assumption (2.14) of Theorem Theorem 2.1 we obtain relation (3.49). Thus the asymptotic formula (3.39) follows. Finally, combining the asymptotic formulae (3.38) and (3.39) we obtain the asymptotic formula (3.37).

As a last and final step of the proof of Lemma 3.4 we insert the asymptotic formula (3.37) into the asymptotic formula (3.34). Then we arrive at conclusion (3.33).

We complete the proof of Theorem 3.1 by showing that at the right endpoints, the approximate solutions $k_{m,j}$ are asymptotic to the functions $\alpha_{m,j} y_m(\sup \mathcal{J}_{m,j})$ of Proposition 3.2. This is the statement of the proposition that follows.

Proposition 3.5. *Let the assumptions and notations of Theorem 3.1 hold. Then,*

$$k_{m,j}(\sup \mathcal{J}_{m,j}) \sim \alpha_{m,j} y_m(\sup \mathcal{J}_{m,j}), \text{ for } j \to \infty. \tag{3.51}$$

To prove Proposition 3.5 we note that according to definition (3.11) and conclusion (3.33) of Lemma 3.4 it is equivalent to,

$$\lim_{j\to\infty} \beta_{m,j} \cdot \frac{z_m(\sup \mathcal{J}_{m,j})}{y_m(\sup \mathcal{J}_{m,j})} = 0. \tag{3.52}$$

We start the proof of relation (3.52) by showing that it holds for the first factor of this product. This is the statement of the lemma that follows.

Lemma 3.6. *Let the assumptions and notations of Theorem 3.1 hold. Then,*

$$\lim_{j\to\infty} \beta_{m,j} = 0. \tag{3.53}$$

We start the proof of Lemma 3.6 by noting that similarly to the way that formula (3.36) and assumptions (2.10) and (2.9) of Theorem 2.1 imply the asymptotic formula (3.34) they also imply the asymptotic formula,

$$\beta_{m,j} \sim -\frac{\pi^{1/2}}{6} \exp\left(\frac{\pi i}{4}\right) \cdot$$
$$\cdot \exp\left[-i \int_{\lambda^{-1/2}\nu}^{\sigma_j} (\lambda - \psi_\nu(\sigma))^{1/2} d\sigma\right] \cdot \left[i y_m (\lambda - \psi_\nu)^{1/4} + y'_m (\lambda - \psi_\nu)^{-1/4}\right](\sigma_j), \quad \text{for } j \to \infty. \quad (3.54)$$

We continue the proof of Lemma 3.6 by asymptotically evaluating the third factor of the asymptotic formula (3.54). More specifically, we claim that for the sequence of definition (2.8),

$$\left[i y_m (\lambda - \psi_\nu)^{1/4} + y'_m (\lambda - \psi_\nu)^{-1/4}\right](\sigma_j) \sim o(1) \cdot \exp\left[-i \int_{\lambda^{-1/2}\nu}^{\sigma_j} (\lambda - \psi_\nu(\sigma))^{1/2} d\sigma\right],$$
$$\text{for } j \to \infty. \quad (3.55)$$

To prove the asymptotic formula (3.55) we asymptotically evaluate $y_m(\sigma_j)$ and $y'_m(\sigma_j)$:

$$y_m(\sigma_j) \sim (4\pi)^{-1/2} \exp(i\frac{\pi}{4}) \cdot (\lambda - \psi_\nu(\sigma_j))^{-1/4} \cdot \exp\left[-i \int_{\lambda^{-1/2}\nu}^{\sigma_j} (\lambda - \psi_\nu(\sigma))^{1/2} d\sigma\right], \quad (3.56)$$

and

$$y'_m(\sigma_j) \sim -i(\lambda - \psi_\nu(\sigma_j))^{1/2} \cdot y_m(\sigma_j), \quad \text{for } j \to \infty. \quad (3.57)$$

To prove the asymptotic formula (3.56) we note that relations (3.41) and (3.40) allow us to apply the asymptotic formula (3.7) to the family of complex numbers $z = -\varphi(\sigma_j)$. Then, combining this application with definition (3.5) we find,

$$y_m(\sigma_j) \sim (4\pi)^{-1/2} \varphi'(\sigma_j)^{-1/2} [-\varphi(\sigma_j)]^{-1/4} \exp\left(-\frac{2}{3}[-\varphi(\sigma_j)]^{3/2}\right), \quad \text{for } j \to \infty. \quad (3.58)$$

Similarly to the way that the asymptotic formula (3.42) implies the asymptotic formula (3.38), we see that the asymptotic formula (3.58) implies the asymptotic formula (3.56). Specifically, inserting the differential equation (3.27) and formulae (3.23), (3.43) into the asymptotic formula (3.58) we obtain the asymptotic formula (3.56).

To prove the asymptotic formula (3.57) we differentiate definition (3.5), which yields,

$$y'_m(\rho) = -\varphi'(\rho)^{1/2} \text{Ai}'(-\varphi(\rho)) - \frac{1}{2}\varphi''(\rho)\varphi'(\rho)^{-3/2} \text{Ai}(-\varphi(\rho)). \quad (3.59)$$

Next we claim that the asymptotic formula (3.57) holds for the first term of formula (3.59) in place of y'_m,

$$-\varphi'(\sigma_j)^{1/2} \text{Ai}'(-\varphi(\sigma_j)) \sim -i(\lambda - \psi_\nu(\sigma_j))^{1/2} \cdot y_m(\sigma_j) \quad \text{for } j \to \infty. \quad (3.60)$$

Indeed, similarly to the way that the asymptotic formula (3.47) and relations (3.41) and (3.40) imply the asymptotic formula (3.48) we see that they also imply

$$\varphi'(\sigma_j)^{1/2} \text{Ai}'(-\varphi(\sigma_j)) \sim -(4\pi)^{-1/2} \varphi'(\sigma_j)^{1/2} [-\varphi(\sigma_j)]^{1/4} \cdot \exp\left[-i \int_{\lambda^{-1/2}\nu}^{\sigma_j} (\lambda - \psi_\nu(\sigma))^{1/2} d\sigma\right]$$
$$(3.61)$$

A Turning Point Problem

Combining the asymptotic formulae (3.61), (3.56) and using the differential equation (3.27) and formula (3.26) we find the asymptotic formula (3.60). We complete the proof of the asymptotic formula (3.57) by noting that relation (3.49) implies that second term of formula (3.59) is asymptotically zero compared to the first one. Combining the asymptotic formulae (3.57) and (3.56), in turn, we obtain the asymptotic formula (3.55).

We complete the proof of Lemma 3.6 by showing that

$$\left| \exp\left[-2i \int_{\lambda^{-1/2}\nu}^{\sigma_j} (\lambda - \psi_\nu(\sigma))^{1/2} d\sigma \right] \right| \leq 1. \quad (3.62)$$

Indeed, in definition (2.11) we have chosen the branch of the square root function to be the principal one and so,

$$\text{Re}(-2i(\lambda - \psi_\nu(\sigma))^{1/2}) = \lim_{\epsilon \to +0} \text{Re}(-2i(\lambda + i\epsilon - \psi_\nu(\sigma))^{1/2}) \geq 0, \quad \text{for } \sigma \in (0, \tau). \quad (3.63)$$

In other words, the real part of the integrand of estimate (3.62) is positive. According to definition (2.8) the upper limit is less than the lower limit and so, estimate (3.62) follows.

Finally, inserting estimate (3.62) and the asymptotic formula (3.55) into the asymptotic formula (3.54) we arrive at conclusion (3.53) and this completes the proof of Lemma 3.6.

We complete the proof of relation (3.52) by showing that the absolute value of second factor of this product is bounded in j. This is the statement of the lemma that follows.

Lemma 3.7. *Let the assumptions of Theorem 3.1 hold and let the approximate solutions z_m and y_m be given by definitions (3.10) and (3.2), respectively. Then there is a family of intervals satisfying assumption (3.1) such that*

$$\limsup_{j \to \infty} \left| \frac{z_m(\sup \mathcal{J}_{m,j})}{y_m(\sup \mathcal{J}_{m,j})} \right| < \infty. \quad (3.64)$$

We start the proof of Lemma 3.7 by recalling relation (3.18) which shows that

$$-\exp\left(\frac{2\pi i}{3}\right) \varphi(\sup \mathcal{J}_{m,j}) \in \mathcal{S}\left(-\frac{\pi}{3}, \frac{\pi}{6} + \varepsilon\right). \quad (3.65)$$

Relations (3.65) and (3.17) show that the sequence of complex numbers $\exp\left(\frac{2\pi i}{3}\right) \varphi(\sup \mathcal{J}_{m,j})$ satisfies assumption (3.8). Hence, similarly to the asymptotic formula (3.38)

$$z_m(\sup \mathcal{J}_{m,j}) \sim (4\pi)^{-1/2} \exp\left(\frac{\pi i}{4}\right) \exp\left(\frac{-\pi i}{6}\right) \cdot (\lambda - \psi_\nu(\sup \mathcal{J}_{m,j}))^{-1/4}$$
$$\cdot \exp\left[i \int_{\lambda^{-1/2}\nu}^{\sup \mathcal{J}_{m,j}} (\lambda - \psi_\nu(\sigma))^{1/2} d\sigma \right], \quad \text{for } j \to \infty. \quad (3.66)$$

We continue the proof of Lemma 3.7 by dividing the absolute value of the previous asymptotic formula by the absolute value of conclusion (3.16) of Lemma 3.3. This yields,

$$\left| \frac{z_m(\sup \mathcal{J}_{m,j})}{y_m(\sup \mathcal{J}_{m,j})} \right| \sim \left| \frac{\exp\left[i \int_{\lambda^{-1/2}\nu}^{\sigma_j} (\lambda - \psi_\nu(\sigma))^{1/2} d\sigma \right]}{\{h_\tau^- - ih_\tau^+\}(\sup \mathcal{J}_{m,j})} \right|, \quad \text{for } j \to \infty. \quad (3.67)$$

We complete the proof of Lemma 3.7 by choosing the sequence $\sup \mathcal{J}_{m,j}$ so that relation (3.28) holds. In other words, we choose this sequence so that the denominator of the asymptotic formula (3.67) equals 2. This completes the proof of conclusion (3.64) and of Lemma 3.7 At the same time it completes the proof of Proposition 3.5 and of Theorem 3.1.

We conclude this section by observing that the proof of Theorem Theorem 3.1 did not use did not use the full force of assumption (2.14). In fact, all that we used was the first half of the inequality of this assumption. This fact allows us to choose $\beta = 1$, $\kappa = 1$, $\alpha_{m,j} = 1$, and $\beta_{m,j} = 0$ in Theorem Theorem 3.1. Then the conclusion reduces to conclusion (3.13) of Proposition 3.2 with $\alpha_{m,j} = 1$. In other words,

$$\liminf_{j \to \infty} |\lambda - \psi_\nu(\sup \mathcal{J}_{m,j})|^{1/4} \cdot |y_m(\sup \mathcal{J}_{m,j})| > 0.$$

Since we have chosen $\kappa = 1$ we see from assumption (3.1) that

$$|\lambda - \psi_\nu(\sup \mathcal{J}_{m,j})| > \frac{1}{2}\lambda,$$

and so,

$$\liminf_{j \to \infty} |y_m(\sup \mathcal{J}_{m,j})| > 0.$$

Now the previous inequality is, essentially, conclusion (5.41) of Theorem 5.1 of [4].

4. COMPLETION OF THE PROOF OF Theorem 2.1

In this section we complete the proof of Theorem 2.1. Recall conclusion (3.1) of Theorem 3.1 which says that conclusion (2.16) of Theorem 2.1 holds for the approximate solutions $k_{m,j}$ in place of the solutions f_m. Hence Theorem 2.1 is implied by the following theorem which says that at the points σ_j the approximate solutions are asymptotic to the solution.

Theorem 4.1. *Let the assumptions and notations of Theorem 2.1 hold and let the approximate solutions $k_{m,j}$ be given by definition (3.11). Then*

$$f_m(\sup \mathcal{J}_{m,j}) \sim k_{m,j}(\sup \mathcal{J}_{m,j}), \text{ for } j \to \infty. \tag{4.1}$$

We start the the proof of Theorem 4.1 by introducing some definitions. First, let $\mathcal{J}$ be a given interval and let w be a given weight function on it. That is to say, w is positive and continuous with the possible exception of finitely many points and at these points the limit of w is infinite. Then, for each $f \in \mathfrak{C}(\mathcal{J})$ define the norm,

$$\|f\|(w) = \sup_{\rho \in \mathcal{J}} |f(\rho)| w(\rho)^{-1}. \tag{4.2}$$

Second, with the help of the functions of definitions (2.6) and (3.2) we define this weight function:

$$w_m(\rho) = |\lambda - \psi_\nu(\rho)|^{-1/4} \cdot \exp(-\mathrm{Re}[i\nu g(\rho \tau^{-1})]). \tag{4.3}$$

We continue the proof of Theorem 4.1 by showing that with respect to this norm, the norm of the difference of the solution and of the approximate solution is small for large j. This is the statement of the technical proposition that follows.

Proposition 4.2. *Let the assumptions of Theorem 2.1 hold and let the approximate solution $k_{m,j}$ be given by definition (3.11). Then, for the weighted norm corresponding to any interval $\mathfrak{J}_{m,j}$ satisfying assumption (3.1)*

$$\limsup_{\substack{j\to\infty \\ \lambda\in\mathfrak{J}}} \|f_m - k_{m,j}\|_m = 0. \tag{4.4}$$

As a first step of the proof of Proposition 4.2 we formulate a crude upper estimate for the norm of conclusion (4.4). This is the statement of Lemma 4.3 which follows. To formulate it, we define the approximate potentials corresponding to the approximate solutions of definitions (3.10) and (3.5) by

$$q_{m,j} = q_{m,j}(\rho) = \frac{1}{2}\{\varphi(\rho),\rho\} - \varphi'(\rho)^2\varphi(\rho), \quad \rho\in\mathfrak{J}_{m,j}. \tag{4.5}$$

Here, as usual, the Schwarzian derivative is given by [1], [6], [12],

$$\{\varphi(\rho),\rho\} = \frac{\varphi'''(\rho)}{\varphi'(\rho)} - \frac{3}{2}\left[\frac{\varphi''(\rho)}{\varphi'(\rho)}\right]^2, \quad \varphi\in\mathfrak{C}^3(\mathfrak{J}_{m,j}),\ (\varphi')^{-1}\in\mathfrak{C}(\mathfrak{J}_{m,j}). \tag{4.6}$$

Then it is an elementary consequence of definitions (4.6), (4.5), (3.10) and (3.5) that for any given φ for which y_m is smooth,

$$y_m'' - q_m y_m = 0,\ \text{and}\ z_m'' - q z_m = 0. \tag{4.7}$$

In other words, the approximate solutions of definitions (3.10) and (3.5) satisfy assumption (A.6) of the Appendix with respect to the approximate potentials of definition (4.5). The corresponding error potential of definition (A.7) of the Appendix, with e_m in place of e, is given by

$$e_m(\rho) = -\frac{1}{2}\{\varphi(\rho),\rho\} + \varphi'(\rho)^2\varphi(\rho) - (\lambda - p_{0,j})(\rho). \tag{4.8}$$

Lemma 4.3. *Let the assumptions and notations of Proposition 4.2 hold. Next let the error potential e_m be given by definition (4.8). Then there is a constant γ such that the norm of conclusion (4.4) of Proposition 4.2 has the upper bound*

$$\|f_m - k_{m,j}\|_m \leq \gamma \cdot \int_{\mathfrak{J}_{m,j}} |e_m(\rho)||\lambda - \psi_\nu(\rho)|^{-1/2}d\rho$$

$$\cdot \exp(\gamma\int_{\mathfrak{J}_{m,j}} |e_m(\rho)||\lambda - \psi_\nu(\rho)|^{-1/2}d\rho)$$

$$\cdot \|k_{m,j}\|_m. \tag{4.9}$$

We start the proof of Lemma 4.3 by formulating an integral equation for the solution f_m over the interval $\mathfrak{J}_m$. More specifically, with the help of the approximate solutions of definitions (3.10) and (3.2) and the error potential (3.10) we define the kernel,

$$Q_m(\xi,\eta) = \begin{cases} (y_m(\eta)z_m(\xi) - z_m(\eta)y_m(\xi))W(y_m,z_m)^{-1}e_m(\eta), & \eta<\xi \\ 0, & \eta>\xi \end{cases} \tag{4.10}$$

Next define the corresponding Volterra operator

$$Q_m f(\xi) = \int_{\inf\mathfrak{J}_{m,j}}^{\xi} Q_m(\xi,\eta)f(\eta)d\eta,\ \xi\in\mathfrak{J}_{m,j}, \tag{4.11}$$

and Volterra equation

$$f_m = k_{m,j} + Q_m f_m. \tag{4.12}$$

We continue the proof of Lemma 4.3 by showing that Corollary A.2 of the Appendix applies to the Volterra equation (4.12). Specifically, define

$$w_{red}(\rho) = |(\lambda - \psi_\nu(\rho)|^{-1/4}, \tag{4.13}$$

and

$$v(\rho) = -\text{Re}[i\nu g(\rho\tau^{-1})]. \tag{4.14}$$

Then we show that Corollary A.2 holds for the weight function w_m of the definition (4.3) in place of w, for the reduced weight function w_{red} of definition (4.13) in place of the one of assumption (A.9) and for the function v of definition (4.14) and for the operator Q_m of definition (4.10) in place of Q of definition (A.8). To see that assumption (A.9) holds we insert definitions (4.14) and (4.13) into definition (4.3). This yields,

$$w_m(\rho) = w_{red}(\rho) \exp(v(\rho)). \tag{4.15}$$

To see that assumption (A.10) holds we need that the estimate [4],[(6.16)] implies in our present notation,

$$|y_m(\rho)| \leq \gamma |\lambda - \psi_\nu(\rho)|^{-1/4} \exp(-\text{Re}[i\nu g(\rho\tau^{-1})]), \quad \rho \in \mathcal{R}^+. \tag{4.16}$$

Inserting definitions (4.14) and (4.13) into estimate (4.16) we find

$$|y_m(\rho)| \leq \gamma w_{red}(\rho) \exp(v(\rho)), \quad \rho \in \mathcal{R}^+. \tag{4.17}$$

That is to say assumption (A.10) of the Appendix holds for y_m in place of y. To see that assumption (A.11) of the Appendix holds we note that the estimate [4],[(6.20)] implies in our present notation,

$$|z_m(\rho)| \leq \gamma |\lambda - \psi_\nu(\rho)|^{-1/4} \exp(\text{Re}[i\nu g(\rho\tau^{-1})]), \quad \rho \in \mathcal{R}^+. \tag{4.18}$$

Inserting definitions (4.14) and (4.13) into estimate (4.18), dividing the resulting estimate by the absolute value of $W(y_m, z_m)$ and using formula (3.35), we find

$$|z_m(\rho) W(y_m, z_m)|^{-1}| \leq 2\pi\gamma w_{red}(\rho) \exp(v(\rho))), \quad \rho \in \mathcal{R}^+. \tag{4.19}$$

That is to say assumption (A.11) of the Appendix holds for z_m in place of z and for the constant $2\pi\gamma$ in place of that constant. To see that

$$k_{m,j} \in \mathfrak{B}(w_m), \tag{4.20}$$

recall definition (3.11) which together with the triangle inequality shows that

$$\|k_{m,j}\|_m \leq \|\alpha_{m,j} y_m\|_m + \|\beta_{m,j} z_m\|_m \tag{4.21}$$

To estimate the first term of inequality (4.21), we combine estimate (4.16) with definitions (4.3) and (4.2). This yields,

$$\|\alpha_{m,j} y_m\|_m \leq \gamma |\alpha_{m,j}|. \tag{4.22}$$

A Turning Point Problem

To estimate the second term of the inequality (4.21) we combine estimate (4.18) with definition (4.3). This yields,

$$|\beta_{m,j} z_m(\rho)| \leq \gamma |\beta_{m,j}| w_m(\rho) \exp(2\operatorname{Re}[i\nu g(\rho\tau^{-1})]). \tag{4.23}$$

We see from definitions (3.1) and (2.5) that $\tau^{-1} \mathcal{J}_{m,j}$ is a compact interval which does not contain the point 1. Hence, according to definition (3.2) the function $2\operatorname{Re}[i\nu g(\rho\tau^{-1})]$ attains its maximum for ρ in the interval $\tau^{-1} \mathcal{J}_{m,j}$. Combining this fact with estimate (4.23) we see that to each j there is a constant $\gamma(j)$ such that

$$\|\beta_{m,j} z_m\|_m \leq \gamma(j) |\beta_{m,j}| \tag{4.24}$$

Inserting estimates (4.24), (4.22) and formula (3.35) into inequality (4.21) we arrive at estimate (4.20). Thus the Volterra equation (4.12) satisfies the assumptions of Corollary A.2 with y_m in place of y, z_m in place of z and with $k_{m,j}$ in place of r.

We complete the proof of Lemma 4.3 by applying conclusion (A.13) of Corollary A.2 of the Appendix with e_m in place of e. Clearly, this yields conclusion (4.9). Incidentally, note that although estimates (4.17), (4.19) and definition (4.13) did involve the constants $|\alpha_{m,j}|$, conclusion (4.9) does not.

As a second step of the proof of Proposition 4.2 we show that the second factor on the right of conclusion (4.9) is small for large j. This is the statement of the lemma that follows.

Lemma 4.4. *Let the assumptions of Proposition 4.2 hold and let the error potential e_m be given by formula (4.8). Then*

$$\limsup_{j \to \infty} \int_{\mathcal{J}_{m,j}} |e_m(\rho)| |\lambda - \psi_\nu(\rho)|^{-1/2} d\rho = 0. \tag{4.25}$$

We start the proof of Lemma 4.4 with a simple formula for the error potential,

$$e_{m,j}(\eta) = e_m(\eta) = p_0(\eta) - \frac{1}{2}\{\varphi(\eta), \eta\} - \frac{1}{4}\eta^{-2}. \tag{4.26}$$

Indeed, inserting the differential equation (3.27) and definition (2.2) into formula (4.8) we find formula (4.26).

We continue the proof of Lemma 4.4 by showing that conclusion (4.25) holds for the first term of formula (4.26):

$$\limsup_{j \to \infty} \int_{\mathcal{J}_m} |p_0(\rho)| |\lambda - \psi_\nu(\rho)|^{-1/2} d\rho = 0. \tag{4.27}$$

To see estimate (4.27) note that estimate (2.7) yields,

$$\sup_{\lambda \in \mathcal{J}} \int_{\mathcal{J}_m} |p_0(\rho)| |\lambda - \psi_\nu(\rho)|^{-1/2} d\rho = 0(\nu^{-\beta}) \cdot \int_{\mathcal{J}_m} |\lambda - \psi_\nu(\rho)|^{-1/2} \psi_\nu^{\frac{\beta}{2}}(\rho) d\rho. \tag{4.28}$$

It is an elementary fact that assumption (3.1) yields,

$$\int_{\mathcal{J}_{m,j}} |\lambda - \psi_\nu(\rho)|^{-1/2} \psi_\nu^{\frac{\beta}{2}}(\rho) d\rho \leq \lambda^{\omega_1 + \omega_2 - 1/2} \cdot \nu \cdot \int_{1-\nu^{\kappa-1}}^{1+2\nu^{\kappa-1}} |\sigma^2 - 1|^{-1/2} \sigma^{1-\beta} d\sigma, \tag{4.29}$$

and this also follows from the scaling lemma of [10, Lemma 3.3] applied to $\omega_1 = -1/2$, $\omega_2 = \beta/2$, and $\mathcal{J} = \mathcal{J}_{m,j}$. An elementary integration gives,

$$\int_{1-\nu^{\kappa-1}}^{1+\nu^{\kappa-1}} |1-\sigma|^{-1/2} \cdot |1+\sigma|^{-1/2} \sigma^{1-\beta} d\sigma < 4 \cdot \nu^{1/2 \cdot (\kappa-1)}. \tag{4.30}$$

Combining the previous three estimates we find

$$\sup_{\lambda \in \mathcal{J}} \int_{\mathcal{J}_{m,j}} |p_0(\rho)| \cdot |\lambda - \psi_\nu(\rho)|^{-1/2} d\rho = O(\nu^{-\beta+1/2 \cdot (\kappa+1)}), \text{ for } \nu \to \infty. \tag{4.31}$$

We see from the second half of assumption (2.14) that the exponent of ν in estimate (4.31) is strictly negative, and so estimate (4.27) follows.

We complete the proof of Lemma 4.4 by showing that conclusion (4.25) also holds for the second term of formula (4.26). This is implied by estimate (10.33) of [4], which says that,

$$\int_{\mathcal{J}_{m,j}} |\{\varphi(\eta),\eta\} - \frac{1}{2}\eta^{-2}|\lambda - \psi_\nu(\eta)|^{-1/2} d\eta \leq \lambda^{-1/2} \int_0^\delta \frac{d\sigma}{|\sigma^2-1|^{1/2}\sigma(|\log\sigma|+1)^2}. \tag{4.32}$$

As a third and last step of the proof of Proposition 4.2 we show that the fourth term of conclusion (4.9) remains bounded. This is the statement of the lemma that follows.

Lemma 4.5. *Let the assumptions and notations of Proposition 4.2 hold. Then*

$$\limsup_{j \to \infty} \|k_{m,j}\|_m < \infty. \tag{4.33}$$

We start the proof of Lemma 4.5 by showing that

$$\limsup_{j \to \infty} \|\alpha_{m,j} y_m\|_m < \infty. \tag{4.34}$$

Indeed, combining conclusion (3.33) of Lemma 3.4 with estimate (4.22) we find estimate (4.34).

We continue the proof of Lemma 4.5 by showing that

$$\lim_{j \to \infty} \|\beta_{m,j} z_m\|_m = 0. \tag{4.35}$$

To prove estimate (4.35), first we show that

$$|\beta_{m,j} z_m(\rho)| \leq \gamma |\beta_{m,j}| w_m(\rho) \exp(2\operatorname{Re}[i\nu g(\sigma_j \tau^{-1})]). \tag{4.36}$$

Indeed, combination of the proof of estimate (3.62) with formula (3.43) shows that the function $-2\operatorname{Re}[i\nu g(\rho\tau^{-1})]$ is increasing and so, the function $2\operatorname{Re}[i\nu g(\rho\tau^{-1})]$ is decreasing. Hence,

$$2\operatorname{Re}[i\nu g(\rho\tau^{-1})] \leq 2\operatorname{Re}[i\nu g(\sigma_j \tau^{-1})] \text{ for } \sigma_j \leq \rho. \tag{4.37}$$

Inserting the previous estimate into estimate (4.23) we obtain estimate (4.36). To prove estimate (4.35), second we show that

$$\lim_{j \to \infty} \beta_{m,j} \exp(2\operatorname{Re}[i\nu g(\sigma_j \tau^{-1})]) = 0. \tag{4.38}$$

A Turning Point Problem

To see estimate (4.38) we need a stronger version of conclusion (3.53) of Lemma 3.6. More specifically, we need that

$$\lim_{j\to\infty} \beta_{m,j} \exp\left[2i \int_{\lambda^{-1/2}\nu}^{\sigma_j} (\lambda - \psi_\nu(\sigma))^{1/2} d\sigma\right] = 0. \tag{4.39}$$

Indeed, inserting estimate (3.62) and the asymptotic formula (3.55) into the asymptotic formula (3.54) we get relation (4.39). Inserting relation (3.43), in turn, into relation (4.39) we obtain relation (4.38). Then, combining relations (4.38) and (4.36) we arrive at estimate (4.35).

We complete the proof of Lemma 4.5 by inserting estimates (4.35) and into estimate (4.21).

Finally, inserting the conclusions of Lemmas 4.5 and 4.4 into conclusion (4.9) of Lemma 4.3 and using that Lemma 4.4 implies that the third factor of conclusion (4.9) is bounded, we arrive at conclusion (4.4) of Proposition 4.2. Hence the proof of Proposition 4.2 is complete.

We complete the proof of Theorem 4.1 by showing that norm estimates imply pointwise estimates. More specifically, we show that there is a j_0 such that

$$|f_m(\sup \mathfrak{I}_{m,j}) - k_{m,j}(\sup \mathfrak{I}_{m,j})| \le |k_{m,j}(\sup \mathfrak{I}_{m,j})| \cdot \|f_m - k_{m,j}\|_m, \; for\, j > j_0. \tag{4.40}$$

To prove estimate (4.40), first we note that

$$|f_m(\sup \mathfrak{I}_{m,j}) - k_{m,j}(\sup \mathfrak{I}_{m,j})| \le w_m(\sup \mathfrak{I}_{m,j}) \cdot \|f_m - k_{m,j}\|_m. \tag{4.41}$$

Indeed, according to assumption (3.1) the point $\sup \mathfrak{I}_{m,j}$ is in $\mathfrak{I}_{m,j}$ and so, we can apply definition (4.2) to w_m in place of w, to $f_m - k_{m,j}$ in place of f and to $\sup \mathfrak{I}_{m,j}$ in place of ρ. Then this application yields estimate (4.41). To prove estimate (4.40), second we show that

$$|k_{m,j}(\sup \mathfrak{I}_{m,j})| \sim 2|\lambda - \psi_\nu(\sup \mathfrak{I}_{m,j}))^{-1/4}|, \; for\, j \to \infty. \tag{4.42}$$

To see estimate (4.42), insert conclusion (3.16) of Lemma 3.3 and conclusion (3.33) of Lemma 3.4 into conclusion (3.51) of Proposition 3.5. After taking absolute values, this yields

$$|k_{m,j}(\sup \mathfrak{I}_{m,j})| \sim |(\lambda - \psi_\nu(\sup \mathfrak{I}_{m,j}))^{-1/4}||\{h_\tau^- - ih_\tau^+\}(\sup \mathfrak{I}_{m,j})|, \; for\, j \to \infty.$$

Next, we choose the family of intervals so that in addition to assumption (3.1) of Theorem 3.1 they satisfy relation 3.28). Then, inserting relation (3.28) into the previous asymptotic formula we find the asymptotic formula (4.42). According to definition (3.2)

$$-\text{Re}\left[i\nu g(\rho\tau^{-1})\right] = 0, \; for\, \rho\tau^{-1} > 1,$$

and so, by definition (4.3)

$$w_m(\sup \mathfrak{I}_{m,j}) = |(\lambda - \psi_\nu(\sup \mathfrak{I}_{m,j}))|^{-1/4}. \tag{4.43}$$

Inserting formula (4.43) into inequality (4.41) and using the asymptotic formula (4.42) we obtain estimate (4.40). Finally, inserting conclusion (4.4) of Proposition 4.2 into estimate (4.40) we arrive at conclusion (4.1) of Theorem 4.1. Hence, Theorem 4.1 follows and so does the main Theorem 2.1 as well.

We conclude this section by motivating definition (3.4). We see from definition (4.5), from formula (4.8) and from the differential equation (4.10) that if we would choose φ so that

$$\lambda - p_{0,j}(\rho) - \varphi'(\rho)^2 \varphi(\rho) + \frac{1}{2}\{\varphi(\rho), \rho\} = 0, \tag{4.44}$$

then the corresponding function of definition (3.5) would be an exact solution to the basic equation (2.3). However, there is no guarantee that with this choice of φ we could get conclusion (2.16) of Theorem 2.1. Therefore, we have solved an approximate equation to (4.27). First, we added p_0 to the left of equation (4.27) and then, in order to get the same approximate solution as in [4] we replaced the Schwarzian derivative on the right by $\frac{1}{2}\rho^{-2}$. Recalling definitions (2.4) and (2.2), we see that this leads to the differential equation (3.27). We have seen that the function of definition (3.4) satisfies this non-linear differential equation. At the same time it follows that

$$\lim_{z \to \tau} \varphi(z) = 0. \tag{4.45}$$

In other words, the function φ maps the approximate turning point τ into the turning point of the Airy equation (3.6).

A. APPENDIX, A LEMMA ON VOLTERRA OPERATORS

In this Appendix we isolate some general facts about the solutions of the basic differential equation (2.3). The first one is a corollary of a resolvent estimate for Volterra operators acting on weighted spaces. This resolvent estimate is a generalization of a result of Love [5], Erdelyi [2] and Olver [6], which was also proved in [10]. To describe it we need some notations. Let $\mathfrak{J}$ be a given interval and let w be a given weight function in the sense of Section 4. That is to say, w is positive and continuous with the possible exception of finitely many points and at these points the limit of w is infinite. Then, we denote by $\mathfrak{B}(w)$ the completion of $\mathfrak{C}(\mathfrak{J})$ with reference to the norm $\|f\|_w$ given by definition (4.2). Let $\mathfrak{B}(\mathfrak{B}(w))$ denote the algebra of bounded operators on $\mathfrak{B}(w)$. Next let V be a given Volterra operator with kernel $V(\xi, \eta)$ acting on the space $\mathfrak{B}(w)$. In other words,

$$Vf(\xi) = \int_0^\xi V(\xi, \eta) f(\eta) d\eta \quad , \quad \xi \in \mathfrak{J}, \quad f \in \mathcal{B}(w). \tag{A.1}$$

Then, we define the Love–Erdelyi–Olver bound of V with respect to w by

$$\|V\|(LEO, w) = \int_{\mathfrak{J}} \sup_{\xi > \eta} w(\xi)^{-1} w(\eta) |V(\xi, \eta)| d\eta. \tag{A.2}$$

Lemma A.1. *Let w be a given weight function, let V be a given Volterra operator acting on $\mathfrak{B}(w)$ and let the Love–Erdelyi–Olver bound be given by definition (A.2). Suppose that*

$$\|V\|(LEO, w) < \infty. \tag{A.3}$$

Then

$$V \in \mathfrak{B}(\mathfrak{B}(w)) \text{ and } (I - V)^{-1} \in \mathfrak{B}(\mathfrak{B}(w)). \tag{A.4}$$

Furthermore,

$$\|V\|(w) \leq \|V\|(LEO, w) \text{ and } \|(I - V)^{-1}\|(w) \leq \exp(\|V\|(LEO, w)). \tag{A.5}$$

A Turning Point Problem

We shall use this LEO-Lemma A.1 in the proof of our Theorem 2.1 via the corollary that follows. It is an adaptation to the case of a Volterra operator which is built up from two approximate solutions of the basic equation (2.3). More specifically, let $q \in \mathfrak{C}(\mathfrak{J})$ be a given approximate potential and let y and z be two given linearly independent solutions of the homogeneous differential equation,

$$y'' - qy = 0 \text{ and } z'' - qz = 0. \tag{A.6}$$

With the help of such an approximate potential we define the *error potential*,

$$e = e_\lambda = e_{\lambda,j} = -q - (\lambda - p_{0,j}) \tag{A.7}$$

and kernel

$$Q(\xi, \eta) = \begin{cases} (y(\eta)z(\xi) - z(\eta)y(\xi))W^{-1}e(\eta) &, \eta < \xi \\ 0 &, \eta > \xi \end{cases} \tag{A.8}$$

Corollary A.2. *Let Q, be a given Volterra operator and let its kernel, $Q(\xi, \eta)$, be of the form (A.8) and let w be a given weight function. Suppose that to this weight function there is a reduced weight function, w_{red}, and an increasing function v, such that,*

$$w(\rho) = w_{red}(\rho) \exp[v(\rho)] \tag{A.9}$$

Next suppose that there is a constant γ such that,

$$|y(\rho)| \leq \gamma w_{red}(\rho) \exp[v(\rho)] \tag{A.10}$$

and

$$|z(\rho)W^{-1}| \leq \gamma w_{red}(\rho) \exp[-v(\rho)]. \tag{A.11}$$

Then, for each $r \in \mathfrak{B}(w)$ the Volterra equation

$$f = r + Qf, \tag{A.12}$$

admits a unique solution $f \in \mathfrak{B}(w)$ and f is such that

$$\|f - r\|_w \leq \gamma^2 \int_{\mathfrak{J}} |e(\rho)|w_{red}(\rho)^2 \, d\rho \cdot \exp\left(\gamma^2 \int_{\mathfrak{J}} |e(\rho)|w_{red}(\rho)^2 \, d\rho\right) \cdot \|r\|_w. \tag{A.13}$$

Furthermore, if r satisfies the differential equation of relation (A.6), then f satisfies the basic differential equation (2.3).

The proof of Corollary A.2 is based on the key estimate

$$\|Q\|(LEO, w) \leq \int_{\mathfrak{J}} |e(\rho)|w_{red}(\rho)^2 \, d\rho. \tag{A.14}$$

We start the proof of the key estimate (A.14) by showing that

$$w(\eta) \cdot |y(\xi)z(\eta)W^{-1}| \leq \gamma^2 w(\xi)w_{red}(\eta)^2. \tag{A.15}$$

To prove estimate (A.15) we apply assumption (A.10) to $\rho = \xi$, assumption (A.11) to $\rho = \eta$ and multiply the results together. This yields,

$$|y(\xi)z(\eta)W^{-1}| \leq \gamma^2 w_{red}(\xi)w_{red}(\eta)\exp[v(\xi) - v(\eta)]. \qquad (A.16)$$

We see from the assumption (A.9) that

$$w(\eta) \cdot w_{red}(\xi)\exp[v(\xi) - v(\eta)] = w(\xi) \cdot w_{red}(\eta), \qquad (A.17)$$

and so, multiplying estimate (A.16) by $w(\eta)$ we obtain estimate (A.15).

We continue the proof of the key estimate (A.14) by showing that

$$w(\eta) \cdot |y(\eta)z(\xi)W^{-1}| \leq \gamma^2 w(\xi)w_{red}(\eta)^2, \text{ for } \eta < \xi. \qquad (A.18)$$

To prove estimate (A.18), note that interchanging the variables ξ and η in estimate (A.16) we find

$$|y(\eta)z(\xi)W^{-1}| \leq \gamma^2 w_{red}(\xi)w_{red}(\eta)\exp[v(\eta) - v(\xi)]. \qquad (A.19)$$

We see from the assumption (A.9) that similarly to formula (A.17),

$$w(\eta) \cdot w_{red}(\xi)\exp[v(\eta) - v(\xi)] = w(\xi) \cdot w_{red}(\eta)\exp[2v(\eta) - 2v(\xi)]. \qquad (A.20)$$

By assumption the function v is increasing, and so,

$$\exp[2v(\eta) - 2v(\xi)] \leq 1, \text{ for } \eta < \xi. \qquad (A.21)$$

Inserting this inequality into formula (A.20) we obtain the inequality,

$$w(\eta)w_{red}(\xi)\exp[v(\eta) - v(\xi)] \leq w(\xi)w_{red}(\eta), \text{ for } \eta < \xi. \qquad (A.22)$$

Now, multiplying estimate (A.19) by $w(\eta)$ and using the inequality (A.22) we arrive at estimate (A.18).

We complete the proof of the key estimate (A.14) by inserting estimates (A.15) and (A.18) into assumption (A.8). This yields,

$$w(\eta) \cdot |Q(\xi, \eta,)| \leq 2\gamma^2 w(\xi) \cdot w_{red}(\eta)^2 |e(\eta)|, \text{ for } \eta < \xi. \qquad (A.23)$$

Since the constant γ^2 is independent of ξ, this, in turn, yields,

$$\sup_{\xi > \eta} w(\xi)^{-1} w(\eta) \cdot |Q(\xi, \eta)| \leq 2\gamma^2 w_{red}(\eta)^2 |e(\eta)|. \qquad (A.24)$$

Integrating estimate (A.24) with respect to η over the interval $\mathcal{J}$ and using definition (A.2) we find the key estimate (A.14).

Having established the key estimate (A.14), we can easily complete the proof of Corollary A.2. In fact, combining estimate (A.14) with the first half of conclusion (A.5) of the LEO-Lemma A.1 and with the Volterra equation (A.12) we obtain

$$\|f - r\|_w \leq \int_{\mathcal{J}} |e(\rho)|w_{red}(\rho)^2 \, d\rho \cdot \|f\|_w \qquad (A.25)$$

Then, using the second half of conclusion (A.5) of the LEO-Lemma A.1 and using the Volterra equation (A.12) again we arrive at conclusion (A.13) of Corollary A.2.

Finally we show that each solution $f \in \mathfrak{B}(w)$ of the Volterra equation (A.12) satisfies the basic differential equation (2.6). For this purpose we need that the assumption $f \in \mathfrak{B}(w)$ implies that f is continuous with the possible exception of the points where w is not continuous. Hence with the possible exception of these points, we can perform the usual differentiation operations on the integral of the Volterra equation (A.12). This yields,

$$(Qf)'' = ef + qQf. \qquad (A.26)$$

Combining this formula with the Volterra equation (A.12), with definition (A.7) and with the assumption that r satisfies the differential equation of relation (A.6) we find that f satisfies the basic differential equation (2.3). More specifically, f satisfies this basic differential equation (2.3) with the possible exception of the points where w is not continuous. Since the coefficients of the the basic differential equation (2.3) are continuous, f satisfies this equation on all of $\mathcal{J}$.

REFERENCES

1. Erdelyi A., *Asymptotic expansions*, Dover New York, 1956.
2. ———, *Asymptotic solutions of differential equations with transition points or singularities. see section 4.*, J. Math. Phys. **1** (1960), 16–26.
3. Devinatz Allen and Rejto Peter., *A limiting absorption principle for Schrödinger operators with oscillating potentials I*, J. Diff. Equations **49** (1983), 85–104.
4. Devinatz Allen, Moeckel Richard, and Rejto Peter., *A limiting absorption principle for Schrödinger operators with von Neumann–Wigner type potentials*, Integral Equations and Operator Theory **14** (1991), 13–68.
5. Love C.E., *Singular integral equations of the volterra type*, Trans. Amer. Math. Soc. **15** (1914), 1467–476.
6. Olver F.W.J., *Asymptotics and special functions*, Academic Press, 1974.
7. Rejto P. and Taboada M., *Limiting absorption principle for Schrödinger operators with generalized von Neumann–Wigner potentials I. (Construction of approximate phase.)*, J. Math. Anal. Appl. **208** (1991), 85–108.
8. ———, *Limiting absorption principle for Schrödinger operators with generalized von Neumann–Wigner potentials II.(The Proof)*, J. Math. Anal. Appl. **208** (1997), 311–336.
9. Langer R.E., *The asymptotic solutions of ordinary linear differential equations of the second order with special reference to a turning point.*, Trans. Amer. Math. Soc. **67** (1949), 461–490.
10. P. Rejto and Taboada M., *Weighted resolvent estimates for Volterra operators on unbounded intervals*, J. Math. Anal. Appl. **160** (1991), 223–235.
11. Yasutaka Sibuya, *Global theory of a second order linear differential equation with a polynomial coefficient*, North-Holland Mathematics Studies 18, North-Holland/American Elsevier, 1975.
12. W. Wasow, *Asymptotic expansions for ordinary differential equations*, Interscience Pure and Applied Mathematics Vol. XIV, Wiley-Interscience, 1966.
13. H. F. Weinberger, *Partial differential equations, a first course*, Blaisdell Publishing, 1965, See Section 38, Oscillation Theorem.

10

EIGENVALUE PROBLEMS FOR SEMILINEAR EQUATIONS

Martin Schechter*

Department of Mathematics
University of California, Irvine
Irvine, CA

ABSTRACT

We present some useful methods of solving eigenvalue problems for semilinear equations. Each of the methods is described fully, and applications are given to boundary value problems for partial differential equations.

1. INTRODUCTION

We shall study semilinear elliptic equations of the form

$$Au = f(x,u), \quad u \in D(A) \tag{1.1}$$

where A is a positive selfadjoint operator on $L^2(\Omega), \Omega \subset \mathbb{R}^n$, and $f(x,t)$ is a Caratheodory function or $\bar{\Omega} \times \mathbb{R}$. We shall make no asymptotic assumptions on $f(x,t)$ other than it is subcritical (cf. below). Moreover, we shall not make any assumptions on $f(x,t)$ in bounded regions. We shall show that problem (1.1) is intimately connected with the eigenvalue problem

$$Au = \lambda f(x,u), \quad \in D(A) \tag{1.2}$$

not only because (1.1) is a special case of (1.2), but also because the absence of three solutions of (1.1) implies the existence of a rich family of eigenfunctions of (1.2) and, conversely, the absence of at least four rich families of eigenfunctions of (1.2) implies the existence of solutions of (1.1).

To describe our hypotheses, let Ω be a domain in $\mathbb{R}^n$, and let A be a selfadjoint operator on $L^2(\Omega)$ satisfying $A \geq \lambda_0 > 0$ and

$$C_0^\infty(\Omega) \subset D := D(A^{1/2}) \subset H^{m,2}(\Omega)) \tag{1.3}$$

---
*Research supported in part by a NSF grant.

Spectral and Scattering Theory, edited by Ramm,
Plenum Press, New York, 1998

for some $m > 0$ (we do not require m to be an integer). Let q be any number satisfying

$$2 \leq q \leq 2n/(n-2m), \quad 2m < n$$
$$2 \leq q < \infty \quad n \leq 2m \quad (1.4)$$

and let $f(x,t)$ satisfy

$$|f(x,t)| \leq V(x)^q |t|^{q-1} + V(x)W(x) \quad (1.5)$$

and

$$f(x,t)/V(x)^q = o(|t|^{q-1}) \text{ as } |x| \to \infty \quad (1.6)$$

where $V(x) > 0$ is a function in $L^q(\Omega)$. If $q = 2$, we require that multiplication by V be a compact operator from D to $L^2(\Omega)$. We also assume that there is a $\delta > 0$ such that

$$2F(x,t) \leq \lambda_0 t^2, \quad |t| \leq \delta \quad (1.7)$$

and that there is a $u_0 \in D \setminus \{0\}$ such that

$$\|u_0\|_D^2 \leq 2 \int F(x, u_0(x)) dx, \quad (1.8)$$

where

$$F(x,t) = \int_0^t f(x,s) ds.$$

Our first results are

Theorem 1.1. *Under the above hypotheses, the following alternative holds: either (a) there are at least two nontrivial solutions of (1.1) or (b) for each $R > \|u_0\|_D$ there is at least one solution of (1.2) satisfying*

$$\|u\|_D = R, \quad 0 < \lambda < 1. \quad (1.9)$$

Theorem 1.2. *Under the same hypotheses, the following alternative holds: either (a) there is at least one nontrivial solution of (1.1) or (b) for each $R > \|u_0\|_D$ there are at least two solutions of (1.2) satisfying (1.10).*

Theorem 1.3. *In addition to the hypotheses of Theorem 1.1 assume that*

$$f(x, u(x)) \not\equiv 0 \text{ for all } u \in D. \quad (1.10)$$

Then the following alternative holds: either (a) there are at least three nontrivial solutions of (1.1) or (b) for each $R > \|u_0\|_D$ there is at least one solution of (1.2) satisfying

$$\|u\|_D = R. \quad (1.11)$$

Theorem 1.4. *Under the same hypotheses, the following alternative holds: either (a) there are at least two nontrivial solutions of (1.1) or (b) for each $R > \|u_0\|_D$ there are at least two solutions of (1.2) satisfying (1.12).*

Theorem 1.5. *Under the same hypotheses, the following alternative holds: either (a) there is at least one nontrivial solution of (1.1) or (b) for each $R > \|u_0\|_D$ there are at least four solutions of (1.12) satisfying (1.2).*

Theorem 1.6. *Assume that there is a $u_0 \in D$ such that*

$$\int_\Omega F(x,u_0)dx > 0 \tag{1.12}$$

and let

$$\mu_0 = \|u_0\|_D^2 / 2 \int_\Omega F(x,u_0)dx. \tag{1.13}$$

Then for each $\mu > \mu_0$ there are at least two solutions of

$$Au = \lambda f(x,u), 0 < \|u\|_D \le \|u_0\|_D, 0 < \lambda < \mu. \tag{1.14}$$

Theorem 1.7. *The following alternative holds. Either (a) there is an infinite number of solutions of*

$$Au = \lambda f(x,u), 0 < \|u\|_D \le \|u_0\|_D, \mu_0 < \lambda < \mu_0 + \epsilon \tag{1.15}$$

for any $\epsilon > 0$, or (b) there are at least two solutions of

$$Au = \lambda f(x,u), 0 < \|u\|_D \le \|u_0\|_D, 0 < \lambda \le \mu_0. \tag{1.16}$$

The theorems of this section will be proved in Section 8. Some of the results of this section were announced in [1].

2. A BOUNDED SADDLE POINT THEOREM

We now present the theory which is used in proving the results of Section 1. Let E be a Hilbert space and let Φ be the set of all continuous maps $\Gamma(t)$ from $E \times [0,1]$ to E such that

a) $\Gamma(0) = I$

b) there is a $u_0 \in E$ such that $\Gamma(t)u \to u_0$ as $t \to 1$ uniformly on bounded subsets of E

c) for each $t \in [0,1), \Gamma(t)$ is a homeomorphism of E onto itself with Γ^{-1} continuous on $E \times [0,1)$.

We let Φ_R denote the set of these mappings $\Gamma(t) \in \Phi$ which map $\bar{B}_R$ into itself for each $t \in [0,1]$, where

$$B_R := \{u \in E : \|u\| < R\}. \tag{2.1}$$

Definition 2.1. A subset A of $\bar{B}_R$ links a subset B of $\bar{B}_R$ with respect to the family Φ_R if

$$A \cap B = \phi \tag{2.2}$$

and for each $\Gamma \in \Phi_R$ there is a $t \in [0,1]$ such that

$$\Gamma(t)A \cap B \ne \phi. \tag{2.3}$$

We have

Theorem 2.1. *Let G be a C^1 functional on $\bar{B}_R$ and let $A, B \subset \bar{B}_R$ be such that A links B with respect to Φ_R and*

$$a_0 := \sup_A G \leq b_0 := \inf_B G. \tag{2.4}$$

Assume

$$a_R := \inf_{\Gamma \in \Phi_R} \sup_{\substack{0 \leq s \leq 1 \\ u \in A}} G(\Gamma(s)u) < \infty \tag{2.5}$$

and there are constants $\theta < 1, \delta > 0$ such that

$$(G'(u), u) + \theta R \|G'(u)\| \geq 0 \tag{2.6}$$

for all $u \in \partial B_R$ satisfying

$$|G(u) - a_R| \leq \delta. \tag{2.7}$$

Then there is a sequence $\{u_k\} \subset \bar{B}_R$ such that

$$G(u_k) \to a_R, \, G'(u_k) \to 0. \tag{2.8}$$

If $a_R = b_0$, then we can also require

$$d(u_k, B) \to 0. \tag{2.9}$$

Proof: Assume first that $a_0 < a_R$. If a sequence satisfying (2.8) did not exist, there would be positive constants δ, m such that

$$\|G'(u)\| \geq m \tag{2.10}$$

holds for all u in the set

$$Q = \{u \in \bar{B}_R : |G(u) - a_R| \leq 3\delta\}.$$

By reducing δ if necessary, we may assume that (2.6) holds for u in the set $\tilde{Q} = \partial B_R \cap Q$. We write

$$Q_0 = \{u \in \bar{B}_R : |G(u) - a_R| \leq 2\delta\}$$
$$Q_1 = \{u \in \bar{B}_R : |G(u) - a_R| \leq \delta\}$$

and

$$Q_2 = \bar{B}_R \setminus Q_0, \, \eta(u) = d(u, Q_2) / [d(u, Q_1) + d(u, Q_2)].$$

Note that

$$\eta(u) = 1, u \in Q_1, \eta(u) = 0, u \in \bar{Q}_2, 0 < \eta(u) < 1, \text{ otherwise}.$$

Eigenvalue Problems for Semilinear Equations

By Theorem 10 of [2], for each $\alpha < 1 - \theta$ there is a locally Lipschitz continuous map $Y(u)$ of $\hat{B}_R = \{u \in \bar{B}_R : G'(u) \neq 0\}$ into E such that

$$\|Y(u)\| \leq 1, \alpha \|G'(u)\| \leq (G'(u), Y(u)), u \in \hat{B}_R \tag{2.11}$$

and

$$(Y(u), u) > 0, u \in \tilde{Q}. \tag{2.12}$$

Let

$$W(u) = -\eta(u)Y(u), u \in \hat{B}_R. \tag{2.13}$$

Then $W(u)$ is locally Lipschitz on the whole of $\bar{B}_R$. We extend it to be locally Lipschitz continuous on the whole of E by setting

$$W(u) = W(Ru/\|u\|), \|u\| > R.$$

Let $\sigma(t)$ be the flow generated by W on E. We note that $\sigma(t)u$ does not exist $\bar{B}_R$ for any $u \in \bar{B}_R$. For we have

$$d\|\sigma(t)u\|^2/dt = 2(\sigma', \sigma) = -2\eta(\sigma)(Y(\sigma), \sigma). \tag{2.14}$$

If $u_1 \in \partial B_R$ is in $\tilde{Q}$, then (2.12) holds in a neighborhood of u_1. If it is not in $\tilde{Q}$, then η vanishes near u_1. Hence the right hand side of (2.14) is ≤ 0 in the neighborhood of each point of ∂B_R. Consequently, $\sigma(t)u$ does not exist $\bar{B}_R$ for $t \geq 0$. Now

$$\begin{aligned} dG(\sigma(t)u)/dt &= (G'(\sigma(t)u), \sigma'(t)u) \\ &= -\eta(\sigma(t)u)(G'(\sigma(t)u), Y(\sigma(t)u)) \\ &\leq -\eta(\sigma(t)u)\alpha m \leq 0 \end{aligned} \tag{2.15}$$

by (2.10) and (2.11). Write a for a_R and let u be any element of $E_{a+\delta}$, where

$$E_c := \{u \in \bar{B}_R : G(u) \leq c\}. \tag{2.16}$$

Take $T > 2\delta/\alpha m$. If there is a $t_1 \in [0, T]$ such that $\sigma(t_1)u \notin Q_1$, then

$$G(\sigma(T)u) \leq G(\sigma(t_1)u) < a - \delta.$$

Hence $\sigma(T)u \in E_{a-\delta}$. On the other hand, if $\sigma(t)u \in Q_1$ for all $t \in [0, T]$, then $\eta(\sigma(t)u) \equiv 1$, and (2.15) implies

$$G(\sigma(T)u) \leq G(u) - \alpha m T$$
$$< a + \delta - 2\delta.$$

Thus

$$\sigma(T)E_{a+\delta} \subset E_{a-\delta}. \tag{2.17}$$

Now by (2.5) there is a $\Gamma \in \Phi_R$ such that

$$\Gamma(s)A \subset E_{a+\delta}, 0 \leq s \leq 1. \tag{2.18}$$

Let

$$\Gamma_1(s) = \sigma(2sT), 0 \leq s \leq \frac{1}{2}$$
$$= \sigma(T)\Gamma(2s-1), \frac{1}{2} < s \leq 1. \quad (2.19)$$

It is easily checked that $\Gamma_1 \in \Phi_R$. Also

$$\Gamma_1(s)A = \sigma(2sT)A \subset E_{a_0} \subset E_{a-\delta}, 0 \leq s \leq \frac{1}{2}$$
$$= \sigma(T)\Gamma(2s-1)A \subset \sigma(T)E_{a+\delta} \subset E_{a-\delta}, \frac{1}{2} < s \leq 1.$$

Thus

$$G(\Gamma_1(s)A) < a - \delta, \quad 0 \leq s \leq 1.$$

This contradicts (2.5) and proves the theorem for the case $a_0 < a$. Now assume $b_0 = a$. If there did not exist a sequence satisfying both (2.8) and (2.9), there would be positive numbers m, δ, T such that $2\delta < \alpha mT$ and (2.10) holds whenever

$$u \in Q = \{u \in \bar{B}_R : d(u,B) \leq 4T, |G(u) - a| \leq 3\delta\}.$$

Let

$$Q_0 = \{u \in \bar{B}_R : d(u,B) \leq 3T, |G(u) - a| \leq 2\delta\}$$
$$Q_1 = \{u \in \bar{B}_R : d(u,B) \leq 2T, |G(u) - a| \leq \delta\}.$$

Since $a = b_0$, we see that $Q_1 \neq \phi$. Define Q_2, η as before with respect to the new Q_j, and let $\sigma(t)$ be the flow generated with respect to the new $W(u)$. Let u be any elements in $E_{a+\delta}$. If there is a $t_1 \leq T$ such that $\sigma(t_1)u \notin Q_1$, then either

$$G(\sigma(t_1)u) < a - \delta \quad (2.20)$$

or

$$d(\sigma(t_1)u, B) > 2T. \quad (2.21)$$

Since $\|\sigma(t)u - \sigma(t')u\| \leq |t - t'|$, we see that (2.21) implies

$$d(\sigma(t)u, B) > T, \quad 0 \leq t \leq T. \quad (2.22)$$

On the other hand, if $\sigma(t)u \in Q_1$ for all $t \in [0,T]$, then

$$G(\sigma(T)u) \leq G(u) - \alpha mT \leq a + \delta - 2\delta.$$

Thus we have either

$$G(\sigma(T)u) < a - \delta \quad (2.23)$$

or (2.22) holds. In either case

$$\sigma(T)E_{a+\delta} \cap B = \phi. \quad (2.24)$$

Moreover
$$\sigma(t)A \cap B = \phi, 0 \leq t \leq T. \qquad (2.25)$$

For we have by (2.15)
$$G(\sigma(t)u) \leq a_0 - \alpha m \int_0^t \eta(\sigma(\tau)u)d\tau, u \in A.$$

If $\sigma(t)u \in B$, we must have $G(\sigma(t)u) \geq b_0 \geq a_0$. The only way this can happen is if $\eta(\sigma(\tau)u) \equiv 0$ for $0 \leq \tau \leq t$. But this implies $\sigma(\tau)u \in \tilde{Q}_2$ for such τ, and this would mean that either
$$d(\sigma(\tau)u, B) > 3T, \quad 0 \leq \tau \leq t$$
or
$$G(\sigma(\tau)u) < a - 2\delta, \quad 0 \leq \tau \leq t.$$

In either case, we cannot have $\sigma(t)u \in B$. Thus (2.25) holds. Let $\Gamma \in \Phi_R$ satisfy (2.18), and define Γ_1 by (2.19). Then $\Gamma_1 \in \Phi_R$. But (2.24) and (2.25) imply that $\Gamma_1(s)A \cap B = \phi$ for all $s \in [0, 1]$, contradicting the fact that A links B relative to Φ_R. Hence there is a sequence satisfying both (2.8) and (2.9). $\square$

Theorem 2.2. *Assume that*
$$m_R := \inf_{B_R} G > -\infty \qquad (2.26)$$

and that there are constants $\delta > 0, \theta < 1$ such that (2.6) holds for all $u \in \partial B_R$ satisfying
$$G(u) < m_R + \delta. \qquad (2.27)$$

Then there is a sequence $\{u_k\} \subset \bar{B}_R$ such that
$$G(u_k) \to m_R, G'(u_k) \to 0. \qquad (2.28)$$

Proof: If such a sequence did not exist, there would be positive constants δ, m such that (2.10) would hold for all u in the set
$$Q = \{u \in \bar{B}_R : G(u) < m_R + 3\delta\}$$

By shrinking δ if necessary, we may assume that (2.6) holds for all u in $\tilde{Q} = Q \cap \partial B_R$. Let
$$Q_0 = \{u \in \bar{B}_R : G(u) < m_R + 2\delta\}$$
$$Q_1 = \{u \in \bar{B}_R : G(u) < m_R + \delta\}$$

and define $Q_2, \eta(u), Y(u), W(u), \sigma(t)$ as before with respect to the new sets Q_j. As we saw in the proof of Theorem 2.1, $\sigma(t)u$ does not exist $\bar{B}_R$ for $t \geq 0$. By (2.26), Q_1 is not empty. Let u be any element of Q_1, and take $T > 2\delta/\alpha m$. Then (2.15) guarantees that $\sigma(t)u \in Q_1$ for all $t \geq 0$. But this means that
$$G(\sigma(T)u) \leq m_R + \delta - \alpha mT < m_R - \delta$$

contradicting (2.26). Thus a sequence satisfying (2.28) exists. $\square$

3. THE MOUNTAIN PASS ALTERNATIVE

An important hypothesis in Theorem 2.1 was (2.6). Without it we cannot be assured that the mapping $\sigma(t)$ stays in $\bar{B}_R$. Consequently, we cannot assert that Γ_1 is in Φ_R. In this section we examine what happens when hypothesis (2.6) is dropped. Of course we cannot expect to obtain a Palais–Smale sequence. However, all is not lost. We do obtain a sequence which can lead not to a critical point, but instead to an eigenvalue. We present the theory here. Assume that G is a C^1 functional on $\bar{B}_R$ and that $A, B \subset \bar{B}_R$ are subsets such that A links B with respect to Φ_R and (2.4) holds. Assume that a_R given by (2.5) is finite and that

$$-\nu(u) \leq K_R, u \in \partial B_R \tag{3.1}$$

where we define

$$\nu(u) := (G'(u), u), \beta(u) := \nu(u)/\|u\|^2$$
$$X(u) := G'(u) - \beta(u)u, \quad u \neq 0. \tag{3.2}$$

We have

Theorem 3.1. *Under the above hypotheses, the following alternative holds. Either* **(a)** *there is a sequence satisfying (2.8) (and (2.9) if $a_R = b_0$) or* **(b)** *there is a sequence $\{u_k\} \subset \partial B_R$ such that*

$$G(u_k) \to a_R, X(u_k) \to 0, \nu(u_k) \leq 0. \tag{3.3}$$

Proof: Suppose there is no sequence satisfying (3.3). Then there are positive constants m, δ such that

$$\|X(u)\| \geq m \tag{3.4}$$

whenever

$$|G(u) - a_R| \leq \delta, \nu(u) \leq 0, u \in \partial B_R. \tag{3.5}$$

If the set described by (3.5) is empty, then (2.6) holds for all $u \in \partial B_R$ satisfying (2.7), and option (a) follows form Theorem 2.1. Let θ be such that

$$0 < \theta^{-2} - 1 < m^2 R^2 / K_R^2. \tag{3.6}$$

Then if u satisfies (3.5), we have

$$\nu(u)^2(\theta^{-2} - 1) < R^2 \|X(u)\|^2 \nu(u)^2 / K_R^2$$

and consequently

$$\theta^{-2}\nu(u)^2 \leq R^2 \|X(u)\|^2 + \nu(u)^2 = R^2 \|G'(u)\|^2.$$

Hence u satisfies (2.6). Thus (2.6) holds for all u satisfying (3.5). It also holds trivially for all u such that $\nu(u) > 0$. Hence, (2.6) holds for all $u \in \partial B_R$ satisfying (2.7). Now all of the hypotheses of Theorem 2.1 are satisfied, and option (a) follows. □

Assume next that (2.6) holds. We have

Corollary 3.2. *Under the hypotheses of Theorem 3.1 there is a sequence $\{u_k\} \subset \bar{B}_R$ such that*

$$G(u_k) \to a_R, X(u_k) \to 0, \nu(u_k) \to \nu \leq 0. \tag{3.7}$$

Proof: By Theorem 3.1, either there is a sequence satisfying (2.8) or one satisfying (3.3). If (2.8) is satisfied, then $X(u_k) \to 0$ since

$$\|G'(u)\|^2 = \|X(u)\|^2 + \|u\|^{-2}\nu(u)^2, u \in E. \tag{3.8}$$

Moreover, (2.8) and $\|u_k\| \leq R$ imply $\nu(u_k) \to 0$. Thus (3.7) holds with $\nu = 0$. If (3.3) holds, then (3.1) implies that there is a subsequence such that $\nu(u_k) \to \nu \leq 0$. This implies (3.7). □

Theorem 3.3. *If (3.1) and (2.26) hold, then there is a sequence $\{u_k\} \subset \bar{B}_R$ such that*

$$G(u_k) \to m_R, X(u_k) \to 0, \nu(u_k) \to \nu \leq 0. \tag{3.9}$$

Proof: If no such sequence existed, then for each $M > 0$ there would be positive constants m, δ such that (2.10) holds for all $u \in E$ satisfying

$$G(u) < m_R + \delta, -M \leq \nu(u) \leq 0, u \in \bar{B}_R \tag{3.10}$$

(this set may be empty). Take $M = K_R$ given in (3.1) and let θ satisfy (3.6). Then it follows from (3.6) that (2.6) holds for all $u \in \partial B_R$ satisfying (3.10). Moreover, (2.6) holds trivially when $\nu(u) \geq 0$. Hence it holds for all $u \in \partial B_R$ satisfying (2.27). The hypotheses of Theorem 2.2 are now satisfied, and we can conclude that there is a sequence such that (2.28) holds. But such a sequence satisfies (3.9), and the theorem is established. □

4. A COMPACTNESS CONDITION

We now introduce a compactness condition, similar to the Palais–Smale condition, but much weaker.

(I) If $\mu \in \mathbb{R}, \nu \leq 0, \{u_k\} \subset \bar{B}_R$ and

$$G(u_k) \to \mu, X(u_k) \to 0, \nu(u_k) \to \nu, \tag{4.1}$$

then $\{u_k\}$ has a convergent subsequence.

Note that condition (I) requires the sequence to be bounded unlike the Palais–Smale condition. We have

Theorem 4.1. *If we add condition (I) to the hypotheses of Theorem 3.1, we can solve*

$$G(u) = a_R, G'(u) = \beta u, \beta \leq 0, u \in \bar{B}_R. \tag{4.2}$$

Moreover, we can take $u \in \partial B_R \cup \bar{B}$ if $a_R = b_0$ and $u \in \partial B_R$ if $\beta \neq 0$.

Proof: By Theorem 3.1, we have either a sequence satisfying (2.8) (and (2.9) if $a_R = b_0$) or a sequence on ∂B_R satisfying (3.3). In either case the sequence satisfies (4.1). By condition (I), it has a convergent subsequence. Thus there is a renamed subsequence such that

$$u_k \to u, G'(u_k) \to G'(u) \text{ in } E \tag{4.3}$$

$$G(u_k) \to G(u), v(u_k) \to v(u), \beta(u_k) \to \beta(u) \text{ in } \mathbb{R}. \tag{4.4}$$

In the case of (2.8) (and (2.9)) we obtain

$$G(u) = a_R, G'(u) = 0 \tag{4.5}$$

with $u \in \bar{B}$ if $a_R = b_0$. If (3.3) holds, we obtain

$$G(u) = a_R, X(u) = 0, v(u) \leq 0, u \in \partial B_R. \tag{4.6}$$

Since $X(u)$ is given by (3.2), we obtain (4.2) in both cases with $u \in \partial B_R \cup \bar{B}$ if $a_R = b_0$ and $u \in \partial B_R$ if $\beta \neq 0$. □

Theorem 4.2. *If (3.1), (2.26) and condition (I) hold, then there is a solution of*

$$G(u) = m_R, G'(u) = \beta u, \beta \leq 0, u \in \bar{B}_R \tag{4.7}$$

with $u \in \partial B_R$ if $\beta \neq 0$.

Proof: By Theorem 3.3 there is a sequence satisfying (3.9). Thus a renamed subsequence satisfies (4.3), (4.4). This implies that there is a solution of

$$G(u) = m_R, X(u) = 0, \quad v(u) \leq 0, u \in \bar{B}_R$$

which in turn implies (4.7). Again we can take $u \in \partial B_R$ if $\beta \neq 0$. □

Theorem 4.3. *If we add (2.26) to the hypotheses of Theorem 4.1, then there are at least two solutions of*

$$G'(u) = \beta u, \beta \leq 0, u \in \bar{B}_R. \tag{4.8}$$

If there are two points $u_0, u_1 \in A \cap B_R$ such that $G(u_1) \leq G(u_0)$ and $u_0 \notin \bar{B}$, then we can take both solutions unequal to u_0.

Proof: By Theorems 4.1 and 4.2 there are solutions of (4.2) and (4.7). If $m_R \leq a_0 < a_R$, then we have two different solutions of (4.8). Since $G(u_1) \leq G(u_0)$, we can take both solutions not equal to u_0. If $m_R < a_0 = a_R$, then $a_R = b_0$, and there are solutions of (4.2) and (4.7), with the former in $\partial B_R \cup \bar{B}$. Again both solutions do not equal u_0. If $m_R = a_0 = a_R$, then every point of A is a solutions of (4.7) and there is a solution of (4.2) in $\partial B_R \cup \bar{B}$. Thus two of the solutions will not equal u_0. □

Corollary 4.4. *Under the same hypotheses, the following alternative holds. Either* **(a)** *there are at least two solutions of*

$$G'(u) = 0, \quad u \in \bar{B}_R \tag{4.9}$$

or **(b)** *there is at least one solution of*

$$G'(u) = \beta u, \quad \beta < 0, \quad u \in \partial B_R. \tag{4.10}$$

If points u_0, u_1 described in Theorem 4.3 exist, then the two solutions of (4.9) can be taken unequal to u_0.

Eigenvalue Problems for Semilinear Equations

Proof: If no solutions of (4.10) exists, then the two solutions of (4.8) given by Theorem 4.3 must satisfy (4.9). □

Corollary 4.5. *Under the same hypotheses, the following alternative holds. Either* **(a)** *there is at least one solutions of (4.9) or* **(b)** *there are at least two solutions of (4.10). If u_0, u_1 described in Theorem 4.3 exist, one can take the solution of (4.9) not equal to u_0.*

Proof: If (4.9) has no solutions, then the two solutions of (4.8) given by Theorem 4.3 must satisfy (4.10). □

5. DUAL SITUATIONS

If we replace G by $-G$ and interchange the sets A and B, we obtain dual theorems of those of Sections 2–4. We state them here.

Theorem 5.1. *Let G be a C^1 functional on $\bar{B}_R$ and let $A, B \subset \bar{B}_R$ be such that B links A with respect to Φ_R and (2.4) holds. Assume that*

$$b_R := \sup_{\Gamma \in \Phi_R} \inf_{\substack{0 \le s \le 1 \\ v \in B}} G(\Gamma(s)v) > -\infty \qquad (5.1)$$

and that there are constants $\delta > 0, \theta < 1$ such that

$$(G'(u), u) \le \theta R \|G'(u)\| \qquad (5.2)$$

holds for all $u \in \partial B_R$ satisfying

$$|G(u) - b_R| \le \delta. \qquad (5.3)$$

Then there is a sequence $\{u_k\} \subset \bar{B}_R$ such that

$$G(u_k) \to b_R, \, G'(u_k) \to 0. \qquad (5.4)$$

If $b_R = a_0$, we can also require that

$$d(u_k, A) \to 0. \qquad (5.5)$$

Theorem 5.2. *Assume that*

$$M_R := \sup_{B_R} G < \infty \qquad (5.6)$$

and that there are constants $\delta > 0, \theta < 1$ such that (5.2) holds for all $u \in \partial B_R$ satisfying

$$G(u) > M_R - \delta. \qquad (5.7)$$

Then there is a sequence $\{u_k\} \subset \bar{B}_R$ such that

$$G(u_k) \to M_R, \, G'(u_k) \to 0. \qquad (5.8)$$

Theorem 5.3. *Assume that B links A, (2.4) and (5.1) hold and*

$$\nu(u) \leq K_R, \quad u \in \partial B_R. \tag{5.9}$$

Then the following alternative holds. Either **(a)** *there is a sequence satisfying (5.4) (and (5.5) if $b_R = a_0$) or |bf (b) there is a sequence $\{u_k\} \subset \partial B_R$ such that*

$$G(u_k) \to b_R, X(u_k) \to 0, \nu(u_k) \geq 0. \tag{5.10}$$

Corollary 5.4. *Under the above hypotheses, there is a sequence $\{u_k\} \subset \bar{B}_R$ satisfying*

$$G(u_k) \to b_R, X(u_k) \to 0, \nu(u_k) \to \nu \geq 0. \tag{5.11}$$

Theorem 5.5. *Assume (5.6) and (5.9) hold. Then there is a sequence $\{u_k\} \subset \bar{B}_R$ such that*

$$G(u_k) \to M_R, X(u_k) \to 0, \nu(u_k) \to \nu \geq 0. \tag{5.12}$$

Now we use the following counterpart of condition (I).

(II) If $\mu \in \mathbb{R}, \nu \geq 0, \{u_k\} \subset \bar{B}_R$ and (4.1) holds, then $\{u_k\}$ has a convergent subsequence.

Theorem 5.6. *In addition the hypotheses of Theorem 5.3, assume that condition (II) holds. Then we can solve*

$$G(u) = b_R, G'(u) = \beta, \beta \geq 0, u \in \bar{B}_R. \tag{5.13}$$

If $b_R = a_0$, we can take $u \in \partial B_R \cup \bar{A}$; if $\beta \neq 0$, we can take $u \in \partial B_R$.

Theorem 5.7. *In addition to the hypotheses of Theorem 5.5, assume that condition (II) holds. Then we can solve*

$$G(u) = M_R, G'(u) = \beta u, \beta \geq 0, u \in \bar{B}_R \tag{5.14}$$

with $u \in \partial B_R$ if $\beta \neq 0$.

Theorem 5.8. *If we assume (2.4), (5.1), (5.6), (5.9) and condition (II), then there are at least two solutions of*

$$G'(u) = \beta u, \beta \geq 0, u \in \bar{B}_R. \tag{5.15}$$

If there are two points $v_0, v_1 \in B \cap B_R$ such that $G(v_0) \leq G(v_1)$ and $v_0 \notin \bar{A}$, then we can take both solutions $\neq v_0$.

Corollary 5.9. *Under the same hypotheses, the following alternative holds. Either* **(a)** *there are at least two solutions of (4.9), or* **(b)** *there is at least one solution of*

$$G'(u) = \beta u, \quad \beta > 0, \quad u \in \partial B_R. \tag{5.16}$$

If points v_0, v_1 described in Theorem 5.8 exist, then the solutions of (4.9) can be taken unequal to v_0.

Corollary 5.10. *Under the same hypotheses, the following alternative holds. Either* **(a)** *there is at least one solution of (4.9) or* **(b)** *there are at least two solutions of (5.16). If v_0, v_1 described above exist, then we can take the solution of (4.9) unequal to v_0.*

6. COMBINED RESULTS

We now show what can be obtained by combining the results of Sections 4 and 5. Again we assume that G is a C^1 functional on $\bar{B}_R$ and that A and B are subsets of $\bar{B}_R$ which link each other with respect to the family Φ_R. We assume that (2.4) holds and that

$$|v(u)| \leq K_R, u \in \partial B_R \tag{6.1}$$

and

$$\sup_{B_R} |G(u)| < \infty. \tag{6.2}$$

We assume a compactness condition which combines (I) and (II).

(III) If $\mu, v \in \mathbb{R}, \{u_k\} \subset \bar{B}_R$ and (4.1) holds, then $\{u_k\}$ has a convergent subsequence.

Note that our assumptions include (2.26), (3.1), (5.6) and (5.9). Since A and B link each other with respect to Φ_R and (2.4) holds, we have

$$m_R \leq b_R \leq a_0 \leq b_0 \leq a_R \leq M_R. \tag{6.3}$$

We have

Theorem 6.1. *The following alternative holds. Either* **(a)** *there are at least four solutions of*

$$G'(u) = \beta u, \beta \in \mathbb{R}, u \in \bar{B}_R \tag{6.4}$$

with $u \in \partial B_R$ if $\beta \neq 0$, or **(b)** *there are at least three solutions of (6.4) with at least one of them a solution of*

$$G'(u) = 0, u \in \bar{B}_R \tag{6.5}$$

or **(c)** *there are at least two solutions of (6.5).*

Proof: If

$$m_R < b_R < a_R < M_R \tag{6.6}$$

then it follows from Theorems 4.1, 4.2, 5.6 and 5.7 that we can solve (6.4) with $G(u)$ equal to each of the numbers in (6.6). This gives option (a). If

$$m_R = b_R < a_R < M_R \tag{6.7}$$

we have at least three distinct solutions with the solutions of (4.7) and (5.13) possibly coinciding. However, if these two equations have the same solution, then we must have $\beta = 0$ since (4.7) requires $\beta \leq 0$ and (5.13) requires $\beta \geq 0$. Thus the alternative holds in this case. If

$$m_R < b_R = a_R < M_R \tag{6.8}$$

then again we have at least three distinct solutions of (6.4) with solutions of (4.2) and (5.13) possibly coinciding. But then again we must have $\beta = 0$ for this solution since (4.2) requires $\beta \leq 0$ while (5.13) stipulates $\beta \geq 0$. The same reasoning applies if

$$m_R < b_R < a_R = M_R \tag{6.9}$$

since (4.2) and (5.14) coincide only if $\beta = 0$. If

$$m_R = b_R = a_R < M_R \tag{6.10}$$

then $a_0 = m_R$, and each point of A is a minimum point for G in $\bar{B}_R$. Since A must have at least two points in order to link B, we obtain at least three solutions of (6.4). If a solution of (5.13) coincides with a solution of (4.2) or (4.7), we must have $\beta = 0$. Thus the theorem is proved in this case as well. The same argument applies when

$$m_R < b_R = a_R = M_R. \tag{6.11}$$

If

$$m_R = b_R < a_R = M_R \tag{6.12}$$

and we only have two distinct solutions of (6.4) because the solutions of (4.7) and (5.13) coincide and the solutions of (4.2) and (5.14) coincide, then in each case we must have $\beta = 0$ because in the case of (4.7) and (4.2) we have $\beta \leq 0$ and in the case of (5.13) and (5.14) we have $\beta \geq 0$. This gives option (c). Finally, if

$$m_R = b_R = a_R = M_R \tag{6.13}$$

then G is constant on $\bar{B}_R$ and (6.5) has an infinite number of solutions. □

Corollary 6.2. *If there are no solutions of*

$$G'(u) = \beta u, \beta \neq 0, u \in \partial B_R \tag{6.14}$$

then (6.5) has at least two solutions.

Corollary 6.3. *The following alternative holds. Either* **(a)** *there are at least two solutions of (6.5) or* **(b)** *there are at least two solutions of (6.14).*

Corollary 6.4. *The following alternative holds. Either* **(a)** *there is at least one solution of (6.5) or* **(b)** *there are at least four solutions of (6.14).*

Remark 6.1. We can improve the results of Theorem 6.1 and Corollary 6.2 if we assume that $\bar{A}$ cannot link a singleton (a set consisting of a single point) with respect to Φ_R. For the only case in which option (c) of Theorem 6.1 arose is when (6.12) holds. This implies

$$\sup_{\substack{u \in A \\ 0 \leq s \leq 1}} G(\Gamma(s)u) = M_R \tag{6.15}$$

for every $\Gamma \in \Phi_R$. If there is only one point $w \in \bar{B}_R$ such that $G(w) = M_R$, then there is a $\Gamma \in \Phi_R$ such that

$$\Gamma(s)\bar{A} \cap \{w\} = \phi, s \in [0,1].$$

But this would contradict (6.15). Hence option (c) can be eliminated from Theorem 6.1 in this case. Similarly, in Corollary 6.2 one obtains at least three solutions of (6.5) when (6.14) has none. It should be noted that a set consisting of two points cannot link a singleton with respect to Φ_R. For if $A = \{u_0, u_1\}$ and $B = \{w\}$, we can find a point $v \in \bar{B}_R$ such that the line segments connecting the u_j with v do not contain w. If we take $\Gamma(s)u = (1-s)u + sv$, then

$$\Gamma \in \Phi_R \text{ and } \Gamma(s)A \cap B = \phi, 0 \leq s \leq 1.$$

Eigenvalue Problems for Semilinear Equations

Theorem 6.5. *Assume, in addition, that $\bar{A}$ cannot link a singleton with respect to Φ_R and that A contains at least two points u_0, u_1 not in $\partial B_R \cup \bar{B}$ with $G(u_1) \leq G(u_0)$. Then (6.4) has at least three solutions not equal to u_0. If $\beta \neq 0$ for these solutions, then (6.4) has at least four solutions $\neq u_0$.*

Proof: If

$$m_R < G(u_0) < a_R < M_R$$

then Theorems 4.1, 4.2 and 5.7 produce three solutions unequal to u_0. If $a_R = M_R$, then the preceding remark shows that there are at least two points where $G(u) = M_R$. This gives three again. If

$$m_R = G(u_0) < a_R < M_R$$

we see that u_1 is a solution of (4.7). Again we obtain three solutions of (6.4) unequal to u_0. If

$$m_R < G(u_0) = a_R < M_R,$$

then $b_0 = a_R$, and (4.2) has a solution in $\partial B_R \cup \bar{B}$ by Theorem 4.1. Two others are obtained from Theorems 4.2 and 5.7. If

$$m_R = G(u_0) = a_R < M_R$$

we see that u_1 is a solution of (4.7) and that $b_0 = a_R$. Hence (4.2) has a solution in $\partial B_R \cup \bar{B}$ and this solution does not equal u_0 or u_1. A third solution comes via Theorem 5.7. Thus in all cases we obtain three solutions not equal to u_0. If $\beta \neq 0$ in all three cases, we note that Theorems 5.6 and 5.7 cannot give the same solution unless $\beta = 0$. Thus Theorem 5.6 provides a fourth solution. This fourth solution cannot be a solution of (4.2) if $\beta \neq 0$ even when $a_R = b_R$. □

7. THE HAMPWILE THEOREM

In this section we state and prove the Hampwile theorem which can be described as a half mountain pass lemma with loose end. It will be used to prove Theorem 1.6.

Again we consider a C^1 functional G on $\bar{B}_R$ and assume that there is an element $e \in \bar{B}_R$ and a $\delta < \|e\|$ such that

$$G(e) < \inf_{\partial B_\delta} G = b_0. \tag{7.1}$$

This implies (2.4) if we take $A = \{e\}$ and $B = \partial B_\delta$. However, in this case A does not link B with respect to Φ_R. For this reason we need special techniques for proving the theorems below. Let

$$B_{R,\delta} := \{u \in \bar{B}_R : \delta \leq \|u\| \leq R\}. \tag{7.2}$$

Let S denote the set of continuous mappings φ from $[0,1]$ to $\bar{B}_R$ such that $\|\varphi(0)\| \leq \delta, \varphi(1) = e$. Define

$$c = \inf_{\varphi \in S} \max_{0 \leq s \leq 1} G(\varphi(s)). \tag{7.3}$$

First we have

Theorem 7.1. *In addition to (7.1) assume that there are constants $\epsilon > 0, \theta < 1$ such that (2.6) holds for all $u \in \partial B_{R,\delta}$ satisfying*

$$|G(u) - c| \leq 3\epsilon. \tag{7.4}$$

Then there is a sequence $\{u_k\} \subset \bar{B}_{R,\delta}$ such that

$$G(u_k) \to c, G'(u_k) \to 0. \tag{7.5}$$

Proof: If there is no sequence satisfying the conclusions of the theorem, then for ϵ sufficiently reduced there is an $m > 0$ such that (2.10) holds for all $u \in \partial B_{R,\delta}$ satisfying (7.4). Reduce ϵ so that $3\epsilon < b_0 - G(e)$. Let

$$Q_0 = \{u \in \bar{B}_R : |G(u) - c| \leq 2\epsilon\}$$
$$Q_1 = \{u \in Q_0 : |G(u) - c| \leq \epsilon\}$$

and define Q_2 and $\eta(u)$ as before. Let α be any number satisfying $0 < \alpha < 1 - \theta$. Then there is a locally Lipschitz continuous map $Y(u)$ of $\hat{B}_R$ to E such that (2.11) and (2.12) hold, where $\hat{Q}$ is the set of all $u \in \partial B_{R,\delta}$ satisfying (7.4). By (7.3) there is a $\varphi \in S$ such that

$$G(\varphi(s)) < c + \epsilon, \quad 0 \leq s \leq 1. \tag{7.6}$$

We may assume that

$$\|\varphi(s)\| \geq \delta, \quad 0 \leq s \leq 1. \tag{7.7}$$

Otherwise we let s_1 be the smallest value of s such that

$$\|\varphi(s)\| \geq \delta, \quad s_1 \leq s \leq 1.$$

Since $\|\varphi(1)\| = \|e\| > \delta$, we must have $s_1 < 1$. Take

$$\varphi_1(s) = \varphi(s_1 + (1 - s_1)s) \quad 0 \leq s \leq 1.$$

Then φ_1 is in S and satisfies both (7.6) and (7.7). Let $W(u)$ be defined by (2.13) and let $\sigma(t)$ be the flow generated by W on E. As before, $\sigma(t)u$ cannot exist $\bar{B}_R$ as t increases. For the same reason, it cannot exist $\bar{B}_\delta$. Let $\sigma(t,s) = \sigma(t)\varphi(s), s \in [0,1], t > 0$. In view of (2.15), we have

$$dG(\sigma(t,s))/dt \leq -\alpha m, \quad \sigma(t,s) \in Q_1 \cap B_{R,\delta}. \tag{7.8}$$

Let T satisfy $2\epsilon < \alpha m T$, and let $s \in [0,1]$ be such that $\sigma(t,s) \in B_{R,\delta}$ for $t \in [0,T]$. If there is a $t_1 \in [0,T]$ such that $\sigma(t_1,s)$ is not in Q_1, then

$$G(\sigma(T,s)) \leq G(\sigma(t_1,s)) < c - \epsilon \tag{7.9}$$

since it cannot be greater than $c + \epsilon$ by (7.6). On the other hand, if $\sigma(t,s)$ is in Q_1 for each $t \in [0,T]$, then

$$G(\sigma(T,s)) \leq G(\varphi(s)) - \alpha m T < c - \epsilon$$

by (7.8). Thus (7.9) holds in this case as well. Let

$$s_1 = \inf\{s : \sigma(t,s) \in B_{R,\delta}, t \in [0,T]\}.$$

Since $G(\varphi(1)) = G(e) < b_0 - 3\epsilon \leq c - 3\epsilon$, we see that $\sigma(t,1) = e, t \in [0,T]$. Consequently $s_1 < 1$ and $\|\sigma(T,s_1)\| = \delta$. Let

$$\varphi_1(s) = \sigma(T, s_1 + (1-s_1)s), \quad 0 \leq s \leq 1.$$

Then $\varphi_1(s)$ is in S and

$$G(\varphi_1(s)) < c - \epsilon, \quad 0 \leq s \leq 1$$

by (7.9). This contradicts (7.3) and proves the theorem. □

Theorem 7.2. *Assume (7.1) and (3.1). Then there is a sequence $\{u_k\} \subset \bar{B}_{R,\delta}$ such that*

$$G(u_k) \to c, X(u_k) \to 0, \limsup \nu(u_k) \leq 0. \tag{7.10}$$

Proof: We essentially follow the proof of Theorem 3.1. If such a sequence did not exist, then for ϵ sufficiently small there would be an $m > 0$ such that (3.4) holds when $u \in \bar{B}_{R,\delta}$ satisfies (7.4) and $\nu(u) \leq 0$. Following the proof of Theorem 3.1 we see that this leads to (2.6) holding for such u with $\theta < 1$ suitably chosen (with R replaced by δ). We can now apply Theorem 7.1 to conclude that there is a sequence in $\bar{B}_{R,\delta}$ satisfying (7.5). But this implies (7.10). □

Corollary 7.3. *If we add (2.26) and Condition (I) of Section 4 to the hypotheses of Theorem 7.2, then there are at least two distinct solutions of*

$$G'(u) = \beta u, \quad \beta \leq 0, \quad u \in \bar{B}_R. \tag{7.11}$$

If $G(e) < 0$, then both solutions are nontrivial.

Proof: By Theorem 7.2 there is a sequence in $\bar{B}_{R,\delta}$ satisfying (7.10). In particular, there is a subsequence satisfying (4.1) with $\mu = c$. Hence, it has a convergent subsequence. Thus $u_k \to u$ in $\bar{B}_{R,\delta}, \beta(u_k) \to \nu/\|u\|^2 \equiv \beta \leq 0$ and $G'(u_k) = X(u_k) + \beta(u_k)u_k \to \beta u$. This gives a solution of (7.11) with $G(u) = c, u \in \bar{B}_{R,\delta}$. Theorem 4.2 provides a solution of (4.7). Since

$$m_R \leq G(e) < b_0 \leq c,$$

the solutions are distinct. If $G(e) < 0, m_R \neq 0$, and both solutions are nontrivial. □

8. THE APPLICATIONS

We now give the proofs of the theorems of Section 1.

Proof of Theorems 1.1 and 1.2: Let

$$G(u) = \|u\|_D^2 - 2\int_\Omega F(x,u)dx, \quad u \in D. \tag{8.1}$$

It is well known that under hypotheses (1.3)–(1.7), G is a C^1 functional on D and satisfies

$$(G'(u), v)_D = 2(u,v)_D - 2(f(u), v), u, v \in D \tag{8.2}$$

(cf., e.g., [2]). Take $0 < \rho < \|u_0\|_D$, and let A be the set $\{0, u_0\}$ and B the set ∂B_ρ. It is well known that A links B (cf., e.g., [3,4]). By hypothesis

$$\int |F(x,u)| dx \le C(\|\nabla u\|_q^q + \|\nabla u\|_q \|W\|_{q'})$$
$$\le C'(\|u\|_D^q + \|u\|_D), u \in D$$
$$\int_\Omega |uf(x,u)| dx \le C''(\|u\|_D^q + \|u\|_D), u \in D$$

Thus

$$|G(u)| \le \|u\|_D^2 + 2C'(\|u\|_D^q + \|u\|_D), u \in D$$

and

$$|\nu(u)| \le 2\|u\|_D^2 + C''(\|u\|_D^q + \|u\|_D), u \in D.$$

Let $R > \|u_0\|_D$. Then these inequalities imply (5.6) and (6.1). I claim that hypothesis (I) of Section 4 is satisfied. To see this let $\{u_k\}$ be a sequence satisfying (4.1) with $\nu \le 0$. Since the u_k are in $\bar{B}_R$, there is a renamed subsequence such that $u_k \to u$ weakly in D and such that

$$\int_\Omega f(x, u_k) v \, dx \to \int f(x, u) v \, dx, v \in D \tag{8.3}$$

and

$$\int_\Omega f(x, u_k) u_k \, dx \to \int f(x, u) u \, dx.$$

We may assume that $\|u_k\|_D \to r$. If $r = 0$, then (I) is established. Otherwise (4.1) implies

$$2(u_k, v)_D - 2(f(u_k), v) - \beta(u_k)(u_k, v)_D \to 0, v \in D \tag{8.4}$$

$$2\|u_k\|_D^2 - 2(f(u_k), u_k) - \beta(u_k)\|u_k\|_D^2 \to 0. \tag{8.5}$$

Thus in the limit

$$2(u, v) - 2(f(u), v) - \nu r^{-2}(u, v)_D = 0. \tag{8.6}$$

and

$$2r^2 - 2(f(u), v) - \nu = 0. \tag{8.7}$$

Setting $v = u$ in (8.6) and comparing with (8.7), we obtain

$$(2 - \nu r^{-2}) \|u\|_D^2 = 2r^2 - \nu. \tag{8.8}$$

Since $\nu \le 0$, this implies $\|u\|_D = r$, showing that $u_k \to u$ strongly in D. Thus (I) is established. It now follows that all of the hypotheses of Theorem 4.3 are satisfied. The theorems follow from Corollaries 4.4 and 4.5, respectively if we make use of the fact that a solution of (4.8) is a solution of (1.2) if we take $\lambda = 2/(2-\beta)$. □

Proof of Theorems 1.3–1.5: If we can verify that hypothesis (III) of Section 6 is a satisfied, then the theorems will follow from Theorems 6.1 and 6.5 together with Corollaries 6.3–6.4. To this end, let $\{u_k\} \subset \bar{B}_R$ be a sequence satisfying (4.1) with $\nu > 0$. As we saw in the preceding proof, this implies that there is a renamed subsequence such that $u_k \to u$ weakly in D, $\|u_k\|_D \to r$, and (8.2)–(8.8) hold. By (8.6)

$$(2 - \nu r^{-2})Au = f(x,u). \tag{8.9}$$

If (1.11) holds, we see that $u \neq 0$ and $\nu \neq 2r^2$. We can now conclude from (8.8) that $\|u\|_D = r$. Consequently, $u_k \to u$ strongly in D and hypothesis (III) is verified. The proof is complete. □

Proof of Theorem 1.6: Let $\mu > \mu_0$ and take

$$G(u) = \|u\|_D^2 - 2\mu \int_\Omega F(x,u)dx. \tag{8.10}$$

Then $G \in C^1(D, \mathbb{R})$ and

$$(G'(u), v) = 2(u,v)_D - 2\mu(f(u), v), u, v \in D. \tag{8.11}$$

Thus if u is a solution of (7.11), then

$$(2 - \beta)Au = 2\mu f(x,u). \tag{8.12}$$

Consequently, u is a solution of (1.2) with $\lambda \leq \mu$. Now

$$G(u_0) = \|u_0\|_D^2 - 2\mu \int_\Omega F(x, u_0)dx < 0$$

by (1.14). Since $G(0) = 0$, we see by continuity that there is a $\delta > 0$ such that (7.10) holds with $e = u_0$. Moreover (3.1) holds for any $R > 0$. We take $R = \|u_0\|_D$. As shown in the proof of Theorem 1.1 and 1.2, condition (I) of Section 4 is also satisfied. Finally (2.26) is satisfied as well. We may now apply Corollary 7.3 to reach the desired conclusion. □

REFERENCES

1. M. Schechter, The Mountain Cliff Theorem, Differential Equations and Mathematical Physics (Birmingham, AL, 1990) 263–279, Math. Sci. Engineering 186 Academic Press, Boston, MA, 1992.
2. M. Schechter, A bounded mountain pass lemma without the (PS) condition and applications, Trans. Amer. Math. Soc. 331(1992) 681–703.
3. P. H. Rabinowitz, Minimax methods in critical point theory with applications to differential equations, Conf. Board of Math. Sci. Reg. Conf. Ser. in Math. No. 65, Amer. Math. Soc. 1986.
4. M. Schechter and K. Tintarev, Pairs of critical points produced by linking subsets with applications to semilinear elliptic problems, Bull. Soc. Math. Belg. 44(1992) 249–261.
5. A. Ambrosetti and P. H. Rabinowitz, Dual variational methods in critical point theory and applications, J. Func. Anal. 14(1973) 349–381.
6. H. Brezis and L. Nirenberg, Remarks on finding critical points, Comm. Pure Appl. Math. 44(1991) 939–964.
7. J. Mawhin and M. Willem, Critical Point Theory and Hamiltonian Systems, Springer, NY, 1989.
8. L. Nirenberg, Variational an topological methods in nonlinear problems, Bul. Amer. Math. Soc. 4(1981) 267–302.
9. M. Schechter, The Hampwile theorem for nonlinear eigenvalues, Duke Mathematical Journal, 59(1989) 325–335.
10. M. Schechter, The Hampwile alternative, Comm. on Applied Nonlinear Analysis, 1(1994)13–6.

SPECTRAL OPERATORS GENERATED BY 3-DIMENSIONAL DAMPED WAVE EQUATION AND APPLICATIONS TO CONTROL THEORY

Marianna A. Shubov

Texas Tech University
Lubbock, Texas

ABSTRACT

We formulate our results on the spectral analysis for a class of nonselfadjoint operators in a Hilbert space and on the applications of this analysis to the control theory of linear distributed parameter systems. The operators, we consider, are the dynamics generators for systems governed by 3-dimensional wave equation which has spacially nonhomogeneous spherically symmetric coefficients and contains a first order damping term. We consider this equation with a one-parameter family of linear first order boundary conditions on a sphere. These conditions contain a damping term as well. Our main object of interest is the class of operators in the energy space of 2-component initial data which generate the dynamics of the above systems. Our first main result is the fact that these operators are spectral in the sense of N. Dunford. This result is obtained as a corollary of two groups of results: (i) asymptotic representations for the complex eigenvalues and eigenfunctions, and (ii) the fact that the systems of eigenvectors and associated vectors form Riesz bases in the energy space. We also present an explicit solution of the controllability problem for the distributed parameter systems governed by the aforementioned equation using the spectral decomposition method.

1. INTRODUCTION

In the present paper, we consider a class of nonselfadjoint operators in a Hilbert space which are generated by the 3-dimensional damped wave equation with spherically symmetric coefficients and with linear dissipative boundary conditions on a sphere centered at the origin. It follows from our results, that these operators provide a class of nontrivial examples of spectral operators. As is well known, an abstract theory of nonselfadjoint spectral operators has been developed a long time ago [7,9]. However, there is still a problem of finding specific examples of such operators.

In fact, in this work, we do not need the most general definition of a spectral operator. The operators, we consider here, are Riesz spectral operators with discrete spectrum. So, we accept the following definition.

Definition 1.1. An operator $\mathcal{L}$ in a complex separable Hilbert space $\mathcal{H}$ is called Riesz spectral if it has the following properties:

(i) $\mathcal{L}$ is either bounded or closed unbounded operator defined on a dense domain $D(\mathcal{L}) \subset \mathcal{H}$;

(ii) $\mathcal{L}$ has a discrete spectrum;

(iii) only a finite number of the eigenvectors have finite chains of associated vectors;

(iv) the system of root vectors (eigenvectors and associated vectors together) forms a Riesz basis (a linear isomorphic image of an orthonormal basis) in $\mathcal{H}$.

Definition 1.1 can be reformulated as follows. $\mathcal{L}$ is an operator whose matrix in an appropriate basis has a Jordan normal form with a finite number of nontrivial Jordan cells of a finite length each. Definition 1.1 is also equivalent to the following one.

Definition 1.2. a) Let $\mathcal{R}(\mathcal{H})$ be the class of bounded linear operators in $\mathcal{H}$. A sequence of vectors $\{\psi_n\}_{n=1}^{\infty} \subset \mathcal{H}$ is called a Riesz basis if there exists an operator $A \in \mathcal{R}(\mathcal{H})$ such that $A^{-1} \in \mathcal{R}(\mathcal{H})$ and the system of vectors $\{A\psi_n\}_{n=1}^{\infty}$ forms an orthonormal basis in $\mathcal{H}$. A is called an orthogonalizer of $\{\psi_n\}_{n=1}^{\infty}$. If A is an orthogonalizer, then all other orthogonalizers have the form UA, where U is an arbitrary unitary operator. The norm $\|A\|$ is uniquely defined by the basis. For any Riesz basis $\{\psi_n\}_{n=1}^{\infty} \subset \mathcal{H}$ there is a unique biorthogonal basis $\{\psi_n^*\}_{n=1}^{\infty}$ defined by the relations: $(\psi_n, \psi_m^*) = \delta_{nm}$.

b) Let $\{\psi_n^*\}_{n=1}^{\infty}$ be a Riesz basis in $\mathcal{H}$ and let $\{\lambda_n\}_{n=1}^{\infty}$ be a sequence of complex numbers. Define an operator S in $\mathcal{H}$ by the formula:

$$S\varphi = \sum_{n=1}^{\infty} \lambda_n (\varphi, \psi_n^*) \psi_n \tag{1.1}$$

on the dense domain

$$D(S) = \left\{ \varphi \in \mathcal{H} : \sum_{n=1}^{\infty} |\lambda_n|^2 |(\varphi, \psi_n^*)|^2 < \infty \right\}. \tag{1.2}$$

The operators of type (1.1), (1.2) are called scalar operators.

c) An operator $\mathcal{L}$ in $\mathcal{H}$ is called spectral operator if it can be represented in the form:

$$\mathcal{L} = S + N \tag{1.3}$$

where S is a scalar operator and N is a bounded finite rank nilpotent operator (i.e., there exists k such that $N^k = 0$). N commutes with S.

(In a more general definition of a spectral operator [7], spectral representation (1.1) may be continuous, i.e., it involves integration with respect to a spectral measure, and N may be a quasinilpotent operator, i.e., it is bounded and its spectrum $\mathfrak{S}(N) = \{0\}$.)

In Section 2 below, we give a precise definition of the aforementioned operators generated by the 3-dimensional damped wave equation and formulate our results concerning their spectral properties. In Section 3, we describe applications of these results to control theory of distributed parameter systems. The controllability problem, discussed in Section 3, can be considered as an example of an inverse ill-posed problem. In Section 3, we give references to the related works in the field of control theory.

The full proofs of the asymptotical and spectral results stated in Section 2 will appear in our works [21–23]. These papers can be considered as a continuation of a series of our works [24–28] devoted to the asymptotical and spectral analysis of nonselfadjoint operators generated by the equation of a damped string with spacially nonhomogeneous damping, modulus of elasticity and density coefficients. We considered this equation with a 2-parameter family of boundary conditions which contained, in particular, the Dirichlet, Newmann and Sommerfeld radiation conditions. In the aforementioned works, we carried out a detailed asymptotic analysis of the spectrum and eigenfunctions of the corresponding dynamics generators and proved the Riesz basis property of their root vectors. Thus, we have shown that the dynamics generators of the above systems are Riesz spectral operators. Even in this 1-dimensional case, the spectral analysis turned out to be rather nontrivial due to the combination of a nonconstant damping and density with nonselfadjoint boundary conditions. The corresponding spectral results were known only for a number of particular cases [6, 10, 13, 14, 29]. We mention here that there exists an extensive literature devoted to the study of eigenfunctions of nonselfadjoint differential operators even for multidimensional problems (see, e.g., [1], [2] and references therein and also [15] and references therein). However, in this general situation, it is possible only to show that the system of the generalized eigenfunctions forms the so-called "basis with brackets." These systems are not unconditional bases and, therefore, they are not Riesz bases. So, the general results are insufficient to claim the spectral property.

The problem, we consider in the present work, is spherically symmetric. This certainly allows us to separate the variables in the spherical coordinates. However, the spectral analysis of the dynamics generator in this case is significantly more complicated than in the aforementioned case of 1-dimensional string equation.

In the conclusion, we mention that at the present level of knowledge in the field of nonselfadjoint operators, it seems to be unrealistic to consider the spectral analysis problem for dynamics generators corresponding to general damped hyperbolic equations on general domains in $\mathbb{R}^n$. However, at the end of Section 2, we formulate our conjecture concerning a class of nonspherically symmetric problems.

2. SPECTRAL PROPERTIES OF DYNAMICS GENERATORS OF SYSTEMS GOVERNED BY SPHERICALLY SYMMETRIC DAMPED WAVE EQUATION

1. We consider the wave equation

$$u_{tt} + 2d(r)u_t + \mathbf{L}u = 0, \tag{2.1}$$

where the differential operator $\mathbf{L}$ is defined by

$$\mathbf{L}\varphi = -\frac{1}{\rho(r)} div\,(p(r)\nabla\varphi) + q(r)\varphi \tag{2.2}$$

for any smooth $\varphi(x)$. Here we use the following notation: $x = (x_1, x_2, x_3) \in \mathbb{R}^3$, $r = |x| = (x_1^2 + x_2^2 + x_3^2)^{1/2}$, $u = u(x,t)$ is defined for $x \in B_a$, where $B_a = \{x \in \mathbb{R}^3 : |x| < a\}, a > 0$. The boundary conditions on the sphere $|x| = a$ are

$$\left.\frac{\partial u}{\partial r} + hu_t\right|_{r=a} = 0, \qquad h \in \mathbb{C} \cup \{\infty\}. \tag{2.3}$$

To $h = \infty$, we formally associate the Dirichlet condition: $u|_{r=a} = 0$.

About the coefficients (damping-$d(r)$, density-$\rho(r)$, modulus of elasticity-$p(r)$ and rigidity of an external harmonic force-$q(r)$), we assume:

$$\begin{aligned}
&\rho \in H^2[0,a]; \qquad p,d \in H^1[0,a]; \qquad q \in L^\infty(0,a); \\
&\rho(r), \ p(r) > 0, \ d(r), \ q(r) \geq 0 \text{ for a.e. } r \in [0,a]; \\
&\sqrt{p(a)/\rho(a)} \neq |h| \text{ if } \operatorname{Im} h = 0.
\end{aligned} \tag{2.4}$$

The energy space $\mathcal{H}$ of this problem is the space of 2-component initial data $U(x) = \begin{pmatrix} u_0(x) \\ u_1(x) \end{pmatrix}$, $x \in B_a$, equipped with the norm:

$$\|U\|_{\mathcal{H}}^2 = \frac{1}{2}\int_{B_a} \left(p(r)|\nabla u_0|^2 + q(r)\rho(r)|u_0|^2 + \rho(r)|u_1|^2\right) dx. \tag{2.5}$$

Problem (2.1)–(2.3) can be represented in the form of the first order evolution equation in the space $\mathcal{H}$:

$$U_t = i\mathcal{L}U, \tag{2.6}$$

where the dynamics generator $\mathcal{L}$ is defined by the matrix differential expression

$$\mathcal{L} = -i \begin{pmatrix} 0 & 1 \\ -\mathbf{L} & -2d(r) \end{pmatrix}, \tag{2.7}$$

on the domain

$$\mathcal{D}(\mathcal{L}) = \left\{ U \in \mathcal{H} : u_0 \in H^2(B_a), u_1 \in H^1(B_a), \left.\frac{\partial u_0}{\partial r} + hu_1\right|_{r=a} = 0 \right\}. \tag{2.8}$$

It is the operator $\mathcal{L}$ which is our main object of interest.

Equation (2.1) with boundary conditions (2.3) defines a strongly continuous semigroup of transformations on the space $\mathcal{H}$. The operator $\mathcal{L}$ is the generator of this semigroup. Note that $\mathcal{L}$ is a closed, nonselfadjoint, maximal, dissipative (if $\operatorname{Re} h \geq 0$) operator in $\mathcal{H}$ whose resolvent is compact and, therefore, the spectrum is discrete.

Our main result about the operator $\mathcal{L}$ consists of the following.

Theorem 2.1. *$\mathcal{L}$ is a spectral operator in the sense of Definition 1.1.*

This theorem is a corollary of the group of the results below which provides a detailed information about the spectral properties of $\mathcal{L}$.

2. To formulate these results, we separate the variables in spherical coordinates in Eq. (2.6). Denote by $U_{\ell m j}(r,t)$ the coefficients in the expansion of $U(x,t)$ from (2.6) with

respect to the spherical harmonics $\{Y_\ell^{mj}(\theta,\varphi); \ell = 0,1,2,\ldots; \quad m = 0,\pm 1,\pm 2 \cdots \pm \ell; \quad j=1$ if $m=0, j=1,2$ if $m \neq 0\}$:

$$U(x,t) = \sum_{\ell,m,j} U_{\ell m j}(r,t) Y_\ell^{mj}(\theta,\varphi). \tag{2.9}$$

Each $U_{\ell m j}$ satisfies the radial wave equation

$$\frac{\partial}{\partial t} U_{\ell m j} = i\mathcal{L}_\ell U_{\ell m j}, \tag{2.10}$$

where

$$\mathcal{L}_\ell = -i \begin{pmatrix} 0 & 1 \\ -L_\ell & -2d(r) \end{pmatrix}, \quad \ell = 0,1,2,\ldots, \tag{2.11}$$

and for any smooth $\varphi(r)$

$$L_\ell \varphi = -\frac{1}{\rho(r)} \left[\frac{1}{r^2} \frac{d}{dr}\left(r^2 p(r) \frac{d\varphi}{dr}\right) - \frac{\ell(\ell+1)}{r^2} p(r)\varphi \right] + q(r)\varphi. \tag{2.12}$$

For each ℓ, the "radial" energy space $\mathcal{H}_\ell$ is defined in a standard way. $\mathcal{H}_\ell$ is the closure of all smooth 2-component Cauchy data $U(r) = \begin{pmatrix} u_0(r) \\ u_1(r) \end{pmatrix}$, such that $u_0(r)$ is equal to zero in a vicinity of $r=0$, in the norm:

$$\|U\|_{\mathcal{H}_\ell}^2 = \frac{1}{2} \int_0^a \left[p(r)|u_0'|^2 + q(r)\rho(r)|u_0|^2 + \frac{\ell(\ell+1)}{r^2} p(r)|u_0|^2 + \rho(r)|u_1|^2 \right] r^2 dr. \tag{2.13}$$

(Note, that due to the Hardy inequality

$$\int_0^a \frac{|u_0(r)|^2}{r^2} dr \leq 4 \int_0^a \left|\frac{du_0}{dr}\right|^2 dr, \quad u_0 \in H^1(0,a), \quad u_0(0)=0,$$

all spaces $\mathcal{H}_\ell$ are metrically equivalent to $\mathcal{H}_0$.)

The domain of $\mathcal{L}_\ell$ in $\mathcal{H}_\ell$ is

$$D(\mathcal{L}_\ell) = \left\{ U = \begin{pmatrix} u_0 \\ u_1 \end{pmatrix} \in \mathcal{H}_\ell : \begin{array}{l} u_0 \in H^2(0,a), \quad u_1 \in H^1(0,a), \quad u_1(0) = \\ 0, \quad (u_0' + h u_1)(a) = 0 \end{array} \right\}. \tag{2.14}$$

Now, we describe the spectral properties of the radial operator $\mathcal{L}_\ell$.

Theorem 2.2. *a) The operator $\mathcal{L}_\ell$ has a countable set of complex eigenvalues which belongs to a strip parallel to the real axis and has only two points of accumulation: $+\infty$ and $-\infty$. For this reason, the spectrum can be represented in the form $\{\lambda_n^\ell, n \in \mathbb{Z}' = \mathbb{Z}\setminus\{0\}\}$, where $\operatorname{Re}\lambda_n^\ell \leq \operatorname{Re}\lambda_{n+1}^\ell$ and $\operatorname{Re}\lambda_n^\ell \to \pm\infty$ as $n \to \pm\infty$. (As will be explained below, it is convenient to exclude $n=0$.)*

b) The spectral asymptotics has the form:

$$|\lambda_n^\ell - \Lambda_n^\ell| \leq C_\ell \frac{\ln|n|}{|n|}, \tag{2.15}$$

where C_ℓ may go to ∞ as $\ell \to \infty$ and

$$\Lambda_n^\ell = \mathcal{M}^{-1}\left[\left(n + \frac{\ell+1}{2}\operatorname{sgn} n\right)\pi + i\left(\mathcal{N} + \frac{1}{2}\ln\frac{\sqrt{p(a)/\rho(a)} + h}{\sqrt{p(a)/\rho(a)} - h}\right)\right]. \quad (2.16)$$

$\mathcal{M}$ and $\mathcal{N}$ are defined by

$$\mathcal{M} = \int_0^a \sqrt{\rho(x)/p(x)}\,dx > 0, \qquad \mathcal{N} = \int_0^a d(x)\sqrt{\rho(x)/p(x)}\,dx. \quad (2.17)$$

Theorem 2.3. *a) All eigenvalues of $\mathcal{L}_\ell$ have geometric multiplicities equal to 1, i.e., for each λ_n^ℓ there is only one linearly independent eigenvector Ψ_n^ℓ. However, a finite number of eigenvalues $\{\lambda_n^\ell, n \in R_\ell \subset \mathbb{Z}'\}$ may have finite algebraic multiplicities m_n^ℓ, i.e., for such λ_n^ℓ there exists a finite chain of associated vectors $\{\Psi_{n,j}^\ell\}_{j=1}^{m_n^\ell - 1}$:*

$$(\mathcal{L}_\ell - \lambda_n^\ell I)\Psi_{n,j}^\ell = \Psi_{n,j-1}^\ell, \qquad \Psi_{n,0}^\ell \equiv \Psi_n^\ell, \qquad \Psi_{n,-1}^\ell \equiv 0.$$

b) Assume that the linearly independent eigenvectors $\{\Psi_n^\ell, n \in \mathbb{Z}'\}$ are selected in such a way that they are almost normalized, i.e., their norms are bounded from above and below by positive constants. Then the whole set of root vectors (eigenvectors and associated vectors together) of $\mathcal{L}_\ell$ forms a Riesz basis in $\mathcal{H}_\ell$.

c) There exists ℓ_0 such that all operators $\mathcal{L}_\ell$ with $\ell \geq \ell_0$ have only eigenvalues with algebraic multiplicities 1, i.e., there are no associated vectors.

The following statement is important for the solution of the moment problem which appears as a step in the spectral decomposition solution of the control problem.

Corollary 2.4. *The system of nonharmonic exponentials and exponential-polynomial functions $\bigcup_{n \in R_\ell}\{t^k e^{i\lambda_n^\ell t}\}_{k=1}^{m_n^\ell - 1} \cup \{e^{i\lambda_n^\ell t}\}_{n \in \mathbb{Z}'}$ forms a Riesz basis in a subspace V_h^ℓ of the space $H = L^2(0, 2\mathcal{M})$. Moreover, $\dim H(\mod V_h^\ell) = \begin{cases} \ell, & \text{if } |h| \neq \infty \text{ and } h \neq 0, \\ \ell + 1, & \text{if } |h| = \infty \text{ or } h = 0 \end{cases}$.*

Corollary 2.5. *All $\mathcal{L}_\ell$ are spectral, and for $\ell \geq \ell_0$ they are even scalar in the sense of Definition 1.2. The operators S_ℓ and N_ℓ from Definition 1.2 (we equipped them with a subindex ℓ) can be written in the form*

$$S_\ell \varphi = \sum_{n \in \mathbb{Z}' \setminus R_\ell} \lambda_n^\ell (\varphi, \Psi_n^{\ell*})\Psi_n^\ell + \sum_{n \in R} \lambda_n^\ell \sum_{j=0}^{m_n^\ell - 1}(\varphi, \Psi_{n,j}^{\ell*})\Psi_{n,j}^\ell, \quad (2.18)$$

$$N_\ell \varphi = \sum_{n \in R_\ell} \sum_{j=1}^{m_n^\ell - 1}(\varphi, \Psi_{n,j}^{\ell*})\Psi_{n,j-1}^\ell, \qquad \varphi \in \mathcal{D}(\mathcal{L}_\ell),$$

where the asterisk denotes the corresponding vector from the biorthogonal basis.

We conclude the list of spectral properties of $\mathcal{L}_\ell$ with the information about the root vectors of the nonselfadjoint quadratic operator pencil $P_\ell(\lambda)$ naturally related to $\mathcal{L}_\ell$. Note that the equation $\mathcal{L}_\ell \Psi = \lambda \Psi$ for $\Psi(r) = \begin{pmatrix} \psi_0(r) \\ \psi_1(r) \end{pmatrix}$ implies that $\psi_1 = i\lambda \psi_0$ and the function $\psi = \psi_0$ satisfies:

$$P_\ell(\lambda)\psi \equiv -L_\ell \psi - 2i\lambda d(r)\psi + \lambda^2 \psi = 0 \quad (2.19)$$
$$\psi(0) = 0, \qquad (\psi' + i\lambda h\psi)(a) = 0. \quad (2.20)$$

Spectral Operators Generated by 3-Dimensional Damped Wave Equation

Equations (2.19), (2.21) define a nonselfadjoint quadratic operator pencil which does not belong to any class of operator pencils for which the spectral analysis has already been developed [12].

The spectrum and the set of root vectors of the pencil $P_\ell(\lambda)$ can be split into two parts:

$$\mathfrak{F}_+^\ell = \{\psi_n^\ell, \quad n > 0\} \quad \text{and} \quad \mathfrak{F}_-^\ell = \{\psi_n^\ell, \quad n < 0\}. \tag{2.21}$$

This requires an explanation. If $\text{Im}\, h \neq 0$, then there are no imaginary eigenvalues, and the splitting is natural: $n < 0$ for $\text{Re}\, \lambda_n^\ell < 0$ and $n > 0$ for $\text{Re}\, \lambda_n^\ell > 0$. If $\text{Im}\, h = 0$, then the spectrum is symmetric with respect to the imaginary axis; the number of purely imaginary eigenvalues is finite. There can be two possibilities:

(i) This number is even, say $2k$. Then all λ_n^ℓ with $\text{Re}\, \lambda_n^\ell = 0$ must have algebraic multiplicities equal to 1, and we can numerate the corresponding eigenfunctions as $\{\psi_n^\ell, n = \pm 1, \pm 2, \ldots, \pm k\}$, where the order can be arbitrary. The indexes $|n| > k$ can be used to count other root functions in the same way as in the case $\text{Im}\, h \neq 0$.

(ii) This number is odd. In this case, only one of the imaginary eigenvalues may have an algebraic multiplicity greater than 1, and this multiplicity must be even. So, again the number of the corresponding root functions is even, and we can split it into two equal parts in an arbitrary way.

Theorem 2.6. *Each of the sets $\mathfrak{F}_+^\ell$ and $\mathfrak{F}_-^\ell$ forms a Riesz basis in the weighted space $L_\tau^2(0,a)$, where $\tau(r) = \rho(r)/p(r)$. (It is assumed that both sets are almost normalized in this space - see Theorem 2.3b).)*

3. Now we return to the operator $\mathcal{L}$ defined in (2.7), (2.8). The spectrum of $\mathcal{L}$ is just the union of the spectra of $\mathcal{L}_\ell$, i.e., it splits into an infinite sequence of infinite series of eigenvalues. Note that the energy space $\mathcal{H}$ can be represented as an infinite orthogonal sum

$$\mathcal{H} = \bigoplus_{\ell=0}^\infty (\mathcal{H}_\ell \otimes K_\ell), \tag{2.22}$$

where K_ℓ is the $\ell(\ell+1)$-dimensional subspace of $L^2(S^2)$ spanned by the spherical harmonics $\{Y_\ell^{mj}(\theta, \varphi)\}$ with fixed ℓ. (S^2 is the unit sphere in $\mathbb{R}^3$). If $\{\Psi_n^\ell(r), n \in \mathbb{Z}'\}$ is the set of root vectors of $\mathcal{L}_\ell$, then due to Theorem 2.3 the set of 2-component functions $\{\Psi_n^\ell(r) Y_\ell^{mj}(\theta, \varphi)\}$ with fixed ℓ forms a Riesz basis in $\mathcal{H}_\ell \otimes K_\ell$.

Theorem 2.1 is a corollary of Theorem 2.3 and the following result.

Theorem 2.7. *The whole set of functions $\{\Psi_n^\ell(r) Y_\ell^{mj}(\theta, \varphi)\}$ (with any n, ℓ, m, j) forms a Riesz basis in $\mathcal{H}$.*

This statement does not follow from Theorem 2.3. To obtain Theorem 2.7 one should combine Theorem 2.3 with the following nontrivial result. Let A_ℓ be an orthogonalizer of the basis $\{\Psi_n^\ell(r), n \in \mathbb{Z}'\}$ in $\mathcal{H}_\ell$. Recall (see Definition 1.2 a)) that $\|A_\ell\|$ is uniquely defined by the basis.

Theorem 2.8. *There exists two constants $C_1, C_2 > 0$ independent on ℓ, such that*

$$C_1 \leq \|A_\ell\| \leq C_2, \quad \ell = 0, 1, \ldots. \tag{2.23}$$

The proof of this theorem is based on a technically rather complicated asymptotic analysis of the behavior of $\Psi_n^\ell(r)$ when $\ell, |n| \longrightarrow \infty$.

4. We conclude this section with a conjecture about a possible generalization of the above results to the nonspherically symmetric damped wave equation.

Conjecture 2.9. *The operator $\mathcal{L}$ which corresponds to Eq. (2.1) in the ball B_a with nonspherically symmetric coefficients that are constant on the sphere $|x| = a$ is a spectral operator with discrete spectrum [7]. In this case, N (from Definition 1.2) may be a quasinilpotent operator.*

We expect that this conjecture can be proved based on the combination of the approach developed in our works [24–28] with the method of our works [18–20] on resonances for 3-dimensional Schrödinger operator with nonspherically symmetric potential.

Remark 2.1. Once our program of the proof of the above conjecture is carried out, the results can be significantly generalized based on a purely geometrical argument. Let the domain $\Omega \subset \mathbb{R}^3$ (or $\mathbb{R}^2$) be diffeomorphic to a ball. Consider in Ω the equation $u_{tt} + d(x)u_t - Lu = 0$, where L is a formally selfadjoint 2nd order elliptic operator whose coefficients are constants on $\partial \Omega$ (the boundary conditions are (2.3) with u_r replaced by the normal derivative). All of the spectral results concerning the operators mentioned in the conjecture can be immediately extended to this new problem if there exists a change of variables which transforms both Ω into the ball B_a and the above equation into Eq. (2.1) (with nonspherically symmetric coefficients). Such a transformation exists if the following geometric condition is satisfied. The Riemannian manifold (Ω, g) (g is the metric on Ω defined by the principal symbol of L) is isometric to the ball B_a with the conformally flat metric defined by $\tilde{g}_{ij}(x) = p(x)\delta_{ij}$. It is a classical result [3, 8] that the latter is always true in the dimension $n = 2$. We expect that our spectral program for the nonspherically symmetric 3-dimensional damped wave equation mentioned in the above conjecture, with suitable adjustments, can be carried out for $n = 2$. Thus, we will obtain a very general class of nonselfadjoint spectral operators. In the dimension $n = 3$, the above condition is satisfied if and only if the covariant derivative of the Einstein tensor of g ($E_{ij} = R_{ij} - \frac{1}{4}Rg_{ij}$, where R_{ij} is the Ricci tensor and R is the scalar curvature of (Ω, g)) is a symmetric 3-tensor [3, 8]. The latter condition is a complicated differential constraint on g which defines, nevertheless, a wide class of g. Again, we expect to obtain a new class of spectral operators.

3. CONTROLLABILITY PROBLEM FOR 3-DIMENSIONAL SPHERICALLY SYMMETRIC DAMPED WAVE EQUATION

Spectral results described in Section 2 have applications to control theory of distributed parameter systems. First, they are useful in stabilization problems. Indeed, it follows from the Riesz basis property of root vectors that the rate of the energy decay in the considered systems is equal to the spectral abscissa of the corresponding dynamical semigroup. Secondly, the spectral results allow us to solve the controllability problem for systems governed by Eq. (2.1) with boundary conditions (2.3) using the spectral decomposition method. This method was originally suggested by D. Russell [16, 17] for 1-dimensional string equation without damping term. The extension of this method to the damped string equation with dissipative boundary conditions is carried out in [30–32] based on the asymptotical and spectral results obtained in [24–28]. We mention that the controllability results are known for general linear hyperbolic equations on domains in $\mathbb{R}^n$ [4, 5, 11]. However, these general results provide only the existence theorems for the desired control functions. The range of applications of the spectral decomposition method is more restricted. Nevertheless, when applicable, this method provides an explicit construction of the desired control functions in terms of the spectral characteristics of the dynamics generator of the system.

In this section, we consider an example of the controllability problem for a system governed by Eq. (2.1). We formulate the theorem which provides the solution of this problem by the spectral decomposition method.

We consider wave equation (2.1) with a forcing term:

$$u_{tt} + 2d(r)u_t + \mathbf{L}u = g(x,t) \tag{3.1}$$

and the same boundary conditions (2.3). Note, that $g(x,t)$ is not assumed to be in a separable form like in the 1-dimensional case of the string equation [30–32] and is not spherically symmetric. The problem (2.1), (2.3) can be represented in the form

$$U_t = i\mathcal{L}U + G(x,t), \quad U(x,t) = \begin{pmatrix} u_0(x,t) \\ u_1(x,t) \end{pmatrix}, \quad G(x,t) = \begin{pmatrix} 0 \\ g(x,t) \end{pmatrix}, \tag{3.2}$$

where $\mathcal{L}$ is defined in (2.7), (2.8).

The controllability problem can be stated as follows. Let $U(x,0) = U^0(x)$ and $U^0 \in \mathcal{H}$ be given. Let $T > 0$. Construct the control $g(x,t)$ such that $U(x,T) = 0$. In other words, the control problem consists of finding the control which "steers" the given initial state $U^0 \in \mathcal{H}$ to zero in time T.

To formulate the answer, we separate variables in spherical coordinates and obtain the sequence of evolution equations (see (2.9)):

$$\frac{\partial}{\partial t} U_{\ell mj}(r,t) = i\mathcal{L}_\ell U_{\ell mj}(r,t) + \tilde{G}_{\ell mj}(r,t), \quad \ell = 0, 1, 2, \ldots, \tag{3.3}$$

where

$$\tilde{G}_{\ell mj}(r,t) = \begin{pmatrix} 0 \\ \tilde{g}_{\ell mj}(r,t) \end{pmatrix} \tag{3.4}$$

are the coefficients in the expansion of $G(x,t)$ with respect to the spherical harmonics. The initial conditions for (3.3) are:

$$U_{\ell mj}(r,0) = U^0_{\ell mj}(r). \tag{3.5}$$

Assumption. The functions $\tilde{g}_{\ell mj}(r,t)$ have the forms

$$\tilde{g}_{\ell mj}(r,t) = g_{\ell mj}(r) f_{\ell mj}(t). \tag{3.6}$$

So that

$$\tilde{G}_{\ell mj}(r,t) = f_{\ell mj}(t) \begin{pmatrix} 0 \\ g_{\ell mj}(r) \end{pmatrix} \equiv f_{\ell mj}(t) G_{\ell mj}(r). \tag{3.7}$$

Now we reformulate the controllability problem: given the functions $g_{\ell mj}(r)$, find the controls $f_{\ell mj} \in L^2(0,T)$ such that $U(x,T) = 0$ (for given U^0 and T).

The solution of this problem is given by Theorem 3.1 below.

To make the statement of this theorem more observable, we introduce an additional assumption about the operators $\mathcal{L}_\ell$ defined in (2.11)–(2.14). Namely, we assume that all these operators have no associated vectors, i.e., their eigenvalues have algebraic multiplicities equal to 1. (In other words, we assume that ℓ_0 from Theorem 2.3 is equal to zero.) In the presence of associated vectors, the formulas for the control functions are significantly more complicated than (3.11) and (3.12) below. (See [31] for the corresponding formulas in the case of the damped string.)

Theorem 3.1. *Let $T = 2\mathcal{M}$, where $\mathcal{M}$ is defined in (2.17). Let us expand the functions $U^0_{\ell m j}(r)$ (see (3.5)) and $G_{\ell m j}(r)$ (see (3.7)) with respect to the bases $\{\Psi_n^\ell\}_{n\in\mathbb{Z}'}$ of all the eigenvectors of the operators $\mathcal{L}_\ell (\ell = 0, 1, 2, \ldots)$:*

$$U^0_{\ell m j}(r) = \sum_{n\in\mathbb{Z}'} \overset{\circ}{u}{}^n_{\ell m j} \Psi_n^\ell(r), \qquad G_{\ell m j}(r) = \sum_{n\in\mathbb{Z}'} g^n_{\ell m j} \Psi_n^\ell(r). \tag{3.8}$$

Recall that for each ℓ, the system of exponentials $\{e^{i\lambda_n^\ell t}, n \in \mathbb{Z}'\}$ forms a Riesz basis in $V_h^\ell \subset L^2(0,T)$ (Corollary 2.4) and denote by $\{w_n^\ell(t), n \in \mathbb{Z}'\}$ the biorthogonal basis in V_h^ℓ. Assume that the initial function $U^0(x)$ satisfies the condition

$$\sum_{\ell, m, j} \sum_{k\in\mathbb{Z}'} |g^k_{\ell m j}|^2 \sum_{n\in\mathbb{Z}'} \left|\frac{\overset{\circ}{u}{}^n_{\ell m j}}{g^n_{\ell m j}}\right|^2 < \infty, \tag{3.9}$$

which is equivalent to (due to the Bari Theorem [9]):

$$\sum_{\ell, m, j} \|G_{\ell m j}\|^2_{\mathcal{H}_\ell} \sum_{n\in\mathbb{Z}'} \left|\frac{\overset{\circ}{u}{}^n_{\ell m j}}{g^n_{\ell m j}}\right|^2 < \infty. \tag{3.10}$$

(If the force profile functions $g_{\ell m j}(r)$ from (3.6) belong to $L^2(0,a)$, then (3.9) defines a dense subspace $\mathcal{H}_g \subset \mathcal{H}$.) In this case, the unique solution of the controllability problem on the time interval $[0,T]$ is given by:

$$f_{\ell m j}(t) = -\sum_{n\in\mathbb{Z}'} \frac{\overset{\circ}{u}{}^n_{\ell m j}}{g^n_{\ell m j}} \overline{w_n^\ell(t)}, \tag{3.11}$$

and

$$g(x,t) = \sum_{\ell, m, j} g_{\ell m j}(r) f_{\ell m j}(t) Y_\ell^{m j}(\theta, \varphi). \tag{3.12}$$

Remark 3.1. Let us discuss conditions (3.9) or (3.10) and conclusions (3.11), (3.12) of Theorem 3.1. Conditions (3.9) (or (3.10)) can be formulated in two steps.

1. Assume that for each triple (ℓ, m, j)

$$\alpha_{\ell m j} \equiv \sum_{n\in\mathbb{Z}'} \left|\frac{\overset{\circ}{u}{}^n_{\ell m j}}{g^n_{\ell m j}}\right|^2 < \infty. \tag{3.13}$$

(3.13) guarantees that the formula (3.11) makes sense and defines the set of control functions

$$f_{\ell m j} \in V_h^\ell \subset L^2(0,T).$$

However, (3.13) is insufficient for (3.12) to make sense.

2. Assume in addition to (3.13) that

$$\sum_{\ell,m,j} \alpha_{\ell m j} \|g_{\ell m j}\|^2_{L^2(0,a)} < \infty, \tag{3.14}$$

where $g_{\ell m j}$ are the force distribution functions from (3.6). Note that (3.14) is equivalent to (3.10) or (3.9). (3.14) guarantees that (3.12) defines the control function

$$g \in L^2(0,T; L^2(B_a)).$$

ACKNOWLEDGMENT

Partial support by the National Science Foundation grant DMS-9706882 and the Advance Research Program of Texas Grant #003644-124 is gratefully acknowledged.

REFERENCES

1. Agranovich, M. S., Convergence of series in root vectors of operators that are nearly selfadjoint. (Russian) - *Trudy Moskov. Mat. Obshch.*, **41**, pp. 163–180, (1980).
2. Agranovich, M. S., Elliptic operators on closed manifolds. (Russian) - Current Problems in Mathematics, Fundamental directions, Vol. 63, pp. 5–169.
3. Aubin, T., *Non-linear Analysis on Manifolds*, Monge–Ampere Equations, Springer, 1982.
4. Bardos, C., Lebeau, G., and Rauch, J., Controle et Stabilisation dans les Problemes Hyperboliques, Appendix II in J.-L. Lions, controllabilite exacte et Stabilisation de Systems Distributes, Vols. 1,2, Masson, 1990.
5. Bardos, C., Lebeau, G., and Rauch, J., Sharp sufficient conditions for the observations, control and stabilization of waves from the boundary, *SIAM J. Cont. Opt.*, **30** (1992), 1024–1065.
6. Cox, S. and Zuazua, E., The rate at which the energy decays in the string damped at one end, Indiana Univ. Math. J., 44 (1995), 545–573.
7. Dunford, N. and Schwartz, J. T., *Linear Operators, Part III, Spectral Operators*, Wiley Interscience, 1971.
8. Gallot, S., Hulin, D., and Lafontaine, J., *Riemannian Geometry*, Springer-Verlag, 1987.
9. Gohberg, I. Ts. and Krein, M. G., *Introduction to the Theory of Linear Nonselfadjoint Operators in Hilbert Space*, AMS Translations, Vol. 18, (1969).
10. Hruščev, S. V., The Regge problem for strings, unconditionally convergent eigenfunction expansions, and unconditional basis of exponentials in $L^2(-T,T)$, J. Oper. Theory 14 (1985), 57–85.
11. Lions, J.-L., Exact controllability, stabilizability and perturbations for distributed systems, SIAM Rev. **30** (1988), 1–68.
12. Marcus, A. S., *Introduction to the Spectral Theory of Polynomial Pencils*, Translation of Mathematical Monographs, Vol. 71, AMS, Providence, RI, (1988).
13. Pekker, M. A. (M. A. Shubov), Resonances in the scattering of acoustic waves by a spherical inhomogeneity of the density, Soviet Math. Dokl., 15 (1974), 1131–1134.
14. Pekker, M. A. (M. A. Shubov), Resonances in the scattering of acoustic waves by a spherical inhomogeneity of the density, Amer. Math. Soc. Transl. (2), 115 (1980), 143–163.
15. Ramm, A. S., Spectral properties of some nonselfadjoint operators and some applications, Spectral Theory of Differential Operators (Birmingham, Ala., 1981), pp. 349–354, North Holland Math. Stud., **55**.
16. Russell, D. L., Controllability and stabilizability theory for linear partial differential equations: recent progress and open questions, *SIAM Review*, **Vol. 20**, No. 4 (1978), 639–839.
17. Russell, D. L., Nonharmonic Fourier series in the control theory of distributed parameter systems, *Journal of Mathematical Analysis and Applications*, **18**, (1967), 542–560.
18. Shubova, M. A., Construction of resonances states for the three-dimensional Schrödinger operator, *Problems of Mathematical Physics* 9, Leningrad University, (1979) 145–180 (Russian), Editor and Reviewer: M. S. Birman.
19. Shubova, M. A., The resonance selection principle for the three-dimensional Schrödinger operator, *Proc. Steklov Inst. Math.*, **159**, (1984), 181–195.
20. Shubova, M. A., The serial structure of resonances of the three-dimensional Schrödinger operator, *Proc. Steklov Inst. Math.*, **159**, (1984), 197–217.

21. Shubov, Marianna A., Asymptotics of spectrum and eigenfunctions for nonselfadjoint operators generated by radial nonhomogeneous damped wave equation, to appear in *Asymptotic Analysis*.
22. Shubov, Marianna A., Riesz basis property of root vectors of nonselfadjoint operators generated by radial damped wave equations. Exact controllability results. Preprint TTU, (1997).
23. Shubov, Marianna A., Riesz basis property of root vectors of dynamics generator for 3-dimensional damped wave equation. Preprint TTU, (1997).
24. Shubov, Marianna A., Basis Property of Eigenfunctions of Nonselfadjoint Operator Pencils Generated by the Equation of Nonhomogeneous Damped String, *Integral Equations and Operator Theory*, **25**, (1996), pp. 289–328.
25. Shubov, Marianna A., Asymptotics of Resonances and Eigenvalues for Nonhomogeneous Damped String, *Asymptotic Analysis*, **13**, (1996), pp. 31–78.
26. Shubov, Marianna A., Nonselfadjoint Operators Generated by the Equation of Nonhomogeneous Damped String, to appear in *Transactions of American Math. Society*.
27. Shubov, Marianna A., Transformation operators for class of damped hyperbolic equations, Preprint TTU, (1997).
28. Shubov, Marianna A., Spectral operators generated by damped hyperbolic equations, *Int. Eqs. and Oper. Theory*, **28**, (1997), pp. 358–372.
29. Shubov, Marianna A., Asymptotics of resonances and geometry of resonance states in the problem of scattering of acoustic waves by a spherically symmetric inhomogeneity of the density, *Differ. and Integ. Eq.* V.8, No. 5 (1995), 1073–1115.
30. Shubov, Marianna A., Martin, C. F. Dauer, J. P., and Belinskiy, B. P., Exact controllability of damped wave equation, to appear in *SIAM J. on Control and Optimization*.
31. Shubov, Marianna A., Exact boundary and distributed controllability of nonhomogeneous damped string. Preprint TTU, (1996).
32. Shubov, Marianna A., Spectral decomposition method for controlled damped string. Reduction of control time. To appear in *Applicable Analysis*.

INVERTIBILITY OF NONLINEAR OPERATORS AND PARAMETER CONTINUATION METHOD

Vladilen A. Trenogin

Department of Mathematics
Moscow State Steel and Alloys Institute, 117936
Moscow, Leninsky prospect 4, Russia

ABSTRACT

The article considers the invertibility of nonlinear operators which map a metric space or a weak metric space into a Banach space. It is shown that a nonlinear operator close to the invertible nonlinear operator is invertible. The analogous result for an operator distant from invertible operator is proved by parameter continuation method. Some properties of resolvent set in nonlinear case are established. A new approach to resolvent set and spectrum of nonlinear operators is offered.

1. INTRODUCTION

The purpose of this article is to apply basic aspects of nonlinear functional analysis to more profound investigations of nonlinear problems. Sometimes the concepts of normed or metric spaces are insufficient. For example these concepts are bad for describing a growth of nonlinear operator in the infinity or in the vicinity of its domain of definition. Therefore, a great number of results in nonlinear theory are, in fact, linear or semilinear ones. Recently we offered a new approach to nonlinear problems [1,2]. Here we give detailed calculation of some main aspects of the corresponding theory. In Section 2 we consider weakly metric spaces and nonlinear operators which map a weakly metric space in a Banach space. This generalization allows to take the singularities of the studied operator into account. Moreover, having in mind these singularities, we introduce a weak metric. Then we introduce two significant numerical characteristics of nonlinear operators. First is a seminorm of nonlinear operator used for the upper bound proof. Second is a measure of correctness of nonlinear operator. It allows to obtain the lower bound. In Section 3 by use of these concepts we prove the generalization of a well-known Banach theorem: an operator close to an invertible operator is also invertible. We obtain here useful bound relations for inverse operators. On the base of this theorem in Section

4 we develop a generalized variant of the parameter continuation method. This method goes back to Bernstein [3] and Schauder [4] and represents a powerful tool for proofs of the existence theorems for nonlinear elliptic boundary value problems. It gives an effective computational algorithm [6] also. In Section 5 we consider an application of Banach generalized theorem, which shows a close connection of our constructions with the monotone operators theory. In Section 6 we give another application for an investigation of a nonlinear operator-function resolvent. In Section 7 we consider some illustrative examples.

2. PRELIMINARIES

In this section we give the basic definitions and their connections. Let X be abstract set and $\rho(x_1,x_2)$ be a nonnegative identical symmetric function. In this case we call X a weakly metric space and ρ a weak metric on X. For weakly metric spaces all concepts of metric spaces can be formulated correctly. Triangle inequality absence may break the usual relations of concepts which hold in metric spaces. But it will not be important for our needs. Let X be a weakly metric space and let Y be a Banach space. The real or complex scenarios are possible. An operator $F: X \to Y$ is called ρ-bounded if the seminorm

$$\| F \|_\rho = \underset{x_1 \neq x_2}{\text{Sup}} \ \| Fx_1 - Fx_2 \| / \rho(x_1,x_2) \tag{2.1}$$

is finite. The set of all ρ-bounded operators will be denoted by $L_\rho(X,Y)$, where the structures of a linear seminormed space are introduced naturally. These definitions give us many different generalizations of the facts known for linear bounded operators. The following definition plays an important role in our consideration. For any operator $G: X \to Y$ we introduce its scalar characteristic — the degree (measure) of its correctness:

$$\| | G | \|_\rho = \underset{x_1 \neq x_2}{\text{Inf}} \ \| Gx_1 - Gx_2 \| / \rho(x_1,x_2). \tag{2.2}$$

From definitions (2.1) and (2.2) it is not difficult to verify that the following generalized triangle inequality is valid. If $F \in L_\rho(X,Y)$, $G: X \to Y$, the

$$\| | F + G | \|_\rho \leq \| F \|_\rho + \| | G | \|_\rho . \tag{2.3}$$

We shall note some trivial facts connected with the correctness measure. If $\| | G | \|_\rho > 0$ then obviously G maps bijectively X onto $R(G)$ (range of G). Moreover, if $\| | G | \|_\rho > 0$ and $y_0 \in R(G)$ then there is the a priori bound $\rho(x,x_0) \leq \| | G | \|_\rho^{-1} \| y - y_0 \|$ for the solution x of the equation $Gx = y$, where $Gx_0 = y_0$.

We call an operator $G: X \to Y$ ρ-continuously invertible operator if $\| | G | \|_\rho > 0$ and $R(G) = Y$.

Corollary 2.1. *Denote by $J_\rho(X,Y)$ the nonlinear set of all ρ-continuously invertible operators. It is sometimes useful to introduce a metric in X by means of an auxiliary operator A, which establishes one-to-one correspondence between X and a fixed Banach space Z. Let $\rho_A(x_1,x_2) = \| Ax_1 - Ax_2 \|$. Evidently, ρ_A is a metric in X. In this case, for brevity, we use the notations*

$$L_A(X,Y), \| F \|_A, \| | G | \|_A, J_A(X,Y).$$

Corollary 2.2. If $G \in J_A(X,Y)$ then $AG^{-1} \in L_I(X,Y)$ and

$$\| AG^{-1} \|_I = ||| G |||_A^{-1}. \tag{2.4}$$

This equality is a direct consequence of definitions (2.1) and (2.2).

Now we describe the simplest situations. Many authors [6,9] examined the case of X being a Banach space and $\rho(x_1,x_2) = \| x_1 - x_2 \|$. In other words, $Z = X$ and $A = I$ is the identity operator of X. Here $L_I(X,Y)$ is the space of the Lipschitz-continuous operators, which contains the space $L(X,Y)$ of linear bounded operators. Practically, a case of $L_I(X,Y)$ differs a little from a linear case. Our theory comprises more interested possibilities. In a Banach space X we may introduce an additional metric structure as follows: $\rho(x_1,x_2) = \| x_1 - x_2 \| \ \Psi(R)$. Here $R = \sqrt{\| x_1 \|^2 + \| x_2 \|^2}$ and $\Psi(R)$ is a weight function: $\Psi(R) > 0$ for $R \geq 0$, $\Psi(R) \to \infty$ by $R \to \infty$. Now, X is the weakly metric space with weak metric ρ and $L_\rho(X,Y)$ contains all locally Lipschitz-continuous nonlinear operators with the Lipschitzian constant not greater than $\Psi(R)$.

3. GLOBAL INVERTIBILITY OF NONLINEAR OPERATORS

We now consider the ρ-continuously invertible operator A, i.e., $A \in J_\rho(X,Y)$. We are interested in ρ-continuously invertibility of an operator $B : X \to Y$ close to A.

Theorem 3.1. If $B - A \in L_\rho(X,Y)$ and

$$\| B - A \|_\rho < ||| A |||_\rho \tag{3.1}$$

then $B \in J_\rho(X,Y)$, $AB^{-1} \in L_I(Y,Y)$ and the estimates hold

$$\| AB^{-1} \|_I \leq \| A \|_\rho / (||| A |||_\rho - \| B - A \|_\rho) \tag{3.2}$$
$$\| AB^{-1} - I \|_I \leq \| B - A \|_\rho / (||| A |||_\rho - \| B - A \|_\rho). \tag{3.3}$$

Proof: We write the equation $Bx = y$ in the equivalent form $u = \Phi u$, where $u = Ax$ and $\Phi u = u - BA^{-1}u + y$. For any $u_1, u_2 \in Y$ we set

$$\| \Phi u_1 - \Phi u_2 \| = \| (B-A)A^{-1}u_1 - (B-A)A^{-1}u_2 \|$$
$$\leq \| B - A \|_\rho \, \rho(A^{-1}u_1, A^{-1}u_2)$$
$$\leq \| B - A \|_\rho ||| A |||_\rho^{-1} \| u_1 - u_2 \|.$$

We applied definitions (2.1) and (2.2) here. According to (3.1) Φ is contractive operator on Y. Thus, the equation $Bx = y$ has unique solution $x = A^{-1}u(y) = B^{-1}y$. Here $u(y)$ is the fixed point of Φ. Thus $R(B) = Y$. From generalized triangle inequality (2.3) for $G = B, F = A$ we have

$$||| B |||_\rho \geq ||| A |||_\rho - \| B - A \|_\rho > 0.$$

Hence $B \in J_\rho(X,Y)$.

Now we will prove estimates (3.2) and (3.3). By definitions (2.1) and (2.2) for any $y_1, y_2 \in Y$

$$\| AB^{-1}y_1 - AB^{-1}y_2 \| \leq \| A \|_\rho \, \rho(B^{-1}y_1, B^{-1}y_2) \leq \| B - A \|_\rho \|| B \||_\rho^{-1} \| y_1 - y_2 \|.$$

From the bound for $\|| B \||_\rho$ we now obtain estimate (3.2). Analogously, from the inequality

$$\| (AB^{-1} - I)y_1 - (AB^{-1} - I)y_2 \| = \| (B-A)B^{-1}y_1, (B-A)B^{-1}y_2 \|$$
$$\leq \| B - A \|_\rho \, \rho(B^{-1}y_1, B^{-1}y_2)$$

we have the estimate (3.3). The proof of Theorem 3.1 is over.

Remark 3.1. We introduced [2] the condition number of nonlinear operator and obtained the bound of the relative error by replacing the precise equation by an approximate one.

Remark 3.2. If $Y = X$ is a Banach space, $\rho(x_1, x_2) = \| x_1 - x_2 \|$, then $B \in J_I(X, Y)$ and the following estimates hold

$$\| B^{-1} \|_I = \|| B \||_I^{-1}$$
$$\leq (\| A^{-1} \|_I - \| B - A \|_I)^{-1} \| B^{-1} - A^{-1} \|_I$$
$$\leq \| A^{-1} \|_I \| B - A \|_I (\| A^{-1} \|_I - \| B - A \|_I)^{-1}.$$

This statement is proved [9]. Here only the last estimate is new.

4. PARAMETER CONTINUATION METHOD AND SOLVABILITY OF NONLINEAR EQUATIONS

In this section we present a generalized variant of the nonlinear analysis fundamental method. We consider an operator-function $A(t)$, where $A(t) i L_\rho(X, Y)$ for each $t \in [0, 1]$. We define the ρ-correct operator-function $A(t)$ on $[0, 1]$ as follows: there exists $\gamma > 0$ such that $\|| A(t) \||_\rho \geq \gamma$ for any $t \in [0, 1]$.

The operator-function $A(t)$ is said to be ρ-continuous on $[0, 1]$ if it is continuous on $[0, 1]$ in seminorm of $L\rho(X, Y)$.

Theorem 4.1. *Let $A(t)$ be ρ-correct and ρ-continuous on $[0, 1]$. If $A(0) \in J_\rho(X, Y)$ then $A(1) \in J_\rho(X, Y)$.*

Proof: Let a set $M \subset [0, 1]$ contains all $t \in [0, 1]$ such that $A(t) \in J_\rho(X, Y)$. But $0 \in M$, so M is non-empty. We shall demonstrate that M is simultaneously open and closed on $[0, 1]$.

M is open: if $t_0 \in M$ then because of ρ-continuity of $A(t)$ there exists $\delta > 0$ such that on the set $S_0 = (t_0 - \delta, t_0 + \delta) \cap [0, 1]$ we have the following inequality $\| A(t) - A(t_0) \|_\rho < \|| A(t_0) \||_\rho$ Applying Theorem 3.1, we obtain $S_0 \subset M$.

M is closed: if $\{t_n\} \subset M, t_n \to t_0, n \to \infty$ then because of ρ-continuity of $A(t)$ there exists n such that $\| A(t_n) - A(t_0) \|_\rho < \gamma$. However for each n $\|| A(t_n) \||_\rho \geq \gamma$. We have used the ρ-correctness of $A(t)$. These bounds and Theorem 3.1 for the operators $A(t_n) \in J_\rho(X, Y)$ and $A(t_0)$ give us $A(t_0) \in J_\rho(X, Y)$. Thus $M = [0, 1]$.

Corollary 4.2. *Theorem 4.1 gives also the following bounds of $A^{-1}(t)$ on $[0,1]$:*

$$\rho(A^{-1}(t)y_1, A^{-1}(t)y_2) \leq \gamma^{-1} \|y_1 - y_2\|, \forall y_1, y_2 \in Y.$$

If $Y = X$ is a Banach space and $\rho(x_1, x_2) = \|x_1 - x_2\|$ then the simpler bound is valid:

$$\|A^{-1}(t)\|_l \leq \gamma^{-1}, \forall t \in [0,1].$$

Practically, a special case of the parameter continuation method is important.

Theorem 4.3. *Let $A(t) = A + t(B - A)$ be a ρ-correct on $[0,1]$ and $B - A \mathrm{i} L_\rho(X,Y)$. If $A \in J_\rho(X,Y)$ then $B \in J_\rho(X,Y)$.*

Proof: This statement follows from Theorem 4.1 since $A(t)$ is ρ-continuous on $[0,1]$. However we shall present another way of verification. It consists in finite number of steps of Theorem 3.1 applications. We exclude the trivial case when $\|B - A\|_\rho = 0$, i.e., $B - A$ is a constant operator on X. In this case a statement of Theorem 4.3 is obvious.

First we consider operators $A = A(0)$ and $A(t)$. For all $t \in [0, t_1), t_1 = \||A|\|_\rho / \|B - A\|_\rho$ we have the inequality

$$\|A(t) - A(0)\|_\rho < \||A(0)\|\|_\rho.$$

From Theorem 3.1 we obtain $[0, t_1) \subset M$ (see proof of Theorem 4.1). If $t_1 > 1$ then $1 \in M$. If $t_1 = 1$ as $(0,1) \subset M$ in accordance with $A(t)$ coercivity we have a bound $\|A(1) - A(t)\|_\rho = (1-t)\|B - A\|_\rho < \gamma \leq \||A(t)\|\|_\rho$ for $t \in (0,1)$ sufficiently close to 1. Hence $1 \in M$. Here we reach point 1 i the first step.

Let be $t_1 < 1$. As before, $t_1 \in M$. Now we have to study the operator $A(t_1) \in J_\rho(X,Y)$ and the operator $A(t)$. We note that $\|A(t) - A(t_1)\|_\rho = (t - t_1)\|B - A\|_\rho < \gamma \leq \||A(t_1)\|\|_\rho$ for all $t \in [t_1, t_2), t_2 = t_1 + \gamma/\|B - A\|_\rho$. If $t_2 \geq 1$ then $1 \in M$. We reached here point 1 in two steps. By this induction argument we prove the statement of Theorem 4.3 in l steps of Theorem 3.1 applications, where l is the smallest natural number such that $l\gamma \geq \|B - A\|_\rho$ or $\||A\|\|_\rho + (l-1)\gamma > \|B - A\|_\rho$.

5. RELATION TO MONOTONE OPERATORS THEORY

Let a set X and a real Banach space Y be given. Here we suppose that there exists operator $A: X \to Y$, which realizes one-to-one mapping X onto Y. We introduce a metric in X by setting $\rho(x_1, x_2) = \|Ax_1 - Ax_2\|$. We consider jet an operator $C\mathrm{i}L_A(X,Y)$. The analysis of Theorem 4.3 proof shows that the following statement is valid.

Theorem 5.1. *Let the operator-function $A(t) = A + tC$ be a ρ-correct on $[0, +\infty)$ i.e., there exists a constant $\gamma > 0$ such that $\||A(t)\|\|_\rho \geq \gamma$ on $[0, +\infty)$. Then $A(t) \mathrm{i} J_A(X,Y)$ for all $t \geq 0$.*

Remark 5.1. It is evident that here $\gamma \leq 1$.

Remark 5.2. The fact that $A(t)$ is ρ-correct on $[0, \infty)$ is closely connected with the standard definition of monotone operators. For a real Hilbert space Y our definition of ρ-correctness is equivalent to the inequality

$$-\sqrt{1-\gamma}\|Cx_1 - Cx_2\|\|Ax_1 - Ax_2\| \leq (Cx_1 - Cx_2, Ax_1 - Ax_2).$$

If in addition $X = Y, C = I, \gamma = 1$ we obtain the monotone operator standard definition [5,6].

Also we study more interesting problem of the operator C invertibility.

Theorem 5.2. Let $C : X \to Y$. If there exists $t_0 > 0$ such that $R(A(t_0)) = Y$ and $|||A(t_0)|||_A > 1$ then

1. the equation $Cx = y$ has the unique solution $x(y)$ for each $y \in Y$,

2. the equation $t^{-1}Ax + Cx = y$ has the unique solution $x(t,y)$ for each $t > t_0, y \in Y$,

3. $\| Ax(t,y) - Ax(y) \| = O(t^{-1})$ by $t \to +\infty$.

Proof: At first, we apply Theorem 3.1 for the operators $A_1 = t_0^{-1}A + C$ and $B_1 = C$. Since $\| A_1 - B_1 \|_A = t_0^{-1} < |||t_0^{-1}A + C|||_A$ so $C \in J_A(X,Y)$. Next, we apply Theorem 3.1 for the operators $A_2 = t_0^{-1}A + C$ and $B_2 = t^{-1}A + C$ by $t > t_0$. Since $\| A_2 - B_2 \|_A = t_0^{-1} - t^{-1} < |||A_2|||_A$ so $t^{-1}A + C \in J_A(X,Y)$ by $t > t_0$.

Now, because of definition (2.1), we can write

$$\| Ax(t,y) - Ax(y) \| = \| A(t^{-1}A + C)^{-1}y - A(t^{-1}A + C)^{-1}(t^{-1}A + C)C^{-1}y \|$$
$$\times \| A(t^{-1}A + C)^{-1} \|_I \, t^{-1} \| AC^{-1}y \|.$$

On the other hand, the equality (2.4) for $G = t^{-1}A + C$ and the generalized triangle inequality (2.3) imply

$$\| A(t^{-1}A + C)^{-1} \|_I = |||t^{-1}A + C|||_A^{-1} \leq (|||C|||_A - t^{-1})^{-1}.$$

Combining two last inequalities for all $t > |||C|||_A^{-1}$ one has $\| Ax(t,y) - Ax(y) \| \leq \| AC^{-1}y \| / (t|||C|||_A - 1)$.

This concludes the proof of Theorem 5.2.

6. PROPERTIES OF THE RESOLVENT SETS AND RESOLVENT OPERATORS.

Here we show that many properties of the resolvent sets and resolvent operators known in a linear case are valid for our nonlinear situation. We offer too a nonstandard definition of the eigenvalues of nonlinear operator-function in connection with bifurcation approach. Let a weakly metric space X with weak metric $r = r(x_1, x_2)$ and a Banach space Y are given. We consider an operator-function $A(\lambda)$ where for each scalar $\lambda A(\lambda) : X \to Y$. We will be use the following definitions. The point λ_0 is called the regular point of $A(\lambda)$ if $A(\lambda_0) \in J(X,Y)$ that is $A(\lambda_0)$ is r-continuously invertible operator. The set of all regular points is called the resolvent set of $A(\lambda)$ and denoted by $\rho = \rho(A(\lambda))$. The complement of ρ with respect to the whole scalar field is called the spectrum of $A(\lambda)$ and denoted by $\sigma = \sigma(A(\lambda))$. The operator-function $R(\lambda) = A^{-1}(\lambda)$ with domain of definition ρ is called the resolvent operator. Now we demonstrate the properties of a resolvent set of nonlinear operator-function on the base of the Theorems 3.1–4.3. For brevity we will say that $A(\lambda)$ is linear if $A(\lambda) = B - \lambda A$ where $A \in J(X,Y), B : X \to Y$.

Theorem 6.1. *Let $A(\lambda)$ be a r-continuous on the whole scalar field then its resolvent set ρ is open and its spectrum σ is closed*

Proof: If $\lambda_0 \in \rho$ then because of r-continuity $A(\lambda)$ we have the inequality $|||A(\lambda) - A(\lambda_0)|||_r < |||A(\lambda_0)|||_r$ for each λ belonging to a certain neighborhood S_0 of the point λ_0. From Theorem 3.1 for each $\lambda \in S_0$ the operator $A(\lambda) \in J(X, Y)$. Hence $S_0 \subset \rho$.

Corollary 6.2. *Let $A(\lambda)$ be a linear operator-function. If $\lambda_0 \in \rho$ and $r_0 = |||A(\lambda_0)|||_r ||A||_r^{-1}$ then the disk $S_0 = (\lambda | |\lambda - \lambda_0| < r_0) \subset \rho$. Besides for each $\lambda \in S_0$ the estimation $||AR(\lambda)||_I \leq (r_0 - |\lambda - \lambda_0|)^{-1}$ is valid.*

It is necessary to prove only the estimate. Because of (2.3) for $G = A(\lambda), F = (\lambda - \lambda_0)A$ we have the inequality $|||A(\lambda)|||_r \geq |||A(\lambda_0)|||_r - |\lambda - \lambda_0| \cdot ||A||_r$. Applying (2.4) we obtain our estimation.

Theorem 6.3. *Let $A(\lambda)$ be r-continuous and r-correct on the connected scalar set $\omega \subset \rho$. If there exists $\lambda_0 i \rho$ then $\omega \subset \rho$.*

Proof: Let $\lambda_1 \in \omega$. Let us take from ω a continuous curve $\lambda = \lambda(t), t \in [0,1], \lambda(0) = 0, \lambda(1) = 1$. The operator-function $A(\lambda(t))$ is r-continuous and r-correct on $[0,1]$. So applying Theorem 4.3 we obtain $\lambda_1 \in \omega$.

Theorem 6.4. *Let $A(\lambda)$ be an operator bundle: $A(\lambda) = \sum_0^m \lambda^s A_s$ where $A_s \in L_\rho(X,Y)$, $s = 0, 1, ..m - 1$, $A_m \in J_\rho(X,Y)$. Then each λ satisfying the inequality $\sum_0^{m-1} |\lambda|^{-m+s} ||A_s||_r < |||A_m|||_r$ belongs to ρ.*

Proof: It is sufficient to apply Theorem 3.1 for the operators A_m and $A_m + \sum_0^{m-1} \lambda^{-m+s} A_s$.

Corollary 6.5. *Let $A(\lambda)$ be a linear operator-function. If $|\lambda| > ||B||_r |||A|||_r^{-1}$ then $\lambda \in \rho$ and*

$$||AR(\lambda)||_I \leq (|\lambda| - ||B||_r |||A|||_r^{-1})^{-1}.$$

Remark 6.1. In the linear case of $A(\lambda)$ for each $\lambda, \mu \in \rho$ the following generalized Hilbert equation for a resolvent operator $R(\mu) = R(\lambda)(I + (\mu - \lambda)R(\mu))$ is valid. This fact is a consequence of the identity $(B - \lambda A)R(\mu) = I + (\mu - \lambda)R(\mu)$. From these reasons it is not difficult to prove that $R(\lambda)$ satisfies on ρ a local Lipschitz condition.

In conclusion of this section we give an new eigenvalue concept and its connection with the theory of branching of solutions of nonlinear equations [10]. We call λ_0 the eigenvalue of $A(\lambda)$ if there exist $x_1 \neq x_2$ such that $A(\lambda_0)x_1 = A(\lambda_0)x_2$.

Remark 6.2. Our eigenvalue definition is more general then the classical one. That shows the example $A(\lambda)x = x^2 - \lambda x$ in R by $\lambda = 0$.

Now we nevertheless show the old eigenvalue concept in our situation. We consider the equation $A(\lambda_0)x = y_0$ where λ_0 and y_0 are fixed. Let x_0 be the solution of this equation. Every theorem of the existence of the one-parameter or multiparameter families of solutions of the equation $A(\lambda)x = y_0$ for λ close to λ_0 may be reformulated as a theorem about the eigenvalues of $A(\lambda)$. The situation is more easily in a linear case of $A(\lambda)$. Let $Ax_0 = 0$ and $Bx_0 = y_0$. If we take the operator B_0 definite by formula $B_0 x = Bx - y_0$ then the equation $B_0 x - \lambda Ax = 0$ has the trivial solution $x = x_0$ for each λ. Therefore in a linear case we may assume that $A(\lambda)x_0 = 0$ for each λ and define an eigenvalue as following: λ_0 is the eigenvalue of $A(\lambda)$ if there exists $x \neq x_0$ such that $A(\lambda_0)x = 0$. A consideration of the nonlinear equation $A(\lambda)x = 0$ lead here to the bifurcation point definition. Many authors investigated the problems of bifurcation phenomenon by topological, variational, analytical and group-theoretical methods. We indicate only our papers [11–15], where we continue our joint investigations.

7. SOME ILLUSTRATIVE EXAMPLES

For the illustrations of the stated above theory we give here a few trivial examples.

1. Let $X = (0, \infty)$, $Y = R$, $Ax = x - 1/x$, $r(x_1, x_2) = |Ax_1 - Ax_2|$, $Bx = bx - a/x$ by and $A(\lambda) = B - \lambda A$. It is easily to verify that $\rho = (-\infty, a) \cup (b, +\infty)$, $\sigma = [a, b]$ and $|||A(\lambda)||| = 0$ for each $\lambda \in \sigma$. The eigenvalues fill out (a, b) and for them the range of $A(\lambda)R(A(\lambda)) = (y \in Y | y > 2\sqrt{(\lambda - a)(b - \lambda)})$ is a nonlinear set. For each $\lambda \in (a, b)$ there exist two different positive solutions of the equation if $y \in R(A(\lambda))$. The points a and b are not eigenvalues. By $\lambda = a$ or $\lambda = b R(A(\lambda)) = X$ and the equation $A(\lambda)x = y$ has for $y \in X$ unique positive solution. Therefore a and b are the branching points of the operator-function $A(\lambda)$.

2. Let $f(x)$, $f : (c, d) \to R$ be a strictly increasing and continuous function with $f(c) = -\infty, f(d) = +\infty$. The interval (c, d) may be finite or infinite. We consider the space X as the space of all continuous on $[0, 1]$ functions with the values on (c, d). We turn X into metric space by setting $\rho_A(x_1, x_2) = \sup |a(x_1(s)) - a(x_2(s))|$ by $s \in [0, 1]$. Let $Y = C[0, 1]$ — the space of all continuous on $[0, 1]$ functions. The operator of superposition $(Ax)(s) = f(x(s))$ transforms X onto Y bijectively. Now let us introduce the integral operator $(Bx)(s) = \int_0^1 K(s, e, x(e)) de$ where the function $K(s, e, x)$ is continuous of its variables by $s, e \in [0, 1], x \in (c, d)$ and satisfies the condition $|K(s, e, x_1) - K(s, e, x_2)| \leq k(s, e)|a(x_1) - a(x_2)|$. It is not difficult to verify that B is A-continuously invertible operator and $||B||_A \leq k = \max \int_0^1 k(s, e) de$. From Theorem 3.1 by $k < 1$ it follows that integral equation $Ax + Bx = y$ has unique solution $x \in X$ for each $y \in Y$. In other words it means the existence and uniqueness of the solution $x(s)$ with values on (c, d). If $A(t) = A + tB$ is A-correct on $[0, 1]$ the existence and uniqueness theorem is valid (see Theorem 4.1) without the condition $k < 1$. Also the assertions about the resolvent set of the operator-function $B - \lambda A$ are valid.

3. Now we explain a weak metric role. Instead of the metric r_A we may exploit a weak metric r more simple than r_A but equivalent to r_A. This approach permits to simplify an obtaining of the necessary bounds. For example we take the operator $(Ax)(s) = \sum_1^{2m+1} c_k(s) x^k(s)$ with continuous on $[0, 1]$ coefficients $c_k(s)$. Let $c_1(s) = 1$, $c_{2m+1}(s) > 0$ on $[0, 1]$ and $\sum_1^{2m+1} k c_k(s) x^{k-1} \geq \delta > 0$, for all $s i [0, 1]$, $x \in R$. It is easily to show that the metric r_A is equivalent to weak metric $r(x_1, x_2) = \sup |x_1(s) - x_2(s)|(1 + x_1^{2m}(s) + x_2^{2m}(s))|$ by $s \in [0, 1]$.

More profound applications of stated above theory for different concrete classes of nonlinear problems in particular for differential equations will be given in our further publications.

This work is supported by the Russian Fund of Fundamental Researches, project 96-01-00512.

REFERENCES

1. Trenogin V. A. Locally invertibility of nonlinear operators and parameter continuation method. (Russian) Funkt. Anal. i Pril., v. 30 N, 2, 1996., p. 93–95.
2. Trenogin V. A. Global invertibility of nonlinear operators and parameter continuation method. Russian) Dokl. Ros. Akad. Nauk, v. 350, N 4, 1996, p. 1–3
3. Bernstein S. N. Math. Ann., 1904, Bd. 59.
4. Schauder J. Math. Z, 38, 1934.
5. Trenogin V. A. Functional Analysis, (Russian) Nauka, Moscow, 1980 and 1993 (Nauka, Novosibirsk).
6. Gaponenko J. L. Journ. Vych. Math. i Math. Phys., (Russian) v. 26, 8, 1986.

7. Trenogin V. A. Properties of resolvent sets and estimations of resolvent of nonlinear operators. (Russian) Dokl. Ros. Akad. Nauk, 1997, (in print).
8. Trenogin V. A. Nonlinear operators in weak metric spaces and its conjugate operators. (Russian) Works of Intern.conf. to 175 years birthday P. L. Chebyshev, v. 1, 1996, p. 335–337, Moscow Univ.
9. Donden A. Rev. rum. math. pure et appl.,25, 10, 1980.
10. Vaiberg M. M., Trenogin V. A. Branching theory of solutions of nonlinear equations. Nauka, Moscow, 1969. English transl. Noordhoff int. publ., 1974.
11. Trenogin V. A., Sidorov N. A., Loginov B. V. Potentiality, group symmetry and bifurcations in the theory of branching equations. Dif. and int. equations, USA, Ohio Univ. v. 3, N 1, 1990, p. 145–154.
12. Trenogin V. A., Sidorov N. A., Loginov B. V. Bifurcation, Potentiality, Group-Theoretical and Iterative Methods. ZAMM, v. 76, suppl. 2, 1996, p. 245–248.
13. Loginov B. V., Trenogin V. A Group Symmetry of Bifurcation Equation and Dynamic Branching. ZAMM, v. 76, suppl. 2,1996, p. 237–240.
14. Loginov B. V., Trenogin V. A., Velmesov P. A. Bifurcation and Stability and Some Problems of Continua Mechanics. ZAMM, v. 76, suppl. 2,1996 p. 241–244.
15. Loginov B. V., Trenogin V. A. Branching equation of Andronov–Hopf bifurcation under group symmetry conditions. Chaos, 1997 (in print).

13

STURM–LIOUVILLE DIFFERENTIAL OPERATORS WITH SINGULARITIES

V. Yurko

Department of Mathematics
Saratov State University
Astrakhanskaya 83, Saratov 410071, Russia
Telephone: (8452) 515538
Fax: (8452) 240446
E-mail: Yurko@scnit.saratov.su

ABSTRACT

We study the inverse spectral problems for second-order differential equations on a finite interval having regular singularities inside the interval. Necessary and sufficient conditions for the solvability of these inverse problems and a procedure for the solution of these problems are given, and uniqueness theorems are proved.

1. Let us consider the boundary-value problem $L = L(p(x), h_0, h_1)$:

$$\ell y \equiv -y'' + p(x)y = \lambda y, \quad p(x) = \sum_{j=1}^{N} \frac{a_j}{(x - \gamma_j)^2} + q(x), \quad 0 < x < T, \tag{1}$$

$$U_0(y) \equiv y'(0) + h_0 y(0) = 0, \quad U_1(y) \equiv y'(T) + h_1 y(T) = 0, \tag{2}$$

$$0 < \gamma_1 < \ldots < \gamma_N < T$$

Here $q(x)$ is a complex-valued function, and a_j, h_0, h_1 are complex numbers. Let $a_j = \nu_j^2 - 1/4$, and for definiteness, let $\mathrm{Re}\,\nu_j > 0$, $\nu_j \notin \mathbf{N}$. We shall assume that

$$q(x) \prod_{j=1}^{N} |x - \gamma_j|^{1 - 2\mathrm{Re}\,\nu_j} \in \mathcal{L}(0, T).$$

Under these conditions we shall say that $L \in V$.

Spectral and Scattering Theory, edited by Ramm,
Plenum Press, New York, 1998

Differential equations with singularities inside the interval play an important role in various areas of mathematics as well as in applications. Moreover, a wide class of differential equations with turning points can be reduced to equations with singularities. In this paper the inverse problem of recovering L from its spectral characteristics is studied. We prove the uniqueness theorems and obtain necessary and sufficient conditions along with a procedure for the solution of the inverse problem.

Inverse spectral problems for Sturm–Liouville differential operators without singularities have been thoroughly studied (see [1–3] and the references therein). An important role in the spectral theory of Sturm–Liouville operators was played by the transformation operator method. For the case when a singular point lies at the end of the interval ($N=1, \gamma_1 = 0$), the inverse problem was investigated in [4–10] and other works.

The presence of singularities inside the interval produces essential qualitative modifications in the investigation of the inverse problem. The transformation operator method in this case is not suitable for the solution of the inverse problems. In this paper we use another method connected with the contour integral method. We note that for the inverse problems without singularities, the contour integral method was first applied in [11,12]. For studying the boundary value problem (1)–(2) an important role is played by the special fundamental system of solutions (FSS) which gives us sewing together solutions at the singular point and gives us an opportunity to obtain the asymptotic behavior of the corresponding Stokes multipliers and to study the so-called Weyl solutions and the Weyl function of the boundary value problem (1)–(2). In Section 2 we construct the special FSS and study properties of the spectrum of L. In Section 3 the uniqueness theorems are proved, and in Section 4 we provide necessary and sufficient conditions of solvability of the inverse problem and also obtain a procedure for the solution of the inverse problem.

2. Let $\lambda = \rho^2, \rho = \sigma + i\tau$. Consider the functions

$$C_{kj}(x,\lambda) = (x-\gamma_j)^{\mu_{kj}} \sum_{m=0}^{\infty} c_{kmj}(\rho(x-\gamma_j))^{2m}, \quad \mu_{kj} = (-1)^k \nu_j + 1/2, \quad k=1,2, j=\overline{1,N},$$

$$c_{10j}c_{20j} = (2\nu_j)^{-1}, \quad c_{kmj} = (-1)^m c_{k0j} \left(\prod_{s=1}^{m} ((2s+\mu_{kj})(2s+\mu_{kj}-1) - a_j) \right)^{-1}.$$

Here and in the sequel, $z^\mu = \exp(\mu(\ln|z| + i\arg z))$, $\arg z \in (-\pi, \pi]$.

Let $S_{kj}(x,\lambda)$, $k=1,2, j=\overline{1,N}$ be solutions of the following integral equations:

$$S_{kj}(x,\lambda) = C_{kj}(x,\lambda) + \int_{\gamma_j}^{x} g_j(x,t,\lambda)\left(p(t) - \frac{a_j}{(t-\gamma_j)^2}\right) S_{kj}(t,\lambda)\,dt, \quad x \in \omega_j \cup \omega_{j-1},$$

$$\omega_j = (\gamma_j, \gamma_{j+1}), \ \gamma_0 = 0, \ \gamma_{N+1} = T, \ g_j(x,t,\lambda) = (C_{1j}(t,\lambda)C_{2j}(x,\lambda) - C_{1j}(x,\lambda)C_{2j}(t,\lambda)).$$

Functions $S_{kj}(x,\lambda)$ are entire in λ of order 1/2, and form a FSS of equation (1), and

$$\det[S_{kj}^{(m-1)}(x,\lambda)]_{k,m=\overline{1,2}} \equiv 1,$$

$$|S_{kj}^m(x,\lambda)| \leq C|(x-\gamma_j)^{\mu_{kj}-m}|, \ |S_{kj}(x,\lambda) - C_{kj}(x,\lambda)| \leq C|(x-\gamma_j)^{2\nu_j+\mu_{kj}}|, \ |\rho(x-\gamma_j)| \leq 1.$$

Here and below, one and the same symbol C denotes various positive constants in the estimates.

In [8, 9] asymptotic properties of $S_{kj}(x,\lambda)$ and corresponding Stokes multipliers are investigated. In particular,

$$S_{kj}^{(m)}(x,\lambda) = \beta_{kj}\rho^{-\mu_{kj}}((-i\rho)^m \exp(-i\rho(x-\gamma_j))[1]_j$$
$$+ (i\rho)^m \exp(i\pi\mu_{kj}\operatorname{sign}(\gamma_j - x))\exp(i\rho(x-\gamma_j))[1]_j), \quad x \in \omega_j \cup \omega_{j-1},$$
$$(\rho, x) \in \Omega, \tag{3}$$

where

$$\Omega = \{(\rho, x) : |\rho(x - \gamma_j)| \geq 1,$$
$$j = \overline{1, N}\},$$
$$[1]_j = 1 + O((\rho(x - \gamma_j))^{-1}),$$
$$\beta_{1j}\beta_{2j} = (-4i\sin\pi\nu_j)^{-1}.$$

We note that the FSS $\{S_{kj}(x,\lambda)\}$ will be used for sewing together solutions at the singular point. In particular, if $q(x)$ is an analytic function, it corresponds to sewing solutions by the analytic continuation in the upper half-plane $\operatorname{Im} x > 0$.

Let us continue the functions $S_{kj}(x,\lambda)$ on the whole interval $[0,T]$ by

$$S_{kj}(x,\lambda) = A_{kj}^{1s}(\lambda)S_{1s}(x,\lambda) + A_{kj}^{2s}(\lambda)S_{2s}(x,\lambda), \quad x \in \omega_s \cup \omega_{s-1}. \tag{4}$$

Using (3) and (4) we obtain the following asymptotic formulas for $x \in \omega_s$, $(\rho,x) \in \Omega$,

$$S_{kj}^{(m)}(x,\lambda) = \beta_{kj}\rho^{-\mu_{kj}}((-i\rho)^m \exp(-i\rho(x-\gamma_j))[1]_\gamma + (i\rho)^m \exp(-i\pi\mu_{kj})\exp(i\rho(x-\gamma_j))[1]_\gamma)$$
$$-2i(i\rho)^m \sum_{p=j+1}^{s} \cos\pi\nu_p \exp(i\rho(x+\gamma_j - 2\gamma_p))[1]_\gamma, \quad s \geq j, \tag{5}$$

$$S_{kj}^{(m)}(x,\lambda) = \beta_{kj}\rho^{-\mu_{kj}}((-i\rho)^m \exp(-i\rho(x-\gamma_j))[1]_\gamma + (i\rho)^m \exp(-i\pi\mu_{kj})\exp(i\rho(x-\gamma_j))[1]_\gamma)$$
$$+2i(i\rho)^m \sum_{p=s+1}^{j-1} \cos\pi\nu_p \exp(i\rho(x+\gamma_j - 2\gamma_p))[1]_\gamma, \quad s < j, \tag{6}$$

$$[1]_\gamma = 1 + \sum_{j=1}^{N} O((\rho(x-\gamma_j))^{-1}).$$

Let us denote

$$\varphi_k(x,\lambda) = (-1)^{k-1}(S_{2j}^{(2-k)}(0,\lambda)S_{1j}(x,\lambda) - S_{1j}^{(2-k)}(0,\lambda)S_{2j}(x,\lambda)),$$

$$\varphi(x,\lambda) = \varphi_1(x,\lambda) - h_0\varphi_2(x,\lambda), \quad \Delta(\lambda) = U_1(\varphi), \quad \delta(\lambda) = U_1(\varphi_2),$$

$$\psi(x,\lambda) = \delta(\lambda)\varphi(x,\lambda) - \Delta(\lambda)\varphi_2(x,\lambda) = U_1(S_{2j})S_{1j}(x,\lambda) - U_1(S_{1j})S_{2j}(x,\lambda),$$

$$\Phi(x,\lambda) = -(\Delta(\lambda))_1\psi(x,\lambda) = \varphi_2(x,\lambda) + M(\lambda)\varphi(x,\lambda), \quad M(\lambda) = \Phi(0,\lambda) = -(\Delta(\lambda))_1\delta(\lambda).$$

The function $M(\lambda)$ is called the Weyl function for L. Clearly,

$$\varphi_k^{m-1}(0,\lambda) = \delta_{km}, \quad \langle\varphi_1,\varphi_2\rangle = 1, \quad \langle\varphi,\Phi\rangle = 1, \quad \langle\varphi,\psi\rangle = -\Delta(\lambda),$$

$$\delta(\lambda) = \psi(0,\lambda), \quad \varphi(0,\lambda) = \psi(T,\lambda) = 1, \quad U_0(\varphi) = U_1(\Phi) = U_1(\psi) = 0, \quad U_0(\Phi) = 1,$$

where $\langle y,z\rangle = yz' - zy'$, and δ_{km} is the Kronecker delta.

It follows from (5)–(6) that for $x \in \omega_s$, $(\rho,x) \in \Omega$ we have

$$\left.\begin{array}{l}\varphi^{(m)}(x,\lambda) = \frac{1}{2}((i\rho)^m \exp(i\rho x)[1]_\gamma + (-i\rho)^m \exp(-i\rho x)[1]_\gamma \\ \quad -2i(i\rho)^m \sum_{j=1}^s \cos\pi\nu_j \exp(i\rho(x-2\gamma_j))[1]_\gamma), \\ \psi^{(m)}(x,\lambda) = \frac{1}{2}((i\rho)^m \exp(-i\rho(T-x))[1]_\gamma + (-i\rho)^m \exp(i\rho(T-x))[1]_\gamma \\ \quad +2i(i\rho)^m \sum_{j=s+1}^N \cos\pi\nu_j \exp(i\rho(T+x-2\gamma_j))[1]_\gamma).\end{array}\right\} \quad (7)$$

In particular, for $|\rho| \to \infty$

$$\left.\begin{array}{l}\Delta(\lambda) = \frac{i\rho}{2}(\exp(i\rho T)[1] - \exp(-i\rho T)[1] - 2i\sum_{j=1}^N \cos\pi\nu_j \exp(i\rho(T-2\gamma_j))[1]), \\ \delta(\lambda) = \frac{1}{2}(\exp(i\rho T)[1] + \exp(-i\rho T)[1] + 2i\sum_{j=1}^N \cos\pi\nu_j \exp(i\rho(T-2\gamma_j))[1]), \\ [1] = 1 + O(\rho^{-1}).\end{array}\right\} \quad (8)$$

Let us give a definition. A function $y(x)$ is called the eigenfunction of L for $\lambda = \lambda^*$, if there exist numbers a_{1j}, a_{2j} ($|a_{1j}| + |a_{2j}| > 0$) such that $y(x) = a_{1j}S_{1j}(x,\lambda^*) + a_{2j}S_{2j}(x,\lambda^*)$ and $U_0(y) = U_1(y) = 0$. Those λ for which there exist eigenfunctions, are called the eigenvalues of L.

It is obvious that the eigenvalues $\{\lambda_k\}_{k\geq 0}$ of L coincide with zeros of the entire function $\Delta(\lambda)$, and $\varphi(x,\lambda_k), \psi(x,\lambda_k)$ are eigenfunctions, and consequently $\psi(x,\lambda_k) = \beta_k\varphi(x,\lambda_k)$, $\beta_k \neq 0$.

Using (7)–(8), by the well-known methods (see, for example, [13]) one can obtain the following properties of the characteristic function $\Delta(\lambda)$ and eigenvalues $\{\lambda_k\}_{k\geq 0}$ of the boundary value problem L.

1. For $|\rho| \to \infty$, $\Delta(\lambda) = O(|\rho|\exp(|\tau|T))$.

2. There exist $h > 0$, $C_h > 0$ such that $|\Delta(\lambda)| \geq C_h|\rho|\exp(|\tau|T)$ for $|\text{Im}\,\rho| \geq h$. Hence, the eigenvalues $\{\lambda_k\}$ lie in the domain $|\text{Im}\,\rho| < h$.

3. The number N_a of zeros of $\Delta(\lambda)$ in the rectangle $R_a = \{\rho: |\tau| \leq h, \sigma \in [a,a+1]\}$ is bounded with respect to a.

4. Denote $G_\delta = \{\rho: |\rho - \rho_k| \geq \delta\}$, $\lambda_k = \rho_k^2$. Then

$$|\Delta(\lambda)| \geq C_\delta|\rho|\exp(|\tau|T), \quad \rho \in G_\delta. \quad (9)$$

5. There exist numbers $R_n \to \infty$ such that for sufficiently small $\delta > 0$ the circles $|\rho| = R_N$ lie in G_δ for all n.

6. Let $\{\rho_k^0\}$ be zeros of the function

$$\Delta_0(\rho) = \rho(\exp(i\rho T) - \exp(-i\rho T) - 2i\sum_{j=1}^N \cos\pi\nu_j \exp(i\rho(T-2\gamma_j)).$$

Then, for $k \to \infty$

$$\rho_k = \rho_k^0 + O\left(\frac{1}{\rho_k^0}\right).$$

In the sequel, for simplicity, we confine ourselves to the case when all zeros of $\Delta(\lambda)$ are simple. In this case we shall say that $L \in V'$.

Let us denote

$$\alpha_k = \underset{\lambda=\lambda_k}{\text{Res}}\, M(\lambda)$$

Since $M(\lambda) = -(\Delta(\lambda))^{-1}\delta(\lambda)$, $\beta_k = \delta(\lambda_k)$ then

$$\alpha_k = -(\dot\Delta(\lambda_k))^{-1}\beta_k \neq 0, \quad (\alpha_k)^{-1} = O(1).$$

The set $\{\lambda_k, \alpha_k\}_{k \geq 0}$ is called the spectral data of the boundary value problem L.

3. The inverse problem is formulated as follows: given the spectral data $\{\lambda_k, \alpha_k\}_{k \geq 0}$, construct L.

This inverse problem is a generalization of the well-known inverse problem for Sturm–Liouville equations without singularities (see [1–3]).

First, let us prove the uniqueness theorems for the solution of the inverse problem. For this we agree that together with L we consider a boundary value problem $\tilde{L} = L(\tilde{p}(x), \tilde{h}_0, \tilde{h}_1)$. If a symbol α denotes an object related to L, then $\tilde\alpha$ will denote the analogous object related to $\tilde{L}$, and $\hat\alpha = \alpha - \tilde\alpha$.

Theorem 1. *If $M(\lambda) = \tilde{M}(\lambda)$ then $L = \tilde{L}$.*

Indeed, let $P(x, \lambda) = [P_{jk}(x, \lambda)]_{j,k=1,2}$, where

$$P_{jk}(x, \lambda) = (-1)^{k-1}(\varphi^{(j-1)}(x, \lambda)\tilde\Phi^{(2-k)}(x, \lambda) - \Phi^{(j-1)}(x, \lambda)\tilde\varphi^{(2-k)}(x, \lambda)).$$

By virtue of (7)–(9), we get for $(\rho, x) \in \Omega$, $\rho \in G_\delta$,

$$|P_{jk}(x, \lambda) - \delta_{jk}| \leq C_\delta |\rho|^{-1}, j \leq k; \quad |P_{21}(x, \lambda)| \leq C_\delta |\rho|.$$

On the other hand, since

$$\Phi(x, \lambda) = \varphi_2(x, \lambda) + M(\lambda)\varphi(x, \lambda),$$

we get by the equality $M(\lambda) = \tilde{M}(\lambda)$ that for each fixed $x \neq \gamma_1, \ldots, \gamma_N$, the functions $P_{jk}(x, \lambda)$ are entire in λ. Consequently, $P_{11}(x, \lambda) = 1$, $P_{12}(x, \lambda) = 0$, i.e. $\varphi(x, \lambda) = \tilde\varphi(x, \lambda)$, $\Phi(x, \lambda) = \tilde\Phi(x, \lambda)$, $L = \tilde{L}$.

Using the contour integral method one can obtain the following equality

$$M(\lambda) = \sum_{k=0}^{\infty} \frac{\alpha_k}{\lambda - \lambda_k}, \qquad (10)$$

where the series converge "with brackets." The following theorem is obvious corollary of (10) and Theorem 1.

Theorem 2. *If $\lambda_k = \tilde\lambda_k$, $\alpha_k = \tilde\alpha_k$, $k \geq 0$, then $L = \tilde{L}$.*

4. Let problems $L, \tilde{L} \in V'$ be such that

$$a_j = \tilde{a}_j \; (j = \overline{1,N}), \quad \sum_{k=0}^{\infty} |\rho_k \alpha_k| \xi_k < \infty,$$

where $\xi_k = |\rho_k - \tilde{\rho}_k| + |1 - (\alpha_k)^{-1} \tilde{\alpha}_k|$. Denote

$$\lambda_{k0} = \lambda_k, \; \lambda_{k1} = \tilde{\lambda}_k, \; \alpha_{k0} = \alpha_k, \; \alpha_{k1} = \tilde{\alpha}_k, \; \tilde{\varphi}_{kj}(x) = \tilde{\varphi}(x, \lambda_{kj}),$$

$$\tilde{D}_{kj}(x, \lambda) = \frac{\langle \tilde{\varphi}(x, \lambda), \tilde{\varphi}_{kj}(x) \rangle}{\lambda - \lambda_{kj}}, \quad \tilde{P}_{ni,kj}(x) = \tilde{D}_{kj}(x, \lambda_{ni}) \alpha_{kj}$$

$$\tilde{\psi}_{k0}(x) = \xi_k^{-1}(\tilde{\phi}_{k0}(x) - \tilde{\phi}_{k1}(x)), \quad \tilde{\psi}_{k1}(x) = \tilde{\phi}_{k1}(x)$$

$$\tilde{H}_{n0,k0}(x) = \xi_n^{-1} \xi_k (\tilde{P}_{n0,k0}(x) - \tilde{P}_{n1,k0}(x)), \quad \tilde{H}_{n1,k1}(x) = \tilde{P}_{n1,k0}(x) - \tilde{P}_{n1,k1}(x)$$

$$\tilde{H}_{n0,k1}(x) = \xi_n^{-1}(\tilde{P}_{n0,k0}(x) - \tilde{P}_{n1,k0}(x) - \tilde{P}_{n0,k1}(x) + \tilde{P}_{n1,k1}(x)), \quad \tilde{H}_{n1,k0}(x) = \xi_k \tilde{P}_{n1,k0}(x).$$

Functions $\varphi_{kj}(x)$, $\psi_{kj}(x)$, $P_{ni,kj}(x)$, $H_{ni,kj}(x)$ are defined analogously.

Denote $\psi(x) = [\psi_{ni}(x)]$, $\tilde{\psi}(x) = [\tilde{\psi}_{ni}(x)]$, $\tilde{H}(x) = [\tilde{H}_{ni,kj}(x)]$, $(n, k \geq 0, i, j = 0, 1)$ and consider the Banach space m of bounded sequences $a = [a_{ni}]_{n \geq 0, i=0,1}$ with the norm $\|a\| = \sup_{n,i} |a_{ni}|$. Then for each fixed $x \neq \gamma_j$, we have $\psi(x), \tilde{\psi}(x) \in m$, and

$$\tilde{\psi}(x) = (E + \tilde{H}(x)) \psi(x) \tag{11}$$

where E is the identity operator. The operator $E + \tilde{H}(x)$, acting from m to m, is linear bounded operator:

$$\|\tilde{H}(x)\| \leq C \sum_{k=0}^{\infty} \xi_k |\alpha_k| < \infty,$$

and the operator $E + \tilde{H}(x)$ has a bounded inverse operator, i.e. equation (11) is uniquely solvable. Moreover we have

$$p(x) = \tilde{p}(x) - 2b'(x), \; h_0 = \tilde{h}_0 + b(0), \; h_1 = \tilde{h}_1 + b(T) \tag{12}$$

where

$$b(x) = \sum_{k=0}^{\infty} (\alpha_{k0} \tilde{\varphi}_{k0}(x) \varphi_{k0}(x) - \alpha_{k1} \tilde{\varphi}_{k1}(x) \varphi_{k1}(x)). \tag{13}$$

Equation (11) is called the main equation of the inverse problem. Solving (11) we find the vector $\psi(x)$, and consequently, the functions $\varphi_{ni}(x)$. Since $\varphi_{ni}(x) = \varphi(x, \lambda_{ni})$ are the solutions of (1), we can construct the function $q(x)$ and the coefficients h_0 and h_1.

Now let us formulate necessary and sufficient conditions for the solvability of the inverse problem.

Theorem 3. *For numbers* $\{\lambda_k, \alpha_k\}_{k\geq 0}$, $\alpha_k \neq 0$, $\lambda_k \neq \lambda_n$ $(k \neq n)$ *to be the spectral data for a* $L \in V'$, *it is necessary and sufficient that the following conditions hold:*

1. *(asymptotics) there exists* $\tilde{L} \in V'$ *such that*

$$a_j = \tilde{a}_j \ (j = \overline{1,N}), \quad \sum_{k=0}^{\infty} |\rho_k \alpha_k| \xi_k < \infty,$$

2. *for each fixed* $x \neq \gamma_j$ *the linear bounded operator* $E + \tilde{H}(x)$, *acting from* m *to* m, *has a unique inverse operator;*

3. $b'(x) \prod_{j=1}^{N} |x - \gamma_j|^{1-2\operatorname{Re}\nu_j} \in \mathcal{L}(0,T)$, *where* $b(x)$ *is found by (13).*

Under these conditions the boundary value problem L is constructed by (12).

ACKNOWLEDGMENT

This research was supported in part by Grant No 97-01-00566 of Russian Foundation for Basic Research and Grant No 96-1.7-4 of Russian Ministry of General and Professional Education (Grant Center for Natural Sciences).

REFERENCES

1. Marchenko V. A., Sturm–Liouville operators and their applications, Naukova Dumka, Kiev, 1977; English transl., Birkhauser, 1986.
2. Levitan B. M., Inverse Sturm–Liouville problems, Nauka, Moscow, 1984; English transl., VNU Sci. Press, Utrecht, 1987.
3. McLaughlin J. R., Analytical methods for recovering coefficients in differential equations from spectral data, SIAM Rev. 28 (1986), 53–72.
4. Stashevskaya V. V., On inverse problems of spectral analysis for a certain class of differential equations, Dokl. Akad. Nauk SSSR 93 (1953), 409–412.
5. Gasymov M. G., Determination of Sturm–Liouville equation with a singular point from two spectra, Dokl. Akad. Nauk SSSR 161 (1965), 274–276.
6. Carlson R., Inverse spectral theory for some singular Sturm–Liouville problems, J. Diff. Equations 106 (1993), 121–140.
7. Zhornitskaya L. A. and Serov V. S., Inverse eigenvalue problems for a singular Sturm–Liouville operator on (0,1), Inverse Problems 10 (1994), no.4, 975–987.
8. Yurko V. A., Inverse problem for differential equations with a singularity, Differentsialnye Uravneniya 28 (1992), 1355–1362; English transl. in Differential Equations 28 (1992), 1100–1107.
9. Yurko V. A., On higher-order differential operators with a singular point, Inverse Problems, 9 (1993), 495–502.
10. Yurko V. A., On higher-order differential operators with a regular singularity, Mat. Sb. 186 (1995), no.6, 133–160; English transl. in Sbornik; Mathematics 186 (1995), no.6, 901–928.
11. Levinson N., The inverse Sturm–Liouville problem, Math. Tidsskr. 13 (1949), 25–30.
12. Leibenzon Z. L., The inverse problem of spectral analysis for higher-order ordinary differential operators, Trudy Moskov. Mat. Obshch. 15 (1966), 70–144; English transl. in Trans. Moscow Math. Soc. 15 (1966).
13. Bellman R. and Cooke K., Differential-difference equations, Academic Press, New York, 1963.
14. Borg G., Eine Umkehrung der Sturm–Liouvilleschen Eigenwertaufgabe, Acta Math. 78 (1946), 1–96.
15. Privalov I. I., Introduction to the theory of functions of a complex variable, 11th ed., Nauka, Moscow, 1967.

LIST OF CONTRIBUTORS

Tuncay Aktosun
Department of Mathematics
North Dakota State University
Fargo, ND

Matania Ben-Artzi
Institute of Mathematics
Hebrew University
Jerusalem 91904, Israel

Stephen Gustafson
Department of Mathematics
University of Toronto
100 St. George Street
Toronto, Canada

Hitoshi Kitada
Department of Mathematical Sciences
University of Tokyo
Komaba, Meguro, Tokyo 153, Japan
E-mail: kitada@ms.u-tokyo.ac.jp

Martin Klaus
Department of Mathematics
Virginia Polytechnic Institute and State University
Blacksburg, VA

Peter Kuchment
Department of Mathematics and Statistics
Wichita State University
Wichita, KS
E-mail: kuchment@twsuvm.uc.twsu.edu
http://www.math.twsu.edu/Faculty/Kuchment/

G. N. Makrakis
Institute of Applied and Computational Math.
FO.R.T.H
P.O. Box 1527, 71 110, Heraklion, Crete, Greece
and Department of Mathematics
University of Crete
E-mail: makrakg@calderon.iacm.forth.gr

Cornelis van der Mee
Department of Mathematics
University of Cagliari
Cagliari, Italy

Jonathan Nemirovsky
Institute of Mathematics
Hebrew University
Jerusalem 91904, Israel

Yehuda Pinchover
Department of Mathematics
Technion-Israel Institute of Technology
32000 Haifa, Israel
E-mail: pincho@tx.technion.ac.il

Alexander G. Ramm
Department of Mathematics
Kansas State University
Manhattan, KS
E-mail: ramm@math.ksu.edu

Peter Rejto
School of Mathematics
206 Church Street
University of Minnesota
Minneapolis, MN

Martin Schechter
Department of Mathematics
University of California at Irvine
Irvine, CA

Marianna A. Shubov
Department of Mathematics
Texas Tech University
Lubbock, TX

Mario Taboada
Department of Mathematics
Old Dominion University
Norfolk, VA

Vladilen A. Trenogin
Department of Mathematics
Moscow State Steel and Alloys Institute, 117936
Moscow, Leninsky prospect 4, Russia

Boris Vainberg
Department of Mathematics
University of North Carolina at Charlotte
Charlotte, NC
E-mail: brvainbe@uncc.edu

V. A. Yurko
Department of Mathematics
Saratov State University
Astrakhanskaya 83, Saratov 410071, Russia
E-mail: Yurko@scnit.saratov.su

INDEX

Acoustic waveguides, 89

Control theory, 177

Embedded eigenvalues, 67, 106

Ginzburg–Landau equation, 33
Global invertibility, 191

Hampwile theorem, 171

Indefinite-weight elliptic problems, 77
Inverse problems, 16, 203
Inverse scattering, 104, 11

Limiting absorption principle, 131
Local time, 39

Maxwell's equations, 19
Mountain pass alternative, 164

Nonlinear operators, 189

Operators with singularities, 199

Parameter-continuation, 189
Phase shift, 111
Principal eigenvalues, 77

Quantum mechanics, 39

Relativity, 39
Resolvent estimates, 19

Saddle point theorem, 159
Scattering by obstacle, 89
Scattering solutions, 101
Schrödinger operators, 67
Semilinear equations, 157
Spectral operators, 177
Sturm–Liouville operators, 199

Turning points, 131

Wave equation, 177
Wave scattering, 1